# Big Data in Psychiatry and Neurology

# Big Data in Psychiatry and Neurology

Edited by

**Ahmed A. Moustafa**
Department of Psychiatry, Wroclaw Medical University, Wroclaw, Poland

School of Psychology & Marcs Institute for Brain and Behaviour, Western Sydney University, Sydney, NSW, Australia

Academic Press is an imprint of Elsevier
125 London Wall, London EC2Y 5AS, United Kingdom
525 B Street, Suite 1650, San Diego, CA 92101, United States
50 Hampshire Street, 5th Floor, Cambridge, MA 02139, United States
The Boulevard, Langford Lane, Kidlington, Oxford OX5 1GB, United Kingdom

Copyright © 2021 Elsevier Inc. All rights reserved.

No part of this publication may be reproduced or transmitted in any form or by any means, electronic or mechanical, including photocopying, recording, or any information storage and retrieval system, without permission in writing from the publisher. Details on how to seek permission, further information about the Publisher's permissions policies and our arrangements with organizations such as the Copyright Clearance Center and the Copyright Licensing Agency, can be found at our website: www.elsevier.com/permissions.

This book and the individual contributions contained in it are protected under copyright by the Publisher (other than as may be noted herein).

**Notices**
Knowledge and best practice in this field are constantly changing. As new research and experience broaden our understanding, changes in research methods, professional practices, or medical treatment may become necessary.

Practitioners and researchers must always rely on their own experience and knowledge in evaluating and using any information, methods, compounds, or experiments described herein. In using such information or methods they should be mindful of their own safety and the safety of others, including parties for whom they have a professional responsibility.

To the fullest extent of the law, neither the Publisher nor the authors, contributors, or editors, assume any liability for any injury and/or damage to persons or property as a matter of products liability, negligence or otherwise, or from any use or operation of any methods, products, instructions, or ideas contained in the material herein.

**Library of Congress Cataloging-in-Publication Data**
A catalog record for this book is available from the Library of Congress

**British Library Cataloguing-in-Publication Data**
A catalogue record for this book is available from the British Library

ISBN 978-0-12-822884-5

For information on all Academic Press publications
visit our website at https://www.elsevier.com/books-and-journals

*Publisher:* Nikki Levy
*Acquisitions Editor:* Joslyn Chaiprasert-Paguio
*Editorial Project Manager:* Tracy I. Tufaga
*Production Project Manager:* Omer Mukthar
*Cover Designer:* Victoria Pearson Esser

Typeset by SPi Global, India

# Dedication

I like to dedicate this book to all researchers who are
starting to learn more about the applications of
big data methods to healthcare and medicine.
I hope that this book will provide some valuable
information to your research endeavors.

# Contents

| | |
|---|---|
| Contributors | xv |
| Editor's biography | xix |
| Preface | xxi |
| Acknowledgment | xxiii |

## 1. Best practices for supervised machine learning when examining biomarkers in clinical populations

*Benjamin G. Schultz, Zaher Joukhadar, Usha Nattala,*
*Maria del Mar Quiroga, Francesca Bolk,*
*and Adam P. Vogel*

| | | |
|---|---|---|
| 1 | Introduction | 1 |
| 2 | Data formatting | 2 |
| 3 | Statistical assumptions | 5 |
| 4 | Sample size estimation | 8 |
| 5 | Choosing parsimonious models | 9 |
| 6 | Reduction of data dimensionality | 11 |
| | 6.1 Scaling | 12 |
| | 6.2 Variable selection | 12 |
| | 6.3 Principal component analysis | 13 |
| | 6.4 Linear discriminant analysis | 14 |
| 7 | Performance metrics | 15 |
| 8 | Resampling methods | 19 |
| 9 | Data leakage | 23 |
| | 9.1 Partitioning | 23 |
| | 9.2 Data reduction and scaling | 23 |
| | 9.3 Variable selection | 23 |
| | 9.4 Balancing datasets | 24 |
| 10 | Supervised machine learning classifiers | 24 |
| 11 | Deep learning and artificial intelligence | 26 |
| 12 | Limitations and future directions | 28 |
| 13 | Conclusions | 28 |
| | References | 29 |

**viii** Contents

## 2. Big data in personalized healthcare
*Lidong Wang and Cheryl Alexander*

| | | |
|---|---|---:|
| 1 | Introduction | 35 |
| 2 | Characteristics, methods, and software platforms of big data | 36 |
| 3 | Big data in the healthcare area | 39 |
| 4 | Big data and big data analytics in personalized healthcare | 42 |
| 5 | Conclusion | 46 |
| | Acknowledgment | 46 |
| | References | 46 |

## 3. Longitudinal data analysis: The multiple indicators growth curve model approach
*Thierno M.O. Diallo and Ahmed A. Moustafa*

| | | |
|---|---|---:|
| 1 | Introduction | 51 |
| 2 | Multivariate dimension reduction techniques: Principal component analysis and factor analysis | 53 |
| | 2.1 Principal component analysis | 54 |
| | 2.2 Factor analysis | 55 |
| | 2.3 Factor analysis and principal component analysis for multiple indicator growth curve models | 57 |
| 3 | Longitudinal measurement invariance | 58 |
| 4 | Multiple indicators growth curve model | 60 |
| | 4.1 The MILCM equations | 61 |
| | 4.2 Specification details | 63 |
| 5 | Steps in fitting an MILCM | 64 |
| | References | 66 |

## 4. Challenges and solutions for big data in personalized healthcare
*Tim Hulsen*

| | | |
|---|---|---:|
| 1 | Introduction | 69 |
| 2 | Standardization | 71 |
| | 2.1 Interoperability and reusability | 71 |
| | 2.2 Standards for clinical data | 72 |
| | 2.3 Standards for -omics data | 74 |
| | 2.4 Standards for imaging data | 75 |
| | 2.5 Standards for biosample data | 76 |
| 3 | Data sharing and integration | 77 |
| | 3.1 Data ownership | 77 |
| | 3.2 Support for data sharing | 78 |
| | 3.3 Data sharing initiatives | 79 |
| | 3.4 Data integration | 81 |
| 4 | Privacy and ethics | 84 |
| | 4.1 Stricter regulations | 84 |

Contents  **ix**

| | | |
|---|---|---|
| 4.2 | Explicit consent | 85 |
| 4.3 | Privacy and ethics in industry | 85 |
| **5** | **Teaching data science** | 87 |
| 5.1 | Need for more training | 87 |
| 5.2 | Training data science to medical students | 87 |
| 5.3 | Available courses in Clinical Data Science | 88 |
| **6** | **Discussion** | 88 |
| | **Competing interest statement** | 89 |
| | **References** | 90 |

## 5. Data linkages in epidemiology
*Sinéad Moylett*

| | | |
|---|---|---|
| **1** | **Introduction** | 95 |
| **2** | **Linking local and national routinely-collected data** | 96 |
| 2.1 | Development of diagnostic algorithms: Structured data | 98 |
| **3** | **Linking routinely- and non-routinely-collected data** | 99 |
| **4** | **Linking structured and unstructured routinely-collected data** | 100 |
| 4.1 | CRIS and CRATE databases | 101 |
| 4.2 | Development of diagnostic algorithms: Unstructured data | 106 |
| **5** | **Conclusion** | 109 |
| | **References** | 111 |

## 6. Neutrosophic rule-based classification system and its medical applications
*Sameh H. Basha, Areeg Abdalla, and Aboul Ella Hassanien*

| | | |
|---|---|---|
| **1** | **Introduction** | 119 |
| **2** | **Theoretical background** | 122 |
| 2.1 | Neutrosophic logic and neutrosophic set | 122 |
| 2.2 | Neutrosophic rule-based classification system | 123 |
| **3** | **NRCS medical applications** | 126 |
| 3.1 | Neutrosophic rule-based prediction system for toxicity effects assessment of biotransformed hepatic drugs | 126 |
| 3.2 | A predictive model for diabetics using NRCS | 129 |
| 3.3 | A predictive model for seminal quality using NRCS | 130 |
| **4** | **Conclusions and future work** | 133 |
| | **References** | 133 |

## 7. From complex to neural networks
*Nicola Amoroso and Loredana Bellantuono*

| | | |
|---|---|---|
| **1** | **Big data and MRI analyses** | 137 |
| **2** | **Modeling purposes: Complex networks** | 141 |
| **3** | **Learning from data** | 145 |

**x** Contents

4 A multiplex model to diagnose neurodegenerative
diseases and anomalous aging — 150
References — 152

## 8. The use of Big Data in Psychiatry—The role of administrative databases

*Manuel Gonçalves-Pinho and Alberto Freitas*

1 Introduction — 155
2 Big Data, administrative databases, and mental health — 156
3 Pros and cons of administrative databases research
in mental health — 157
4 Conclusions — 162
References — 163

## 9. Predicting the emergence of novel psychoactive substances with big data

*Robert Todd Perdue and James Hawdon*

1 Introduction — 167
2 Internet search queries as data — 169
3 Methods — 171
4 Results — 172
5 Discussion and conclusion — 174
References — 177

## 10. Hippocampus segmentation in MR images: Multiatlas methods and deep learning methods

*Hancan Zhu, Shuai Wang, Liangqiong Qu, and Dinggang Shen*

1 Introduction — 181
2 Patch-based multiatlas labeling for Hippocampus
segmentation — 184
  2.1 Weighted voting label fusion — 185
  2.2 Local learning-based label fusion — 187
  2.3 Supervised metric learning for label fusion — 188
  2.4 An evaluation of different patch-based multiatlas
labeling methods — 190
3 Deep learning-based methods for Hippocampus
segmentation — 194
  3.1 Multiatlas-based deep learning method for
hippocampus segmentation — 196
  3.2 End-to-end dilated residual dense U-net for
hippocampus segmentation — 205
4 Conclusion — 210
Acknowledgments — 211
References — 212

Contents **xi**

## 11. A scalable medication intake monitoring system

*Diane Myung-Kyung Woodbridge and Kevin Bengtson Wong*

| | | |
|---|---|---|
| 1 | Introduction | 217 |
| 2 | Related work | 218 |
| 3 | System architecture | 220 |
| | 3.1 Smartwatch application | 221 |
| | 3.2 Cloud services | 222 |
| | 3.3 Data storage | 222 |
| | 3.4 Distributed data processing | 223 |
| 4 | Algorithms | 223 |
| | 4.1 Distributed preprocessing | 224 |
| | 4.2 Distributed AutoML and machine learning | 226 |
| 5 | Experiment results | 228 |
| | 5.1 Experiment setting | 228 |
| | 5.2 Results | 231 |
| 6 | Conclusion | 236 |
| | Acknowledgments | 238 |
| | References | 238 |

## 12. Evaluating cascade prediction via different embedding techniques for disease mitigation

*Abhinav Choudhury, Shubham Shakya, Shruti Kaushik, and Varun Dutt*

| | | |
|---|---|---|
| 1 | Introduction | 241 |
| 2 | Background | 243 |
| 3 | Method | 245 |
| | 3.1 Dataset | 245 |
| | 3.2 Generating graph embeddings | 245 |
| | 3.3 Cascade prediction | 249 |
| | 3.4 Evaluation metrics | 253 |
| 4 | Results | 253 |
| | 4.1 Cascade prediction using MLP | 253 |
| | 4.2 Cascade prediction using LSTM | 253 |
| | 4.3 Model comparison | 255 |
| 5 | Discussion and conclusions | 255 |
| | Acknowledgment | 259 |
| | References | 259 |

## 13. A two-stage classification framework for epileptic seizure prediction using EEG wavelet-based features

*Sahar Elgohary, Mahmoud I. Khalil, and Seif Eldawlatly*

| | | |
|---|---|---|
| 1 | Introduction | 263 |
| 2 | Materials and methods | 265 |
| | 2.1 Dataset | 265 |
| | 2.2 Two-stage zero-crossings wavelet-based framework | 266 |
| | 2.3 Comparative analysis methods | 271 |

xii Contents

3 Results 272
   3.1 Stage 1: Interictal and preictal binary classification 272
   3.2 Stage 2: Preictal classification into early and late stages 277
4 Discussion 283
5 Conclusions 284
   Disclosure statement 285
   References 285

## 14. Visual neuroscience in the age of big data and artificial intelligence
*Kohitij Kar*

1 Confining the problem space 287
2 Chapter roadmap 288
3 Understanding vision—What do we seek to reveal? 288
   3.1 The first generation of neural network models 290
   3.2 Next generation of neural network models 291
   3.3 Experiments to falsify and improve models 292
4 How to evaluate the current models of vision? 293
   4.1 Prediction 293
   4.2 Control 297
5 The vision community is coming together to combine data and models 297
   5.1 Allen brain observatory 299
   5.2 Brain-score 299
   5.3 The Algonauts project 300
6 Conclusion 301
   References 301

## 15. Application of big data and artificial intelligence approaches in diagnosis and treatment of neuropsychiatric diseases
*Qiurong Song, Tianhui Huang, Xinyue Wang,*
*Jingxiao Niu, Wang Zhao, Haiqing Xu, and Long Lu*

1 Introduction 305
2 Main data sources 307
   2.1 Genomics 307
   2.2 EEG signals 308
   2.3 Eye movement data 308
   2.4 Neuroimaging data 309
   2.5 Wearable equipment data 309
3 Main algorithms 310
4 Applications 311
   4.1 Early warning 311
   4.2 Diagnosis 312
   4.3 Prognosis 316

Contents **xiii**

|  |  |  |
|---|---|---|
| 5 | Challenges and promising solutions | 317 |
| | 5.1 Privacy and security of patient information | 317 |
| | 5.2 Information island | 319 |
| | 5.3 Storage and analysis capabilities | 319 |
| | 5.4 Lack of specialized personnel | 320 |
| 6 | Conclusions | 320 |
| | References | 321 |

## 16. Harnessing big data to strengthen evidence-informed precise public health response

*G.V. Asokan and Mohammed Yousif Mohammed*

|  |  |  |
|---|---|---|
| 1 | Public health | 325 |
| 2 | Global burden of disease | 326 |
| | 2.1 Noncommunicable diseases | 327 |
| | 2.2 Infectious diseases | 327 |
| 3 | Health systems and public health system | 328 |
| | 3.1 Public health surveillance system | 329 |
| 4 | Big data in precision public health | 332 |
| 5 | Case studies | 333 |
| | 5.1 Noncommunicable diseases | 333 |
| | 5.2 Infectious disease: COVID-19 | 334 |
| | References | 335 |

## 17. How big data analytics is changing the face of precision medicine in women's health

*Maryam Panahiazar, Maryam Karimzadehgan,*
*Roohallah Alizadehsani, Dexter Hadley, and*
*Ramin E. Beygui*

|  |  |  |
|---|---|---|
| 1 | Introduction | 339 |
| 2 | The role of big data and deep learning in personalized medicine to empower women's health | 341 |
| 3 | Use case studies | 342 |
| | 3.1 Advanced data analytics on skin conditions from genotype to phenotype | 342 |
| | 3.2 Big data platform to use machine learning on EHR data for personalized medicine in heart failure survival analysis and patient similarity | 344 |
| | 3.3 Large-scale labeling of free-text pathology report for deep learning to improve women's health in breast cancer | 346 |
| 4 | Conclusion | 347 |
| | Acknowledgements | 348 |
| | Author contribution | 348 |
| | References | 348 |

Index 351

# Contributors

*Numbers in parentheses indicate the pages on which the authors' contributions begin.*

**Areeg Abdalla** (119), Faculty of Science, Cairo University, Cairo, Egypt

**Cheryl Alexander** (35), Institute for IT innovation and Smart Health, Vicksburg, MS, United States

**Roohallah Alizadehsani** (339), Institute for Intelligent Systems Research and Innovation, Deakin University, Melbourne, VIC, Australia

**Nicola Amoroso** (137), Dipartimento di Farmacia—Scienze del Farmaco, Università di Bari; Istituto Nazionale di Fisica Nucleare, Sezione di Bari, Bari, Italy

**G.V. Asokan** (325), Public Health Program, College of Health and Sport Sciences, University of Bahrain, Salmanya Medical Complex, Manama, Kingdom of Bahrain

**Sameh H. Basha** (119), Faculty of Science, Cairo University, Cairo, Egypt

**Loredana Bellantuono** (137), Dipartimento Interateneo di Fisica, Università di Bari, Bari, Italy

**Ramin E. Beygui** (339), Department of Surgery, Division of Cardiothoracic Surgery, School of Medicine, University of California San Francisco, San Francisco, CA, United States

**Francesca Bolk** (1), Murdoch Children's Research Institute, The University of Melbourne, Melbourne, VIC, Australia

**Abhinav Choudhury** (241), Indian Institute of Technology Mandi, Kamand, India

**Thierno M.O. Diallo** (51), School of Social Science, Western Sydney University, Sydney, NSW, Australia; Statistiques & M.N., Sherbrooke, QC, Canada

**Varun Dutt** (241), Indian Institute of Technology Mandi, Kamand, India

**Seif Eldawlatly** (263), Computer and Systems Engineering Department, Faculty of Engineering, Ain Shams University; Faculty of Media Engineering and Technology, German University in Cairo, Cairo, Egypt

**Sahar Elgohary** (263), Computer and Systems Engineering Department, Faculty of Engineering, Ain Shams University, Cairo, Egypt

**Alberto Freitas** (155), Department of Community Medicine, Information and Health Decision Sciences, Faculty of Medicine, University of Porto; Center for Health Technology and Services Research (CINTESIS), Porto, Portugal

**xvi** Contributors

**Manuel Gonçalves-Pinho** (155), Department of Community Medicine, Information and Health Decision Sciences, Faculty of Medicine, University of Porto; Center for Health Technology and Services Research (CINTESIS), Porto; Department of Psychiatry and Mental Health, Centro Hospitalar do Tâmega e Sousa, Penafiel, Portugal

**Dexter Hadley** (339), Department of Clinical Sciences, College of Medicine, University of Central Florida, Orlando, FL, United States

**Aboul Ella Hassanien** (119), Faculty of Computers and Information, Cairo University; Scientific Research Group in Egypt (SRGE), Cairo, Egypt

**James Hawdon** (167), Center for Peace Studies and Violence Prevention, Virginia Polytechnic Institute and State University, Blacksburg, VA, United States

**Tianhui Huang** (305), School of Information Management, Wuhan University, Wuhan, PR China

**Tim Hulsen** (69), Department of Professional Health Solutions & Services, Philips Research, Eindhoven, The Netherlands

**Zaher Joukhadar** (1), Melbourne Data Analytics Platform, The University of Melbourne, Melbourne, VIC, Australia

**Kohitij Kar** (287), McGovern Institute for Brain Research and Department of Brain and Cognitive Sciences; Center for Brains, Minds, and Machines, Massachusetts Institute of Technology, Cambridge, MA, United States

**Maryam Karimzadehgan** (339), Independent Researcher, Google

**Shruti Kaushik** (241), Indian Institute of Technology Mandi, Kamand, India

**Mahmoud I. Khalil** (263), Computer and Systems Engineering Department, Faculty of Engineering, Ain Shams University, Cairo, Egypt

**Long Lu** (305), School of Information Management, Wuhan University, Wuhan, PR China

**Mohammed Yousif Mohammed** (325), Medical Laboratory Sciences Program, College of Health and Sport Sciences, University of Bahrain, Salmanya Medical Complex, Manama, Kingdom of Bahrain

**Ahmed A. Moustafa** (51), Department of Psychiatry, Wroclaw Medical University, Wroclaw, Poland; School of Psychology & Marcs Institute for Brain and Behaviour, Western Sydney University, Sydney, NSW, Australia

**Sinéad Moylett** (95), Laboratory of Neuroimmunology, KU Leuven, Leuven, Belgium

**Usha Nattala** (1), Melbourne Data Analytics Platform, The University of Melbourne, Melbourne, VIC, Australia

**Jingxiao Niu** (305), School of Information Management, Wuhan University, Wuhan, PR China

**Maryam Panahiazar** (339), Department of Surgery, Division of Cardiothoracic Surgery; Bakar Computational Health Sciences Institute, School of Medicine, University of California San Francisco, San Francisco, CA, United States

Contributors **xvii**

**Robert Todd Perdue** (167), Department of Sociology & Anthropology, Elon University, Elon, NC, United States

**Liangqiong Qu** (181), Department of Biomedical Data Science, Stanford University, Stanford, CA, United States

**Maria del Mar Quiroga** (1), Melbourne Data Analytics Platform, The University of Melbourne, Melbourne, VIC, Australia

**Benjamin G. Schultz** (1), Centre for Neuroscience of Speech, The University of Melbourne, Melbourne, VIC, Australia

**Shubham Shakya** (241), National Institute of Technology Kurukshetra, Kurukshetra, India

**Dinggang Shen** (181), School of Biomedical Engineering, ShanghaiTech University; Shanghai United Imaging Intelligence Co., Ltd., Shanghai, China; Department of Artificial Intelligence, Korea University, Seoul, Republic of Korea

**Qiurong Song** (305), School of Information Management, Wuhan University, Wuhan, PR China

**Adam P. Vogel** (1), Centre for Neuroscience of Speech, The University of Melbourne; Redenlab, Melbourne, VIC, Australia

**Lidong Wang** (35), Institute for Systems Engineering Research, Mississippi State University, Starkville, MS, United States

**Shuai Wang** (181), Imaging Biomarkers and Computer-Aided Diagnosis Laboratory, Radiology and Imaging Sciences, National Institutes of Health Clinical Center, Bethesda, MD, United States

**Xinyue Wang** (305), School of Information Management, Wuhan University, Wuhan, PR China

**Kevin Bengtson Wong** (217), Data Science, University of San Francisco, San Francisco, CA, United States

**Diane Myung-Kyung Woodbridge** (217), Data Science, University of San Francisco, San Francisco, CA, United States

**Haiqing Xu** (305), Child Health Division, Department of Maternal and Child Health, Maternal and Child Hospital of Hubei Province, Tongji Medical College, Huazhong University of Science and Technology, Wuhan, PR China

**Wang Zhao** (305), Suzhou Zealikon Healthcare Co., Suzhou, PR China

**Hancan Zhu** (181), School of Mathematics Physics and Information, Shaoxing University, Shaoxing, China

# Editor's biography

Dr. Ahmed A. Moustafa is an associate professor in Cognitive and Behavioural Neuroscience at Marcs Institute for Brain, Behaviour, and Development & School of Psychology, Western Sydney University. Ahmed's h index is 41 and total number of citation is 8910 (Google Scholar). Ahmed is trained in computer science, psychology, neuroscience, and cognitive science. His early training took place at Cairo University in mathematics and computer science. Before joining Western Sydney University as a lab director, Ahmed spent 11 years in America studying psychology and neuroscience. Ahmed conducts research on computational and neuropsychological studies of addiction, schizophrenia, Parkinson's disease, PTSD, and depression. He has published over 200 papers in high-ranking journals including *Science*, *PNAS*, *Journal of Neuroscience*, *Brain, Neuroscience and Biobehavioral Reviews*, *Nature (Parkinson's disease)*, *Neuron*, among others. Ahmed has obtained grant funding from Australia, USA, Qatar, UAE, Turkey, and other countries. Ahmed has recently published four books: (a) *Computational Models of Brain and Behavior*, which provides a comprehensive overview of recent advances in the field of computational neuroscience; (b) *Social Cognition in Psychosis*; (c) *Computational Neuroscience Models of the Basal Ganglia*, which provides several models of the basal ganglia; and (d) *Cognitive, Clinical, and Neural Aspects of Drug Addiction*.

**Ahmed A. Moustafa**
Associate Professor, Cognitive and Behavioral Neuroscience,
Western Sydney University, Sydney, NSW, Australia &
Department of Human Anatomy and Physiology, the Faculty of Health Sciences,
University of Johannesburg, South Africa

# Preface

Several researchers, neurologists, psychiatrists, and policy makers have stressed the importance of (a) gathering large datasets on psychiatric and neurological disorders, (b) making these datasets available, and (c) using complex analytics to provide individualized treatments for these diseases. The goal of this book is to provide an up-to-date overview of what has been achieved so far in the field of *Big Data in Psychiatry and Medicine*, including applications of big data methods to psychiatric and neurological disorders. This book will further provide chapters that explain future work in this area.

This book is divided into two sections. In the first section *"Methods and big data analyses,"* several chapters discuss general analytic methods that have relevance to several big data approaches, including supervised machine learning and growth curve model approaches. Further, some of the chapters in this section discuss strategies to collect large datasets to be used for machine learning analytics. In the second section *"Examples of big data applications,"* several chapters provide examples of the applications of big data methods to neuroscience, public health, women's health, as well as clinical health. For example, some of the chapters focus on the applications of big data to neurosciences, such as understanding the hippocampus and visual brain. Other chapters focus on big data applications to health-related issues, including using big data to understand the emergence of psychoactive substances, diagnose and treat epilepsy, enable medication intake monitoring, help develop disease mitigation approaches, as well as provide medication intake monitoring, and provide techniques for disease mitigation. This section also includes a chapter that provides a precision medicine approach to help focus on women's health issues.

The contributors in this book come from interdisciplinary fields, such as medicine, clinical psychology, pharmacology, statistics, neuroinformatics, and computer science. The book is of help to researchers, students, and hospital staff to implement new methods to collect big datasets from several patient populations. Further, this book will explain how to use several algorithms and machine learning methods to analyze big datasets, and thus help provide individualized treatment for psychiatric and neurological patients.

# Acknowledgment

I would like to thank all chapter authors and contributors, including, but not limited to, Lidong Wang, Tim Hulsen, Sinéad Moylett, Sameh Basha, Nicola Amoroso, Manuel Gonçalves-Pinho, Robert Perdue, Dinggang Shen, Diane Woodbridge, Abhinav Choudhury, Varun Dutt, Seif Eldawlatly, Jason Lu, Asokan Vaithinathan, and Maryam Panahiazar. Thanks also to other contributors who were involved in writing the chapters. In particular, I like to thank Ben Schultz, Thierno Diallo, Areeg Abdalla, and Kohitij Kar for discussion on several topics covered in this book.

For all chapter contributors, I have learned a lot from every chapter you provided in this book. Thanks heaps for your efforts.

Chapter 1

# Best practices for supervised machine learning when examining biomarkers in clinical populations

Benjamin G. Schultz[a], Zaher Joukhadar[b], Usha Nattala[b], Maria del Mar Quiroga[b], Francesca Bolk[c], and Adam P. Vogel[a,d]

[a]*Centre for Neuroscience of Speech, The University of Melbourne, Melbourne, VIC, Australia,*
[b]*Melbourne Data Analytics Platform, The University of Melbourne, Melbourne, VIC, Australia,*
[c]*Murdoch Children's Research Institute, The University of Melbourne, Melbourne, VIC, Australia,*
[d]*Redenlab, Melbourne, VIC, Australia*

## 1 Introduction

Machine learning is a powerful tool for predicting outcomes as it can simultaneously consider multiple features to identify and delineate classes (e.g., healthy or unhealthy). This approach has several advantages over traditional univariate statistical approaches (see Bzdok, Altman, & Krzywinski, 2018; Bzdok & Meyer-Lindenberg, 2018), with broad scope for machine learning in health and medicine. Machine learning has been used to identify and distinguish medical conditions using genetic (cf. Libbrecht & Noble, 2015; Pattichis & Schizas, 1996), speech (cf. Hegde, Shetty, Rai, & Dodderi, 2019), neural (Craik, He, & Contreras-Vidal, 2019; Kassraian-Fard, Matthis, Balsters, Maathuis, & Wenderoth, 2016), imaging (Thrall et al., 2018), and movement (cf. Figueiredo, Santos, & Moreno, 2018; Kubota, Chen, & Little, 2016) data. There are several decisions made by researchers undertaking machine learning that are seldom explicitly reported (or known) including sample size estimation, variable screening and data reduction, selection of machine learning algorithm(s), selection of training and test datasets, and parameter adjustment. This chapter discusses current best practice for supervised machine learning applications in health and medicine, and describes some of the common pitfalls that researchers may encounter.

---

Big Data in Psychiatry and Neurology. https://doi.org/10.1016/B978-0-12-822884-5.00013-1
Copyright © 2021 Elsevier Inc. All rights reserved.

## 2 Data formatting

One of the most important aspects of machine learning, especially across multiple experiments, is data formatting. There are very few guidelines for how data should be formatted in this context and data sharing more generally. While data formatting may seem trivial to some, incorrect data formats can provide misleading results. Following standardized formatting conventions facilitates open data sharing and streamlines collaboration (Ellis & Leek, 2018). Correctly formatting data is important for machine learning in medicine where datasets typically consist of data collected from different sites and studies. Here we describe the best practices for data formatting that work across a range of different software with a focus on tidy data formats and the use of headers (or embedded metadata) for case-wise information (Chen, 2017; Ellis & Leek, 2018; Wickham, 2014; Wickham & Grolemund, 2016). Although there are other formatting styles, here we only discuss formats that are readable/importable for most statistical software that can perform machine learning (e.g., Python, R, SPSS, SAS, STATA).

---

**Best practice 1:**
*Use tidy data formats for machine learning and data sharing*

---

Tidy datasets have a specific structure where each column contains one variable, rows contain one observation, and tables contain one observational unit (e.g., sample, group, or experiment) (Wickham, 2014). There are several advantages of tidy datasets relating to standardized practices in data visualization, exploration, screening, transformations, and analysis. For example, Tables 1–3 show three ways to arrange the same data where Tables 1 and 2 are considered "messy data" and Table 3 is considered "Tidy data" (i.e., *long format*). To perform most machine learning applications for data in Tables 1 and 2, information would need to be combined into a format that can be read and interpreted by statistical software. In Table 1, data for each participant is in a different subtable. In order to perform any visualization or analysis, these tables would need to be combined using additional steps, such as melting and casting. Furthermore, note that each subtable contains *header information* that is not, strictly speaking, inside the table itself (i.e., ID and Group) and would require some scripting to extract automatically. Header information can be included in a multitude of different ways. Embedded metadata, often found within the file properties, can provide information that may require specialized scripts or functions to extract. Information may also be contained in the filename itself (e.g., Part01_control, Part02_control, Part03_disease, Part04_disease) and may require string-splitting to extract this information. In more complicated cases, this header information may be missing or formatted in different ways (e.g., control_Part01, Part02_control, Part03_take2_disease, disease_Part04), requiring manual recoding and/or idiosyncratic code to standardize the data

## Best practices for machine learning Chapter | 1    3

**TABLE 1** Example data separated by observational units (participants).

| ID: **Part 1** | | | ID: **Part 2** | | |
|---|---|---|---|---|---|
| **Group:** Control | | | **Group:** Control | | |
| | **Time1** | **Time2** | | **Time1** | **Time2** |
| **Task A** | 81 | 80 | **Task A** | 70 | 69 |
| **Task B** | 75 | 76 | **Task B** | 71 | 72 |
| **ID:** Part 3 | | | **ID:** Part 4 | | |
| **Group:** Disease | | | **Group:** Disease | | |
| | **Time1** | **Time2** | | **Time1** | **Time2** |
| **Task A** | 50 | 43 | **Task A** | 40 | 32 |
| **Task B** | 78 | 79 | **Task B** | 74 | 73 |

**TABLE 2** Example data in "Wide format" as required for within-subject designs for some statistical software (e.g., SPSS).

| ID | Group | TaskA_Time1 | TaskB_Time1 | TaskA_Time2 | TaskB_Time2 |
|---|---|---|---|---|---|
| Part01 | Control | 81 | 75 | 80 | 76 |
| Part02 | Control | 70 | 71 | 69 | 72 |
| Part03 | Disease | 50 | 78 | 43 | 79 |
| Part04 | Disease | 40 | 74 | 32 | 73 |

**TABLE 3** Example data in "long format" in line with "Tidy Data" recommendations (Wickham, 2014).

| (a) | id | Task | Time | Score | (b) | id | ID | Group |
|---|---|---|---|---|---|---|---|---|
| | 1 | A | 1 | 81 | | 1 | Part01 | Control |
| | 2 | A | 1 | 70 | | 2 | Part02 | Control |
| | 3 | A | 1 | 50 | | 3 | Part03 | Disease |
| | 4 | A | 1 | 40 | | 4 | Part04 | Disease |
| | 1 | B | 1 | 75 | | | | |
| | 2 | B | 1 | 71 | | | | |

*Continued*

**4** Big data in psychiatry and neurology

**TABLE 3** Example data in "long format" in line with "Tidy Data" recommendations (Wickham, 2014).—cont'd

| (a) | id | Task | Time | Score | (b) | id | ID | Group |
|-----|----|------|------|-------|-----|----|----|-------|
| | 3 | B | 1 | 78 | | | | |
| | 4 | B | 1 | 74 | | | | |
| | 1 | A | 2 | 80 | | | | |
| | 2 | A | 2 | 69 | | | | |
| | 3 | A | 2 | 43 | | | | |
| | 4 | A | 2 | 32 | | | | |
| | 1 | B | 2 | 76 | | | | |
| | 2 | B | 2 | 72 | | | | |
| | 3 | B | 2 | 79 | | | | |
| | 4 | B | 2 | 73 | | | | |

(e.g., sequences of "if this then that" conditional statements to fix deviant file naming conventions).

Table 2 is also considered "messy" data, which may be surprising to SPSS (or similar software) users where within-subject designs typically require this format (i.e., *wide format*). This can be considered "messy" because the same variable is scattered across different columns representing different conditions. This makes it difficult to perform certain analyses (e.g., checking for a normal distribution) across all conditions without first transforming the data into long format; the format does not imply that the four columns (TaskA_Time1, TaskB_Time1, TaskA_Time2, TaskB_Time2) contain the same variable under different conditions (Tasks A and B, Times 1 and 2). To perform machine learning, these data need to be explicitly defined as belonging to these conditions. Tidy Data is a standardized data structure that "maps the meaning of a dataset to its structure" (Wickham, 2014, p. 4) where the data structure informs the data interpreter of related categorical and continuous variables across conditions. Data for machine learning diverge from one rule of Tidy Data (Wickham, 2014); relational data on a single subject or participant that is used as a feature or the target category (or categories) being classified must be repeated instead of being placed in a separate table (see Table 4).

Best practices for machine learning **Chapter | 1** **5**

**TABLE 4** Example data in "long format" including relational data (i.e., Group, the target to be classified) that are necessary for machine learning.

| ID | Group | Task | Time | Score |
|---|---|---|---|---|
| Part01 | Control | A | 1 | 81 |
| Part02 | Control | A | 1 | 70 |
| Part03 | Disease | A | 1 | 50 |
| Part04 | Disease | A | 1 | 40 |
| Part01 | Control | B | 1 | 75 |
| Part02 | Control | B | 1 | 71 |
| Part03 | Disease | B | 1 | 78 |
| Part04 | Disease | B | 1 | 74 |
| Part01 | Control | A | 2 | 80 |
| Part02 | Control | A | 2 | 69 |
| Part03 | Disease | A | 2 | 43 |
| Part04 | Disease | A | 2 | 32 |
| Part01 | Control | B | 2 | 76 |
| Part02 | Control | B | 2 | 72 |
| Part03 | Disease | B | 2 | 79 |
| Part04 | Disease | B | 2 | 73 |

## 3  Statistical assumptions

Although machine learning is qualitatively different from other parametric tests and linear statistical methods, many approaches have similar statistical assumptions. Only when these assumptions are upheld do machine learning approaches provide reliable outcomes. For example, many data reduction techniques (e.g., principal components analysis) assume linear relationships between variables and deviations from linearity must be normally distributed, as per parametric assumptions. Some machine learning methods can be robust to mild violations of their statistical assumptions. Regardless of the machine learning method used, researchers must assess their model for gross violations of statistical assumptions as these violations can lead to biased, inaccurate, and unreliable predictions. Model assessments provide statistical support for the reliability of results and the appropriateness of the chosen machine learning approach. These assessments may also reveal latent unmodeled information from the data

**6  Big data in psychiatry and neurology**

that could be included in the machine learning model to improve the results. Linear regression is a basic machine learning method and is used to model linear relationships between a dependent variable (output, e.g., class or group) and multiple predictor variables (input features). Here we discuss statistical assumptions in the context of general linear regression models that also apply to other linear machine learning approaches.

The *assumption of linearity* requires that the relationship between the dependent variable and the fitted values of features is linear when the other variables are held constant (Ernst & Albers, 2017). The fitted values, also called the *predicted* values, are the outcomes that are generated when fitting the linear model. Deviations from linearity can undermine the model and render it unsuitable for the data. This can happen when the data contain hidden nonlinear relationships that are unmodeled. The assumption of linearity can be assessed by viewing plots of the observed by predicted values or the residuals by the predicted values; there should be a linear trend between the observed and expected probabilities. Violations of the assumption of linearity can sometimes be mitigated through transformations or normalization (e.g., logarithmic or exponential transforms) of the original input features but may require the use of nonlinear models.

The *assumption of homoscedasticity* (meaning "same variance") requires that the error (i.e., noise) of the relationship between the dependent variable and features is similar across all features. When this assumption is not upheld, the standard errors in the output are not reliable (Yang, Tu, & Chen, 2019). Homoscedasticity can be assessed by examining relationships between observed residuals and predicted values. If the plot shows a relationship or linear trend between residuals and features, then the linear model is not appropriate. Fig. 1 shows four examples of residual plots where Fig. 1A fulfills the assumption of linearity because residuals are randomly dispersed. Nonrandom residual dispersions, such as linear trends (Fig. 1B), U-shaped (Fig. 1C), or inverted U-shaped (Fig. 1D) patterns suggest nonlinear models are more appropriate for the data (e.g., nonlinear regression, see Bates & Watts, 1988). If the assumption of homoscedasticity is violated, heteroscedasticity-corrected errors can be calculated that account for the heteroscedasticity present in the model (see Hoechle, 2007).

The *assumption of independence* requires that values within features are not correlated or derived from the same source. For example, data collected from the same participant over time or in different tasks are connected. If these values are serially correlated, the estimates of their variances will not be reliable. These effects can be mitigated by specifying known dependencies in the model and accounting for these random effects using linear mixed-effects models, or similar (Bates & DebRoy, 2004; Bates & Watts, 1988).

The *assumption of normality* requires that errors are normally distributed. This is sometimes referred to as a weak assumption as, if it is violated, unreliable results will only be obtained if the dataset is small. This assumption does

Best practices for machine learning Chapter | 1  7

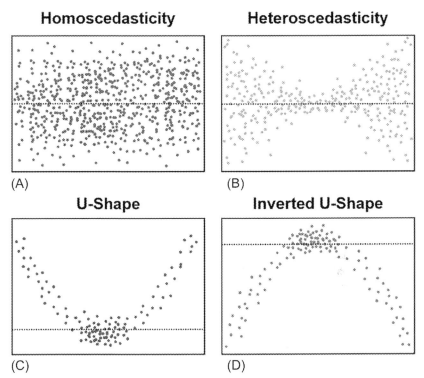

FIG. 1 Examples of residual dispersions that (A) fulfill the assumption of homoscedasticity or (B) violate this assumption through heteroscedasticity, or (C) nonlinear u-shaped or (D) inverted u-shaped distributions.

not need to be upheld when dealing with large datasets (Schmidt & Finan, 2018). Normality can be assessed through visual inspection of normal probability or normal quantile plots to ensure the distribution of errors has a linear relationship with the theoretical standardized residual. Alternatively, normality can be tested using the Anderson-Darling test, the Jarque-Bera test, the Kolmogorov-Smirnov test, or the Shapiro-Wilk test. To mitigate violations in normality, the sample size should be increased to more than 10 observations per variable (Schmidt & Finan, 2018).

The *assumption of linearly unrelated features* (i.e., *absence of multicollinearity*) assumes features are not highly correlated with one another. Multicollinearity prevents a model from accurately associating variance in the dependent variable with the most informative predictor variable(s) and may lead to incorrect interpretations of the model. One can detect multicollinearity by looking at correlations between the input variables and collinearity statistics (e.g., tolerance and variable inflation factor). Multicollinearity can be mitigated through the data reduction techniques discussed later in the chapter (see Section 6).

> **Best practice 2:**
> *Check statistical assumptions for machine learning processes and hidden steps*

Some machine learning approaches do not necessarily need to meet parametric assumptions (e.g., decision trees and neural networks; Breiman, 2001b; Brownlee, 2016). However, some software contain hidden data reduction steps prior to machine learning that still require parametric assumptions to be met (e.g., Gao et al., 2018). We recommend that researchers specify the assumptions required for their chosen approach. If the assumptions are unknown due to proprietary algorithms, then a conservative approach should meet parametric assumptions even if this is not strictly necessary. We further suggest that any procedures used to manipulate the data are reported.

## 4 Sample size estimation

Building supervised machine learning models that are both accurate and generalizable requires large datasets. Health data often contain a large number of variables (i.e., *high-dimensionality*) from a small number of participants (e.g., fewer than 40). It is challenging to collect high-quality (i.e., low-noise, artifact-free) data from large clinical cohorts; patient populations tend to experience fatigue more rapidly leading to poorer performance over time within testing sessions or missing data due to premature session termination. Most clinical studies are acquired from a single site (e.g., clinic or hospital), or the population is rare. There are several initiatives around the world aiming to link health data from multiple locations and experiments to create large corpora of data (*big data*). These initiatives assist in building more reliable machine learning models for health applications (Farinelli, Barcellos De Almeida, & Linhares De Souza, 2015). However, many health studies that use machine learning are still conducted using very small numbers of participants, some less than 10 (Halilaj et al., 2018).

A common question when performing machine learning is "what is an appropriate sample size?" Unfortunately, there is no single answer to this question as there are several factors to consider, including the number of features/variables relative to the number of cases. From a theoretical perspective, training a model with a larger feature set relative to sample size will lead to a more idiosyncratic model that may not accurately classify cases when applied to new data. This will not always eventuate in practice; some variables can hinder classification, some are redundant in the presence of other variables, and others may have a negligible effect on classification accuracy. As discussed later, it is important to assess the usefulness of variables within any model (machine learning or otherwise) and variables that harm the model (or do not help the model) should be discarded.

Sample size calculators have been developed for some machine learning approaches, such as predictive logistic regression (Hsieh, Bloch, & Larsen, 1998; Palazón-Bru, Folgado-De La Rosa, Cortés-Castell, López-Cascales, & Gil-Guillén, 2017). Some have suggested other rules like 10–20 samples per feature (Concato, Peduzzi, Holford, & Feinstein, 1995; Peduzzi, Concato, Feinstein, & Holford, 1995; Peduzzi, Concato, Kemper, Holford, & Feinstein, 1996), or a minimum sample size of between 100 (Palazón-Bru et al., 2017) and 500 (Bujang, Sa'at, & Bakar, 2018) regardless of the number of features. Although models with small-to-moderate sample sizes typically overestimate relationships between variables using categorical classification (i.e., the finite-sample bias; King & Zeng, 2001), even large samples of 150 cases per feature may not completely attenuate this bias (van Smeden et al., 2016). There are certain machine learning approaches that are better suited to small sample sizes (Sharma & Paliwal, 2015) and/or large feature sets that reduce data dimensionality (Jollife & Cadima, 2016) as discussed in the following sections. There are also dynamic resampling techniques that can estimate appropriate sample sizes for training sets and measure classification performance with different sample sizes that are not discussed here (see Byrd, Chin, Nocedal, & Wu, 2012; Figueroa, Zeng-Treitler, Kandula, & Ngo, 2012). However, there are no universally accepted practices for *a priori* determination of sample sizes with the exception that larger sample sizes provide less biased estimates (suggested $N = 75$–$100$, Beleites, Neugebauer, Bocklitz, Krafft, & Popp, 2013). Instead, we discuss the best practices to estimate error variance and model fit, and advocate for data sharing and transparency when reporting results and methods to increase replicability and meta analyses.

---

**Best practice 3:**
*Be cautious when interpreting results from small samples and use appropriate machine learning approaches*

---

## 5  Choosing parsimonious models

It can be difficult to choose a machine learning approach, especially when starting out in the field. Some researchers may choose an approach based on common practice in their field or by what is within their capabilities. This decision-making strategy is understandable given the range of options for performing machine learning. There is no catch-all solution in machine learning and every approach has advantages and disadvantages (i.e., the "no free lunch" theorem; Wolpert, 1996). However, given that machine learning provides a situation where a near-infinite number of variables can be used to discriminate between categories using various approaches, it is prudent to discuss the concept of parsimony in model selection (Vandekerckhove, Matzke, &

**10**  Big data in psychiatry and neurology

Wagenmakers, 2015). Parsimony follows the principle of Ockham's razor "Entities are not to be multiplied beyond necessity" (William of Ockham, c. 1287–1347).

When choosing the parameters and features that are "necessary" in machine learning, it is considered best practice to select the simplest model that best explains the data (Zellner, Keuzenkamp, & McAleer, 2001). There are some metrics that can inform how well a model explains the data relative to the model's complexity including Akaike's information criterion (AIC) and Bayesian information criterion (BIC). These metrics penalize more complex models (i.e., models with more variables and interactions) relative to the goodness of fit, to achieve a tradeoff between simplicity and information loss (Burnham & Anderson, 2004). Lower AIC or BIC values indicate a more parsimonious model relative to other models with the same (or similar) feature sets; BIC penalizes complexity more than AIC and a correction can be applied to the AIC for small sample sizes (AICc). The choice of whether to use AIC or BIC depends on whether the researcher favors information retention or simplicity, respectively. However, AIC and BIC are rarely used when selecting variables and model parameters for machine learning (but see Demyanov, Bailey, Ramamohanarao, & Leckie, 2012).

The absence of parsimony metrics means that researchers must pay attention when selecting/removing variables and adjusting model parameters to ensure that the model is generalizable (i.e., does not overfit the data) and not needlessly complex. We advocate that metrics of parsimony be considered when selecting variables and model parameters, and that these values are reported for baseline and saturated models, in addition to the final model. With parsimony in mind, there are several aspects of machine learning that impact model complexity and generalizability.

The flexibility of the learning algorithm is an important aspect of machine learning. In some instances, a machine learning approach might overfit the sample data and not generalize to other datasets or the population from which the sample was drawn. Most supervised machine learning approaches can be adjusted (automatically or through set parameters) to provide a tradeoff between bias and variance (e.g., GridSearch or Genetic Programming; Nagarajah & Poravi, 2019). This ensures that learning algorithms are not so flexible (i.e., low bias) that they produce a different fit for each training set (i.e., high variance). We recommend that, where applicable, these parameters are reported to improve replicability and so the reader can assess the complexity and level of bias of a model.

---

**Best practice 4:**
*Report adjustments to model parameters and any algorithms used to attain these values*

---

Another aspect to consider is the function complexity and whether there are complex interactions between the input variables. Complex models require more data for the training set and a learning algorithm with low bias. Bayesian variable elimination methods are sensitive to complex interactions. They are useful in identifying whether simple or complex functions exist within a dataset and whether they explain enough variance to be considered parsimonious (Zhang, 2016). It is difficult to predict the function complexity prior to performing machine learning in the absence of preestablished theoretical or data-driven models that describe relationships between features (e.g., path models with mediator variables). While this problem cannot be completely avoided, we advocate that variables be sufficiently described for machine learning applications and that their inclusion is justified even for exploratory analyses.

An additional issue when identifying the most parsimonious model is the number of variables or features. High-dimensional data does not necessarily guarantee a better outcome when using machine learning; some variables may confuse algorithms and decrease classification accuracy. Here we discuss several methods for feature selection/elimination that can identify the relevant and irrelevant features for classifying a target.

## 6  Reduction of data dimensionality

Applying machine learning methods to datasets with small sample sizes and high dimensionality without careful consideration can lead to overfitting. Overfitting occurs when models perform well when predicting the training samples but fail when introduced to new data (e.g., test data or new samples). When building supervised machine learning models, feature engineering and dimension reduction are crucial steps that mitigate the risk of overfitting while maintaining high levels of accuracy. Before applying machine learning methods, it is critical researchers use their domain knowledge and construct feature sets with a justified number of features to reduce the dimensions of data while retaining sufficient relevant information.

Speech data, for example, undergo signal processing to arrive at a set of acoustic features that, theoretically, represent the quality of speech and potential deficits in speech articulators (Noffs et al., 2020; Vogel et al., 2017; Vogel et al., 2020; Vogel, Fletcher, & Maruff, 2010). There are multiple parameters and algorithms that can be used to measure voice attributes, and a plethora of acoustic features that show potential for classifying diseases that affect speech (cf. Hegde et al., 2019). This creates the possibility of high-dimensional data that may contain redundant variables or overfit the data when using machine learning approaches. Similar instances of high dimensionality also occur for other types of data including fMRI (cf. Kassraian-Fard et al., 2016), EEG (cf. Craik et al., 2019; Sun & Zhou, 2014), movement and motion capture

# 12 Big data in psychiatry and neurology

(cf. Figueiredo et al., 2018; Kubota et al., 2016), and genetics (cf. Libbrecht & Noble, 2015). To remedy this, researchers can reduce data dimensions through variable selection and/or automated dimensionality reduction algorithms (Mares, Wang, & Guo, 2016).

## 6.1 Scaling

Before applying data reduction methods and (most) machine learning methods, variables should be scaled so all features have comparable ranges and distributions (Zheng & Casari, 2018). Scaling is particularly important for methods that use distance measures (e.g., Euclidean distance), such as linear regression, $k$-means, and $k$-nearest neighbors. In the absence of scaling, results may be unreliable. The most common scaling methods are $z$-scores (i.e., standard normalization) and range normalization so the values fall into a range between 0 and 1.

## 6.2 Variable selection

One way to reduce the number of variables in high-dimensional data is through variable selection and/or elimination. These methods examine covariation between variables to identify potential redundancies and discard variables that explain the least variance when distinguishing between target categories. For example, there are stepwise variable selection procedures based on parsimony metrics (AIC, BIC) that determine which variables account for the most variance and retain them and/or which variables account for the least variance and eliminate them (see Zhang, 2016). These procedures can also be performed with interactions between variables to determine complex relationships between features that may explain additional variance. However, increasing model parsimony may decrease classification accuracy due to the removal of variables that explain little variance relative to added model complexity.

Other machine learning methods, such as decision trees, may automatically ignore variables that are less informative either in isolation or in combination with other variables (e.g., *XGboost*; Gao et al., 2018). Software for decision trees often provide metrics for the variance explained by each variable. Researchers can decide upon an appropriate variance-explained threshold to remove a variable based on their situation and desired sensitivity and specificity rates. For example, a feature that reliably explains 1% of the variance may seem trivial but this 1% could be the difference between an accurate and false diagnosis (depending on the machine learning application). We recommend that variable selection procedures are reported including the selection methods, thresholds for retaining/removing variables, and the initial feature set with feature weights. This practice aims to increase transparency and inform future research that may use similar variables or features when classifying clinical groups. Accuracy outcomes from machine learning are driven by the sets and subsets of features so it is important that these outcomes are reproducible; describing the feature set is essential to achieving this goal.

---

**Best practice 5:**
*Report feature selection procedures and list variables that were retained or excluded*

---

### 6.3 Principal component analysis

Another way to reduce the number of variables in high-dimensional data is by identifying groups of features that are highly correlated and combining them to create a new variable. Principal component analysis (PCA) is a statistical method used to reduce the dimensionality of large feature sets by determining relationships between features. Dimensionality is reduced through the creation of new factors that are uncorrelated and capture the directions of maximal variance from the original features (see Fig. 2). The goal is to reduce the number of features while retaining as much variability as possible by finding lines of best fit within a feature set. PCA finds several orthogonal lines also known as principal components in $n$-dimensional space to the best fit of the data (James, Witten, Hastie, & Tibshirani, 2000). The first principal component represents the direction of maximum variance in the data, the second principal component has an orthogonal direction where the observations vary second most, and so on. PCA identifies directions of maximum variability to reduce the dimensionality into a smaller number of representative variables.

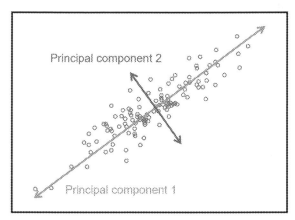

**FIG. 2** Example of principal component analysis (PCA) on simulated data. Principal component 1 (*orange line, lower left to upper right diagonal line* in print version) represents the direction of maximum variation and Principal component 2 (*blue line, upper left to lower right diagonal line* in print version) represents a second direction of maximum variation that is orthogonal to principal component 1.

**14** Big data in psychiatry and neurology

There are several pitfalls that should be avoided when using PCA. First, because PCA is a linear transformation, it assumes that relationships between variables are linear (see Jollife & Cadima, 2016). In most cases, however, linear relationships are only an approximation, especially for large feature sets. When it is known that correlations among variables are nonlinear, researchers should consider using nonlinear dimension reduction methods such as *manifold learning* (Nsimba & Levada, 2019). When using data reduction, it can be difficult to explain the resulting principal components without human intervention, that is, subjectively interpreting the overarching theme of the variables assigned to a component. As PCA assumes linear correlations among the variables, some variables might be assigned to a component where they seem "out of place." PCA cannot explain why a group of features contribute to a principal component (Lee, Lee, & Park, 2012). It is up to the researcher to assess the loadings of variables onto a specific component and assign a meaningful label for each component.

Another pitfall is that PCA is heavily influenced by outliers. Even small numbers of univariate and/or multivariate outliers can produce unreliable principal components. This is because PCA aims to minimize the quadratic norms that are sensitive to outliers that have high magnitudes (Xu, Caramanis, & Sanghavi, 2012). These issues can be attenuated by scaling variables, and identifying and removing outliers using standard statistical distance measures (e.g., Mahalanobis' distance, Cook's distance).

## 6.4 Linear discriminant analysis

Linear discriminant analysis (LDA) can be used to reduce the number of variables in high-dimensional data. LDA is similar to PCA in that it can be used on continuous variables, assumes data are normally distributed, and summarizes groups of features into composite features, with each original feature given a different weight. However, where PCA searches for directions of maximal variability in the data, LDA chooses weights based on the multidimensional space that can best discriminate between classes (i.e., groups; see Fig. 3). Because LDA uses the classes in its estimations, it is a *supervised* dimensionality reduction technique and can be used for machine learning classification problems without deferring to other data reduction or machine learning approaches.

For large feature sets relative to the size of the training set, a form of regularization called *shrinkage* can be used to improve generalization to unseen data (Ledoit & Wolf, 2004). Shrinkage makes as many of the weights as possible to be close to zero, effectively ignoring uninformative features. Benefits of LDA include providing reliable outcomes for small sample sizes, the ability to classify more than two classes, an absence of learning parameters (i.e., hyperparameters) to adjust, and generalizable outcomes in real-world settings (Sharma & Paliwal, 2015).

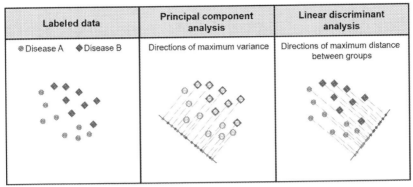

**FIG. 3** Example of LDA using a simulated dataset of patients with two different diseases (Disease A and B; *left panel*). PCA is blind to classes (indicated by a *gray* fill) and a one-dimensional PCA space is incapable of distinguishing the two groups *(middle panel)*. Projecting the data to the one-dimensional LDA subspace *(right panel)* leads to accurate classification of patients with Disease A and B.

## 7 Performance metrics

There are several different metrics that can be used to assess the performance of machine learning models. In general, "overall accuracy" is not as meaningful, particularly when the sample size differs between the target groups/events (as is often the case for uncommon diseases or seizures). Choosing the right metric is crucial when evaluating and comparing different machine learning models. No single metric is the best solution to assess every model and some applications may prefer the reduction of false alarms over increased hit rate (i.e., correct classifications). Depending on the characteristics of the dataset and the classes to be predicted, different metrics may emphasize different aspects of model performance. Researchers should choose the right metric(s) to ascertain whether a certain model is superior to another while being aware of the advantages and disadvantages. Regardless of the metric chosen, model performance should be assessed on data beyond the training dataset (e.g., test data or new unseen data; Kassraian-Fard et al., 2016).

---
**Best practice 6:**
*Consider multiple metrics for model performance*

---

For supervised machine learning classifiers, most metrics are derived from a confusion matrix, which is a tabular representation of the veridical observations against the model predictions (see Table 5). The values on the green *(light gray in print version)* diagonal indicate accurate identification (true positives) and correct rejection (true negatives). Some metrics are unreliable when the number of cases differs between classes (e.g., more healthy controls relative to patients)

## 16 Big data in psychiatry and neurology

**TABLE 5** Possible outcomes for a binary classifier.

| | | Observed | |
|---|---|---|---|
| | | **Yes** | **No** |
| **Predicted** | **Yes** | **True Positives (TP)** | **False Positives (FP)** |
| | **No** | **False Negatives (FN)** | **True Negatives (TN)** |

and there are more than two classes. Machine learning classifiers perform well when they can minimize error, that is, false positives and false negatives. It is difficult to minimize both types of error simultaneously; each error type has its own implications, and neither is a desirable result. A false negative could have severe implications where a person does not receive timely and potentially life-saving treatment or interventions. A false positive may misdiagnose a person who would receive costly and ineffective treatments and, in the worst-case scenario, may receive treatment that causes harm. Metrics that consider these error rates are, thus, an important consideration for any model performance.

*Overall accuracy* (i.e., "accuracy" in the field of machine learning) is one of the simplest metrics that can be calculated from the confusion matrix. It represents the proportion of correct classifications relative to the total number of classifications (see Eq. 1). When the sample sizes of classes are balanced, overall accuracy is a reasonable assessment of model performance for predicting a certain class. However, when classes have unequal sample sizes, performance of a majority class might overshadow that of a minority class or classes. Where the classes are markedly skewed, the performance of the majority class may produce high overall accuracy even if the minority classes were poorly predicted. This can be problematic for the identification of medical conditions because these conditions, by definition, fall into a minority class because they deviate from the population majority, that is, healthy controls. Medical diagnosis can be viewed as a form of anomaly detection where the positive class (i.e., disease or condition) forms a minority population. Therefore, classes are expected to be highly skewed with more controls than patients. A model may produce high overall accuracy but remain insensitive to the presence of the condition of interest. For these reasons, overall accuracy is a poor metric to assess model performance when classes are not uniformly distributed (Valverde-Albacete & Peláez-Moreno, 2014).

$$\text{Overall accuracy} = \frac{TP + TN}{TP + FP + FN + TN} \tag{1}$$

*Precision* measures the proportion of predicted positives that are positive cases relative to the sum of all true and false positives (see Eq. 2a). Although precision can be influenced by unbalanced samples sizes, there are corrections that can be applied based on d' (pronounced "dee-prime"; see Eq. 2b). These metrics aim to maximize the identification of positive cases while penalizing false positives. Precision is strongly affected by borderline cases and tradeoffs between liberal models (predicting more "positive" cases) and conservative models (predicting more "negative" cases) (Tharwat, 2018; Van Rijsbergen, 1979).

$$\text{Precision} = \frac{TP}{TP + FP} \tag{2a}$$

$$\text{Corrected Precision} = \frac{TP}{TP + FN} - \frac{FP}{TN + FP} \tag{2b}$$

*Sensitivity* (also called *Recall*) emphasizes the proportion of the true positives relative to the number of observed cases (see Eq. 3) and is used to measure a classifier's ability to accurately detect all cases that belong to a class. Sensitivity is penalized by false negatives and assesses how sensitive a model is to a specific condition or disease. However, sensitivity does not consider false positives that may become more frequent when a model tends to classify borderline cases as positive. A model with a bias to classify cases as positive would have high sensitivity but remain unable to identify and reject cases without the condition or disease. Thus, there needs to be a tradeoff between the acceptance of positive cases and the rejection of negative cases (Tharwat, 2018; Van Rijsbergen, 1979).

$$\text{Sensitivity} = \frac{TP}{TP + FN} \tag{3}$$

*Specificity* emphasizes the proportion of true negatives relative to the sum of all true negatives and false positives (see Eq. 4) and is used to measure a classifier's ability to identify individuals who do not belong to a specific class. While sensitivity measures a model's ability to accurately detect positive cases, specificity measures model performance in terms of rejecting negative cases. These two metrics are less affected by unbalanced designs and work together to describe model performance and tradeoffs between liberal and conservative estimates when accepting positive cases (Tharwat, 2018; Van Rijsbergen, 1979).

$$\text{Specificity} = \frac{TN}{TN + FP} \tag{4}$$

*Balanced accuracy* is a suitable *global metric* when classes have unequal sample sizes. It is based on the measure of sensitivity (see Eq. 3). Balanced accuracy is the mean sensitivity across all groups. When classes are uniformly

18    Big data in psychiatry and neurology

distributed, overall accuracy and balanced accuracy produce similar (or identical) values. For unbalanced distributions, the minority class(es) bias the metric toward a somewhat representative number (Brodersen, Ong, Stephan, & Buhmann, 2010; Kelleher, Mac Namee, & D'arcy, 2015).

The $F1$ *score* is the harmonic mean of precision and recall that equally weighs precision and recall (see Eq. 5a). To give more weight to either recall or precision, one can create a weighted $f_\beta$ *score* where beta ($\beta$) manages the tradeoff between precision and recall (see Eq. 5b). The default beta value for the $F1$ score is 1.0 which equally weights precision and recall. In some cases, however, both precision and recall are important, but slightly more emphasis on one or the other is desired. For example, when false negatives are deemed more detrimental to clinical outcomes than false positives, one can lower the beta value to give more weight to precision ($\beta < 1$, e.g., $\beta = 0.5$). Similarly, beta can be increased if the researcher prefers to avoid false positives ($\beta > 1$, e.g., $\beta = 2$) (Hand & Christen, 2018; Tharwat, 2018; Van Rijsbergen, 1979).

$$F1 = 2\frac{Precision \times Recall}{Precision + Recall} \tag{5a}$$

$$f_\beta = \left(1 + \beta^2\right)\frac{Precision \times Recall}{\left(\beta^2 \times Precision\right) + Recall} \tag{5b}$$

Precision, recall, $F1$, and $f_\beta$ metrics can be calculated in three different ways: Macro, weighted, and micro (see Van Asch, 2013 for equations). Macro calculations are the simplest and are computed as the mean score of all the class scores. Weighted calculations weight each class according to its sample size and then computes the mean score from these weighted scores. As macro and weighted calculations are averaged calculations, they are susceptible to the shortcomings of averaging, namely, outlier classes will bias the average and possibly obscure the model's true predictive capabilities for the other classes. Micro calculations examine all classes simultaneously to compute the model performance from the positives and negatives in the confusion matrix. Since micro calculations are not computed as an average but as a proportion, it is less biased when the classes have unbalanced sizes. We advocate that model performance for each class should be reported in addition to the overall model performance. This level of transparency allows readers to better evaluate model performance for specific conditions or groups.

---

**Best practice 7:**
*Report classification metrics for each class*

---

Although most machine learning models are assessed in terms of accuracy, sensitivity, specificity, precision, and $F1$ score, an interesting metric to consider for scenarios with unbalanced classes is *Matthew's Correlation Coefficient* (**MCC**). MCC considers all the values of a confusion matrix (TP, FP, TN, FN) to provide a single measure of model performance. MCC computes the correlation between the observed and predicted classes where higher correlations represent better predictions (see Eq. 6). MCC is a symmetric measure in that it is only high if the model predicts both true negatives and true positives correctly. In contrast, the $F1$ score is highly influenced by true and false positives and does not place any emphasis on the detection of true negatives. Medical machine learning applications are usually more concerned with identifying positive cases (e.g., people with a disease or condition). Therefore more interpretable combinations of precision and recall ($F1$ or $f_\beta$) may be a more suitable metric than MCC in certain cases (Boughorbel, Jarray, & El-Anbari, 2017; Chicco & Jurman, 2020; Matthews, 1975).

$$MCC = \frac{TP \times TN - FP \times FN}{\sqrt[2]{(TP+FP) \times (TP+FN) \times (TN+FP) \times (TN+FN)}} \tag{6}$$

## 8  Resampling methods

Before using a trained machine learning model to predict outcomes of unseen real-world data, the reliability of the model needs to be validated. There are several techniques based on resampling that can validate machine learning models and estimate the expected model performance (conversely, the prediction error) when deployed in real-world settings. Resampling is the process of repeatedly taking random (or pseudorandom) samples from a dataset and refitting a machine learning model on every sample. It allows us to evaluate model performance by estimating errors and ensuring model outcomes are not specific to one training and test set. Resampling also helps to achieve a bias-variance tradeoff and decide upon an appropriate level of model complexity.

The bias-variance tradeoff is the point at which adding further complexity decreases model performance across different resamples. *Bias* is the error introduced when using a simple model to describe a complex problem. Bias is measured as the average of all the differences between the true values and the predicted values. Simple models introduce some restrictions or preferences that make them generalize better but also make the difference between the true values and predicted values larger. For example, linear regression is restricted by the assumption that there are linear relationships between the dependent variable and predictor variables; this assumption is rarely fulfilled in ecological settings. These models are considered to have high bias because they are biased toward the restrictions they impose (Mitchell, 1980). *Variance* is the degree to which a model will vary due to small, potentially random, fluctuations in the training set. Ideally, models should not change drastically between resamples (or when encountering new data) in terms of performance metrics, model parameters, and underlying variables achieved through data reduction. Models with high variance are sensitive to small changes in the training set, resulting in increased variability of model performance across resamples. Generally, the more complex the model the higher the variance.

Bias and variance are two competing properties. Simple models have high bias and low variance. Models that have high bias miss important relationships between input and output variables (i.e., underfitting). Increasing the complexity of the model decreases the bias and increases the variance. Models that have high variance tend to *overlearn* small fluctuations and random noise in the training data (i.e., overfitting) leading to decreased accuracy when predicting new, unseen data. When applying machine learning methods on health data, researchers need to achieve an appropriate bias-variance tradeoff with minimal error (see Fig. 4). This can be achieved by measuring the prediction error (also called *test error*) resulting from resampling different training and test splits. Prediction error is the average error of classifying cases that were not used in the training process. Training error is the average error of predicting the model output of the same data used in the training process.

For most machine learning applications, a large dataset is ideal when building and evaluating a classification model (i.e., minimum $N = 75–100$; Beleites et al., 2013). However, this is often unattainable for most real-world problems especially in health research (see Section 4). Resampling techniques allow researchers to build reliable machine learning models when independent large test datasets are not available, by using different configurations of the available data to evaluate the model (Good, 2006). In the absence of an independent test dataset, the available dataset is split into training and test sets. The model learns from the training set and model performance is evaluated using the withheld test

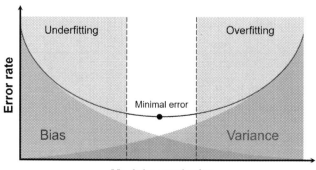

**FIG. 4** Example of tradeoff between bias (*blue shaded area, left dark shaded area* in print version) and variance (*red shaded area, right dark shaded area* in print version) tradeoff in a machine learning model. *Gray*-shaded regions show underfitting *(left light gray shaded area)* and overfitting *(right light gray shaded area)* and the optimal bias-variance tradeoff is the point of minimal error.

set. There are several resampling techniques that determine the way we split data into training and testing sets.

*Monte Carlo cross validation (MCCV)* is one of the simplest resampling methods. Data are randomly divided into a training set and a test set across different resamples. Typical splits range from 50:50 to 80:20 (Training: Validation) and it is often recommended that more cases are used in the training set than the validation set ( Janze, 2017). Some have proposed that a 67:33 split is most appropriate for machine learning approaches (Dobbin & Simon, 2011). The machine learning model is trained using the training set and is validated using the test set. The error across resamples provides an estimate of the prediction error in ecological settings. The simplicity of MCCV makes it an attractive choice for researchers.

Although MCCV has some advantages over the other resampling methods, there are also several drawbacks and pitfalls if not applied carefully, especially for health research. As discussed in previous sections, health datasets are relatively small compared to other big data. MCCV withholds a subset of data to validate the trained model. In the case of small sample sizes, the sizes of training and test subsets are heavily restricted, as are the number of resamples before testing all possible combinations of training and test sets. This may lead to inflation and/or high variability of prediction error across resamples. Although MCCV approaches could exhaustively resample across all possible configurations of training and test sets (i.e., *exhaustive cross validation*), this is rarely performed when validating models (i.e., *nonexhaustive cross validation*). Therefore, resampling may produce different outcomes for different sets and, potentially, select sets that do not accurately reflect the data through overestimation or underestimation of prediction error. We recommend that researchers use the largest number of resamples that is feasible for their sample size to ensure representative prediction error values are obtained. For small sample sizes, it is feasible to test all possible combinations of training and test sets. For big data, it would be prudent to resample until prediction error reaches an asymptote, that is, until the cumulative prediction error does not change substantially when new resamples are performed. Then, the mean and variability of model performance metrics should be reported as indices of model reliability.

---

**Best practice 8:**
*Use cross validation to evaluate models and report the mean and variability of model performance metrics across resamples*

---

A pitfall of the random resampling process relates to how cases are split into training and test sets. Health datasets often have longitudinal data of repeated observations from the same person. If data from an individual appear in both the training and test set, classification accuracy could be artificially increased based on characteristics of the individual person rather than characteristics of the condition or disease (Kassraian-Fard et al., 2016; Neto et al.,

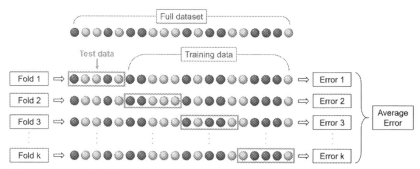

**FIG. 5** Example of k-fold cross validation. Each *circle* represents a case. Classes are distinguished by *dark and light shades inside circles*. Each fold selects a different test set, indicated by *rectangles around groups of circles. Circles outside the rectangles* represent the training sets.

2019). For this reason, researchers must be wary when selecting which cases are assigned to training and test sets. This potential pitfall is applicable to all cross validation methods.

*K-fold cross validation* is an alternative resampling method to MCCV in which data are divided into equally sized groups called *k*-folds. For each iteration, one of the folds is withheld as a test set, and the machine learning method is trained on the remaining folds. The prediction error is calculated *k* times, each using a different fold for the test set (Fig. 5). The final prediction error is the average of all prediction errors calculated across resamples. The main advantage of *k*-fold cross validation is the reduced computational power required when examining a subset of nonexhaustive resamples because the model only resamples *k* times. However, there is no universal agreement on the appropriate number of folds. Some have suggested specific calculations of *k* (Davison & Hinkley, 1997) or have used a fixed number of folds based on common practice in their field (Bengio & Grandvalet, 2004). Regardless of the method used to arrive at *k*, usually only a small number of resamples are performed, which may not accurately reflect model performance.

*Leave-one-out cross validation* splits the data into one set that has all observations but one. The model is trained on the larger set and tested on the single sample that was left out (Fig. 3). Each resample withholds a different case and trains the model on the set containing the remaining observations. This process is repeated for the number of cases. The total validation error is the average of all validation errors calculated across resamples. This method has some advantages compared to nonexhaustive MCCV. First, training can be performed on a larger set of observations which means the model is less likely to overestimate the actual prediction error. Second, there is no random element in assigning cases to training and test sets because each observation is systematically withheld and tested. The leave-one-out method is not recommended when working with health data, especially if data has repeated measures (Barrow & Crone, 2016). This is because there is a higher risk for the model to learn identity characteristics instead of class characteristics. Moreover, it allows outlier cases to inflate model performance in all resamples except the one in which they are withheld.

# 9 Data leakage

Supervised machine learning often serves to predict or classify new, unseen data for ecological (i.e., real-world) applications. Data leakage occurs when the model is trained using information that would not be available for new data in real-world applications. This may lead to superior performance for initial training and test datasets where all outcomes are known, but poor generalizations of the model when tested on unseen data. As previously discussed, the common approach is to separate the available data into *training* and *test* subsamples; the model learns from the training data and is evaluated based on its performance on the unseen test data. Researchers must ensure that information from the test subsample does not permeate (i.e., *leak*) into the training subsample. Data leakage can happen at multiple stages of data preparation and preprocessing.

---

**Best practice 9:**
*Identify confounding data contingencies in training and test sets*

---

## 9.1 Partitioning

When dividing data into training and test sets, researchers must ensure these sets are independent. For example, some problems include multiple data points belonging to the same *unit* or participant (Neto et al., 2019), such as measures of blood pressure from the same patient at different points in time. In these cases, all data belonging to the same patient or unit must be treated as a group, and either all be allocated to the training subsample, or all to the test subsample. Otherwise the model can learn to identify features or traits that belong to the patient but have no relationship to the underlying health condition in question.

## 9.2 Data reduction and scaling

Data preparation steps, such as scaling and dimensionality reduction, must be performed based on the training set only, and later applied as linear transformations to the test set for model evaluation (i.e., reporting performance metrics for a model). However, when building the *final* model for distribution, you may use your whole dataset to obtain the most robust model (Brownlee, 2016).

## 9.3 Variable selection

All features, or independent variables, used to train the model should be available prior to the occurrence of the predicted state/event. For example, consider

**24** Big data in psychiatry and neurology

building a model to identify a disease. The researcher may have access to a measure of disease severity and use this data to train the model. However, disease severity would only be measured in an individual once the disease has already been identified. Therefore, disease severity cannot be included in the training model because a measure of severity would not be available in the absence of a preestablished diagnosis. Including this variable would boost the performance of the model to levels that are unattainable when applying the model to new data. Other examples include age of death and event-dependent variables that cannot be determined prior to the occurrence of some future event.

## 9.4 Balancing datasets

Many applications of classification models refer to datasets that are heavily unbalanced in favor of events or classes that occur more often. For example, if 90% of your data comes from Class A and only 10% of your data from Class B, a very simple model that always predicts "Class A" will be correct 90% of the time. One approach that has been used to force the model to perform well in the classes that occur less often is to *oversample* data in the smaller classes. Oversampling randomly duplicates data in the smaller classes to achieve roughly equal numbers of data points in the different groups (Batista, Prati, & Monard, 2004; He & Garcia, 2009). Oversampling can only be done *after* splitting the dataset into training and test sets to ensure independence between training and test data (Vandewiele et al., 2020). Other approaches include *undersampling*, where data is deleted from the overrepresented class, and the creation of artificial data points (He & Garcia, 2009). Alternatively, some metrics (e.g., corrected precision) can be used to assess model performance for unbalanced datasets (see Section 7).

## 10 Supervised machine learning classifiers

Thus far, we have discussed practices that apply to a wide range of machine learning classifiers. There are several machine learning approaches available, each with specific advantages and disadvantages. For example, some approaches are more suitable for smaller samples sizes and/or large feature sets. This section briefly describes the most common classifiers for supervised machine learning and deep learning approaches.

*Logistic regression* is one of the most common machine learning tools and can be used to predict binary classes. Parametric assumptions must be met when using logistic regression and the classes must have a Bernoulli distribution (Dobson & Barnett, 2018). Although logistic regression is often used to define a baseline model compared to more sophisticated machine learning approaches, it can provide viable models for clinical outcomes (e.g., falls) when using a small set of features (see Gao et al., 2018). There are also regularized variants of logistic regression (e.g., Lasso-regularized and ridge logistic regression) that remove uninformative features and/or identify near-linear relationships between features (i.e., collinearity and multicollinearity; Hoerl & Kennard, 1970; Tibshirani, 1996). The primary disadvantage of logistic regression is that it may require large sample sizes to achieve reliable model performance, particularly in the presence of high-dimensional feature sets.

*Support vector machines (SVM)* aim to construct an optimal decision boundary, called a "hyperplane," that can distinguish classes. The primary advantage of SVMs is that they perform reliably for noisy and correlated feature sets with high dimensionality (Gao et al., 2018). There are also nonlinear SVM approaches using "kernel functions" that can be used if classes are not linearly separable in the original 2-dimensional feature space. Another advantage is that variants of SVM can also be used to predict classes with more than two outcomes (Crammer & Singer, 2001). For these reasons, SVMs have been widely used for neural imaging data (e.g., multivoxel pattern analysis; cf. Mahmoudi, Takerkart, Regragui, Boussaoud, & Brovelli, 2012). The main disadvantage of SVMs is that they require powerful computing resources, especially for big data and high-dimensional feature sets.

*Supervised Naïve Bayes* is a classification technique based on the Bayes theorem. It is *naïve* because it assumes that the effect of each feature on the outcome is independent of the other features (Lewis, 1998; Manning et al., 2012). Naïve Bayes classifiers are simple, fast, and can deal with high dimensional feature sets (Hand & Yu, 2001; Webb, Boughton, & Wang, 2005). A disadvantage of the Naïve Bayes approach is the assumption that the individual influence of each feature on the class is independent of other features. For example, when predicting vulnerability to lung disease one may look at several features including age, sex, medical history, smoking habits, weight, and socioeconomic status. Although we know that these features are interdependent in real life, Naïve Bayes incorrectly assumes these features are independent of each other and measures their influences on the class separately before combining these influences to predict outcomes. When the assumption of independence is upheld, Naïve Bayes can outperform comparable models and can use small training sets with reliable outcomes (Zhang, 2005).

*Decision trees* build a series of if-then-else structures to model the training data and use the same structure to predict unseen data (see Fig. 6; Kotsiantis, 2013; Podgorelec, Kokol, Stiglic, & Rozman, 2002; Quinlan, 1983). In line with other machine learning approaches, more complex decision rules result in models that are more prone to overfitting. To avoid overfitting, *pruning* processes are employed that discard tree *branches* that do not improve model performance. The smallest tree constructed from the entire dataset is selected as the optimal tree, thus, following the principle of parsimony. There are several advantages to decision tree approaches. First, they provide interpretable models that are easy to visualize as a flowchart of categorical decisions and assess the relative contribution of features in the decision process (see Fig. 6). Second, decision trees can be performed on datasets with missing feature data and

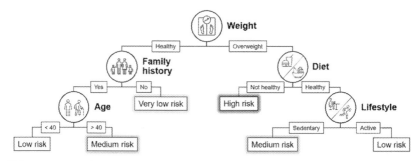

**FIG. 6** Example of a decision tree for high and low risk of diabetes (for illustration purposes only). *Circles* represent features, *lines* represent branches, and *colored rectangles* (*rectangles with gray shadowing* in print version) represent classes.

correlated and/or irrelevant input features. A disadvantage of decisions trees, however, is that they are prone to overfitting and can produce unstable results after small changes to the input features (Deng, Runger, & Tuv, 2011). For this reason, they are often recommended for medium-to-large sample sizes. Machine learning models built upon decision trees (and ensembles consisting of multiple decision trees) outperform other machine learning approaches (Gao et al., 2018). Variants of decision trees are described as follows.

*Random forests* construct an ensemble consisting of multiple decision trees based on different subsets of the training data. Each submodel is called a *weak tree learner* because each is only trained on a small subset of data. Weak tree learners are randomly resampled (called *bagging* or *bootstrap aggregation*) to find an optimal subset of trees that can be averaged over to find the model with the lowest variance. While individual trees might be sensitive to noise, the averaged model is robust to noise (Breiman, 2001a; Ho, 1995).

*Extreme Gradient Boosting (XGBoost)* is an optimized implementation for decision tree-based ensemble machine learning that uses *boosting* instead of random resampling (Chen & Guestrin, 2016; Nielsen, 2016). Boosting iteratively builds weak tree learners and places higher weights on instances with large errors. Thus the model is trained to focus on cases that are difficult to predict so it can learn from previous classification errors. The implementation of XGBoost is further improved by hardware optimizations such as *parallelization*, as well as *caching* and *memory optimization* to deliver fast computing performance. XGBoost has demonstrated better model performance than other decision tree machine learning approaches with performance comparable to more computationally demanding approaches (Gao et al., 2018).

## 11 Deep learning and artificial intelligence

*Deep learning* is a subfield of machine learning based on artificial intelligence (e.g., artificial neural networks; ANNs) (Esteva et al., 2019; Lecun, Bengio, & Hinton, 2015). It aims to mimic biological learning systems in the form of multiple connected *hidden layers* that contain complex interactions between the

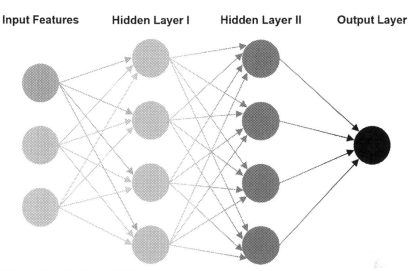

**FIG. 7** Example of an artificial neural network. *Circles* represent nodes and *arrows* represent pathways from the input features to the output layer.

input features that lead to the *output layer* (see Fig. 7). Input features that strongly predict classes are given higher weights and strengthened, whereas uninformative features are progressively attenuated. Each layer the data passes through adjusts the weights of the individual features and combines multiple features to progressively transform the inputs into the output layer. Given enough data, deep learning can find the correct mathematical transformations to represent the linear or nonlinear relationships between the input feature(s) and the output (e.g., classes). Although deep learning approaches perform well for large datasets, they can provide unreliable results for small sample sizes due to overfitting (Cho, Lee, Shin, Choy, & Do, 2015). The layers of abstraction can learn complex relationships between features but, in small datasets, sources of noise may drive the model leading to overfitting and poor generalizability. To prevent overfitting, *dropout* techniques are used. Dropout is a regularization technique in which a percentage of a layer's nodes is ignored in a training round (Srivastava, Hinton, Krizhevsky, Sutskever, & Salakhutdinov, 2014). This disrupts the codependencies between different nodes to identify spurious codependencies that may have arisen due to noise. Dropout, therefore, is used to prevent overfitting by preventing nodes from coadapting to noise.

ANNs are recommended for processing large unstructured datasets (e.g., vast libraries of text, images, and audio) such as those involving image recognition, natural language processing, and pattern mining in bioinformatics. The main disadvantage is that ANNs require significant computational power. ANNs also require fine tuning of various parameters, such as the number of layers and units per layer, with final values derived from trial and error and unspecified decision strategies. This makes it difficult to interpret the properties of the parameters and feature combinations that lead to model outcomes without

28    Big data in psychiatry and neurology

computationally intensive processes like dynamic programming or iterative data perturbations (Ribeiro, Singh, & Guestrin, 2016). In other words, although deep learning may produce a model that outperforms other machine learning approaches, the paths to achieving those outcomes are rarely explained.

---

**Best practice 10:**
*Interpret machine learning outcomes by considering the contribution of individual features and interactions between features*

---

## 12    Limitations and future directions

Machine learning can be incredibly sensitive to features within datasets that are conceptually unrelated to the desired classification outcomes. For example, a dataset may contain data collected from different experiments where the protocol deviated in very slight ways. This may not be problematic for most situations but if Protocol A was used for the majority of one class and Protocol B was used for the majority of a second class, then classes might be distinguished through machine learning solely based on the underlying protocols. Currently, there are few ways to account for these random factors in machine learning. A method that has been proposed for within-subject factors and random effects for binary classification is the generalized mixed effects regression tree (GMERT; Hajjem, Larocque, & Bellavance, 2017). This model is suitable for binary outcomes and count data, and can handle unbalanced clusters but has not yet been widely adopted for machine learning in health applications (but see Beltempo, Bresson, & Lacroix, 2020). Given the high degree of noise in patient data, and the fact that patient data is often collected longitudinally, machine learning approaches that can account for random effects and nested designs are a promising avenue for more reliable outcomes.

The breadth of machine learning options leaves room for analyzing the same dataset using different approaches and data reduction techniques. We advocate that, where possible, data should be shared so researchers can optimize machine learning approaches. By following standard practices in data formatting and reporting all steps of the machine learning process, the scientific community could replicate results with new data or combined datasets from different experiments and/or laboratories. This would also remedy the limitations of small samples that may affect the reliability of machine learning outcomes when using data from a single experiment.

## 13    Conclusions

In this chapter, we presented several supervised machine learning approaches and recommended best practices that should be considered for most applications. We recommend that researchers use Tidy Data formatting (Wickham, 2014) to facilitate data sharing and ensure data is interpreted correctly by machine learning software (Best practice 1). When using machine learning approaches, researchers should check the statistical assumptions required by the chosen model (Best practice 2). If the assumptions are unknown, parametric assumptions should be satisfied to ensure results are reliable. In the case of small sample sizes ($N < 75–100$, Beleites et al., 2013), results should be interpreted with caution and researchers should choose appropriate machine learning approaches (e.g., LDA; Best practice 3). Adjustments to model parameters and algorithms for automatic parameter selection should be described (Best practice 4). To achieve parsimony, dimensionality of large feature sets should be reduced using data reduction approaches, such as feature elimination or component extraction (e.g., PCA); all features (and components) considered in the initial analyses and final models should be reported (Best practice 5).

When evaluating a model, multiple metrics for model performance should be considered based on whether group sizes are balanced, and preferences for liberal (more false positives) or conservative (more false negatives) classification (Best practice 6). Where possible, classification metrics should be reported for each class to assess if some classes are easier to identify relative to others (Best practice 7). Moreover, cross validation techniques should be used to ensure outcomes are reliable and to estimate the margin of error for real-world deployment (Best practice 8). As cross validation is not immune to confounding data contingencies, such as the role of identity for repeated measures (Neto et al., 2019), we recommend that researchers ensure that training and test sets do not contain the same participant (Best practice 9). Finally, we suggest that researchers interpret the final model and consider how features, and interactions thereof, contribute to distinguishing classes (Best practice 10). This might mean using simpler machine learning approaches that are interpretable over more complicated approaches (e.g., deep learning).

These best practices aim to facilitate transparency, data sharing, replicability, and interpretability of machine learning outcomes. Adopting these practices should lead to better understanding between health professionals and data scientists, and facilitate research that has the potential to predict health outcomes. We also discussed some of the common pitfalls and challenges that are encountered during the machine learning process, including appropriate sample sizes and the potential to overfit data. Although these pitfalls are sometimes unavoidable, we suggest several methods of cross validation that can be used to estimate the margin of error. Through careful validation techniques, the reliability of machine learning outcomes can be ascertained for a broad range of health applications that use big data.

# References

Barrow, D. K., & Crone, S. F. (2016). Cross-validation aggregation for combining autoregressive neural network forecasts. *International Journal of Forecasting, 32*(4), 1120–1137.

Bates, D. M., & DebRoy, S. (2004). Linear mixed models and penalized least squares. *Journal of Multivariate Analysis, 91*(1), 1–17.

**30** Big data in psychiatry and neurology

Bates, D. M., & Watts, D. G. (1988). Nonlinear regression analysis and its applications. *1988*. John Wiley & Sons, Inc.

Batista, G. E. A. P. A., Prati, R. C., & Monard, M. C. (2004). A study of the behavior of several methods for balancing machine learning training data. *SIGKDD Explorations Newsletter, 6* (1), 20–29. https://doi.org/10.1145/1007730.1007735.

Beleites, C., Neugebauer, U., Bocklitz, T., Krafft, C., & Popp, J. (2013). Sample size planning for classification models. *Analytica Chimica Acta, 760*, 25–33.

Beltempo, M., Bresson, G., & Lacroix, G. (2020). Using machine learning to predict nosocomial infections and medical accidents in a NICU. *Institute for the Study of Labor discussion paper no. 13099*. Available at SSRN:. https://ssrn.com/abstract=3568304.

Bengio, Y., & Grandvalet, Y. (2004). No unbiased estimator of the variance of k-fold cross-validation. *Journal of Machine Learning Research*, 1089–1105. 5.

Boughorbel, S., Jarray, F., & El-Anbari, M. (2017). Optimal classifier for imbalanced data using Matthews Correlation Coefficient metric. *PLoS One*. https://doi.org/10.1371/journal.pone.0177678.

Breiman, L. (2001a). Random forests. *Machine Learning*. https://doi.org/10.1023/A:1010933404324.

Breiman, L. (2001b). Statistical modeling: The two cultures (with comments and a rejoinder by the author). *Statistical Science, 16*(3), 199–231.

Brodersen, K. H., Ong, C. S., Stephan, K. E., & Buhmann, J. M. (2010). The balanced accuracy and its posterior distribution. In *2010 20th International conference on pattern recognition*. https://doi.org/10.1109/ICPR.2010.764.

Brownlee, J. (2016). *Machine learning mastery with python* (pp. 100–120). Machine Learning Mastery Pty Ltd.

Bujang, M. A., Sa'at, N., & Bakar, T. M. I. T. A. (2018). Sample size guidelines for logistic regression from observational studies with large population: Emphasis on the accuracy between statistics and parameters based on real life clinical data. *The Malaysian Journal of Medical Sciences, 25*(4), 122.

Burnham, K. P., & Anderson, D. R. (2004). Multimodel inference: Understanding AIC and BIC in model selection. *Sociological Methods & Research, 33*(2), 261–304.

Byrd, R. H., Chin, G. M., Nocedal, J., & Wu, Y. (2012). Sample size selection in optimization methods for machine learning. *Mathematical Programming, 134*(1), 127–155.

Bzdok, D., Altman, N., & Krzywinski, M. (2018). *Points of significance: Statistics versus machine learning*. Nature Publishing Group.

Bzdok, D., & Meyer-Lindenberg, A. (2018). Machine learning for precision psychiatry: Opportunities and challenges. *Biological Psychiatry: Cognitive Neuroscience and Neuroimaging, 3*(3), 223–230. https://doi.org/10.1016/j.bpsc.2017.11.007.

Chen, D. Y. (2017). *Pandas for everyone: Python data analysis*. Addison-Wesley Professional.

Chen, T., & Guestrin, C. (2016). XGBoost. https://doi.org/10.1145/2939672.2939785.

Chicco, D., & Jurman, G. (2020). The advantages of the Matthews correlation coefficient (MCC) over F1 score and accuracy in binary classification evaluation. *BMC Genomics, 21*(1), 6.

Cho, J., Lee, K., Shin, E., Choy, G., & Do, S. (2015). *How much data is needed to train a medical image deep learning system to achieve necessary high accuracy?*. ArXiv Preprint ArXiv:1511.06348.

Concato, J., Peduzzi, P., Holford, T. R., & Feinstein, A. R. (1995). Importance of events per independent variable in proportional hazards analysis I. Background, goals, and general strategy. *Journal of Clinical Epidemiology, 48*(12), 1495–1501.

Craik, A., He, Y., & Contreras-Vidal, J. L. (2019). Deep learning for electroencephalogram (EEG) classification tasks: A review. *Journal of Neural Engineering, 16*(3), 31001.

Crammer, K., & Singer, Y. (2001). On the algorithmic implementation of multiclass kernel-based vector machines. *Journal of Machine Learning Research, 2*(Dec), 265–292.

Davison, A. C., & Hinkley, D. V. (1997). *Bootstrap methods and their application.* Issue 1 Cambridge University Press.

Demyanov, S., Bailey, J., Ramamohanarao, K., & Leckie, C. (2012). AIC and BIC based approaches for SVM parameter value estimation with RBF kernels. In *Asian conference on machine learning* (pp. 97–112).

Deng, H., Runger, G., & Tuv, E. (2011). Bias of importance measures for multi-valued attributes and solutions. In *Lecture notes in computer science (including subseries lecture notes in artificial intelligence and lecture notes in bioinformatics).* https://doi.org/10.1007/978-3-642-21738-8_38.

Dobbin, K. K., & Simon, R. M. (2011). Optimally splitting cases for training and testing high dimensional classifiers. *BMC Medical Genomics, 4*(1), 31.

Dobson, A. J., & Barnett, A. G. (2018). *An introduction to generalized linear models.* CRC Press.

Ellis, S. E., & Leek, J. T. (2018). How to share data for collaboration. *The American Statistician, 72*(1), 53–57.

Ernst, A. F., & Albers, C. J. (2017). Regression assumptions in clinical psychology research practice-a systematic review of common misconceptions. *PeerJ.* https://doi.org/10.7717/peerj.3323.

Esteva, A., Robicquet, A., Ramsundar, B., Kuleshov, V., DePristo, M., Chou, K., et al. (2019). A guide to deep learning in healthcare. *Nature Medicine.* https://doi.org/10.1038/s41591-018-0316-z.

Farinelli, F., Barcellos De Almeida, M., & Linhares De Souza, Y. (2015). Linked health data: How linked data can help provide better health decisions. *Studies in Health Technology and Informatics.* https://doi.org/10.3233/978-1-61499-564-7-1122.

Figueiredo, J., Santos, C. P., & Moreno, J. C. (2018). Automatic recognition of gait patterns in human motor disorders using machine learning: A review. *Medical Engineering & Physics, 53,* 1–12.

Figueroa, R. L., Zeng-Treitler, Q., Kandula, S., & Ngo, L. H. (2012). Predicting sample size required for classification performance. *BMC Medical Informatics and Decision Making, 12*(1), 8.

Gao, C., Sun, H., Wang, T., Tang, M., Bohnen, N. I., Müller, M. L. T. M., et al. (2018). Model-based and model-free machine learning techniques for diagnostic prediction and classification of clinical outcomes in Parkinson's disease. *Scientific Reports, 8*(1), 1–21. https://doi.org/10.1038/s41598-018-24783-4.

Good, P. I. (2006). *Resampling methods: A practical guide to data analysis.* https://doi.org/10.1007/0-8176-4444-X.

Hajjem, A., Larocque, D., & Bellavance, F. (2017). Generalized mixed effects regression trees. *Statistics & Probability Letters, 126,* 114–118. https://doi.org/10.1016/j.spl.2017.02.033.

Halilaj, E., Rajagopal, A., Fiterau, M., Hicks, J. L., Hastie, T. J., & Delp, S. L. (2018). Machine learning in human movement biomechanics: Best practices, common pitfalls, and new opportunities. *Journal of Biomechanics.* https://doi.org/10.1016/j.jbiomech.2018.09.009.

Hand, D., & Christen, P. (2018). A note on using the F-measure for evaluating record linkage algorithms. *Statistics and Computing.* https://doi.org/10.1007/s11222-017-9746-6.

Hand, D. J., & Yu, K. (2001). Idiot's Bayes—Not so stupid after all? *International Statistical Review.* https://doi.org/10.1111/j.1751-5823.2001.tb00465.x.

He, H., & Garcia, E. A. (2009). Learning from imbalanced data. *IEEE Transactions on Knowledge and Data Engineering, 21*(9), 1263–1284. https://doi.org/10.1109/TKDE.2008.239.

Hegde, S., Shetty, S., Rai, S., & Dodderi, T. (2019). A survey on machine learning approaches for automatic detection of voice disorders. *Journal of Voice, 33*(6), 947.e11–947.e33. https://doi.org/10.1016/j.jvoice.2018.07.014.

## 32 Big data in psychiatry and neurology

Ho, T. K. (1995). Random decision forests. In *Proceedings of the international conference on document analysis and recognition, ICDAR*. https://doi.org/10.1109/ICDAR.1995.598994.

Hoechle, D. (2007). Robust standard errors for panel regressions with cross-sectional dependence. *The Stata Journal, 7*(3), 281–312.

Hoerl, A. E., & Kennard, R. W. (1970). Ridge regression: Biased estimation for nonorthogonal problems. *Technometrics, 12*(1), 55–67. https://doi.org/10.1080/00401706.1970.10488634.

Hsieh, F. Y., Bloch, D. A., & Larsen, M. D. (1998). A simple method of sample size calculation for linear and logistic regression. *Statistics in Medicine, 17*(14), 1623–1634.

James, G., Witten, D., Hastie, T., & Tibshirani, R. (2000). An introduction to statistical learning. *Current Medicinal Chemistry*. https://doi.org/10.1007/978-1-4614-7138-7.

Janze, C. (2017). Shedding light on the role of sample sizes and splitting proportions in out-of-sample tests: A Monte Carlo cross-validation approach. *Atas Da Conferência Da Associação Portuguesa de Sistemas de Informação, 17*(17), 245–259.

Jollife, I. T., & Cadima, J. (2016). Principal component analysis: A review and recent developments. *Philosophical Transactions of the Royal Society A: Mathematical, Physical and Engineering Sciences, 374*(2065). https://doi.org/10.1098/rsta.2015.0202.

Kassraian-Fard, P., Matthis, C., Balsters, J. H., Maathuis, M. H., & Wenderoth, N. (2016). Promises, pitfalls, and basic guidelines for applying machine learning classifiers to psychiatric imaging data, with autism as an example. *Frontiers in Psychiatry, 7*(Dec). https://doi.org/10.3389/fpsyt.2016.00177.

Kelleher, J. D., Mac Namee, B., & D'arcy, A. (2015). *Fundamentals of machine learning for predictive data analytics: Algorithms, worked examples, and case studies*. MIT Press.

King, G., & Zeng, L. (2001). Logistic regression in rare events data. *Political Analysis, 9*(2), 137–163.

Kotsiantis, S. B. (2013). Decision trees: A recent overview. *Artificial Intelligence Review*. https://doi.org/10.1007/s10462-011-9272-4.

Kubota, K. J., Chen, J. A., & Little, M. A. (2016). Machine learning for large-scale wearable sensor data in Parkinson's disease: Concepts, promises, pitfalls, and futures. *Movement Disorders, 31*(9), 1314–1326.

Lecun, Y., Bengio, Y., & Hinton, G. (2015). Deep learning. *Nature*. https://doi.org/10.1038/nature14539.

Ledoit, O., & Wolf, M. (2004). Honey, I shrunk the sample covariance matrix. *The Journal of Portfolio Management, 30*(4), 110–119. https://doi.org/10.3905/jpm.2004.110.

Lee, Y. K., Lee, E. R., & Park, B. U. (2012). Principal component analysis in very high-dimensional spaces. *Statistica Sinica*. https://doi.org/10.5705/ss.2010.149.

Lewis, D. D. (1998). Naive(Bayes) at forty: The independence assumption in information retrieval. In *Lecture notes in computer science (including subseries lecture notes in artificial intelligence and lecture notes in bioinformatics)*. https://doi.org/10.1007/bfb0026666.

Libbrecht, M. W., & Noble, W. S. (2015). Machine learning applications in genetics and genomics. *Nature Reviews Genetics, 16*(6), 321–332.

Mahmoudi, A., Takerkart, S., Regragui, F., Boussaoud, D., & Brovelli, A. (2012). Multivoxel pattern analysis for FMRI data: A review. *Computational and Mathematical Methods in Medicine, 2012*, 961257.

Manning, C. D., Raghavan, P., Schutze, H., Manning, C. D., Raghavan, P., & Schutze, H. (2012). Text classification and naive Bayes. In *Introduction to information retrieval*. https://doi.org/10.1017/cbo9780511809071.014.

Mares, M. A., Wang, S., & Guo, Y. (2016). Combining multiple feature selection methods and deep learning for high-dimensional data. *Transactions on Machine Learning and Data Mining, 9*(1), 27–45.

Matthews, B. W. (1975). Comparison of the predicted and observed secondary structure of T4 phage lysozyme. *Biochimica et Biophysica Acta (BBA)—Protein Structure*. https://doi.org/10.1016/0005-2795(75)90109-9.

Best practices for machine learning **Chapter | 1** **33**

Mitchell, T. M. (1980). The need for biases in learning generalizations. In *Readings in machine learning*.

Nagarajah, T., & Poravi, G. (2019). A review on automated machine learning (AutoML) systems. In *2019 IEEE 5th international conference for convergence in technology, I2CT 2019*. https://doi.org/10.1109/I2CT45611.2019.9033810.

Neto, E. C., Pratap, A., Perumal, T. M., Tummalacherla, M., Snyder, P., Bot, B. M., et al. (2019). Detecting the impact of subject characteristics on machine learning-based diagnostic applications. *NPJ Digital Medicine, 2*(1), 1–6. https://doi.org/10.1038/s41746-019-0178-x.

Nielsen, D. (2016). *Tree boosting with XGBoost—Why does XGBoost win "every" machine learning competition?*. Norwegian University of Science and Technology. Master's thesis.

Noffs, G., Boonstra, F. M. C., Perera, T., Kolbe, S. C., Stankovich, J., Butzkueven, H., et al. (2020). Acoustic speech analytics are predictive of cerebellar dysfunction in multiple sclerosis. *The Cerebellum, 19*, 1–10.

Nsimba, C. B., & Levada, A. L. M. (2019). Nonlinear dimensionality reduction in texture classification: is manifold learning better than PCA? In *Lecture notes in computer science (including subseries lecture notes in artificial intelligence and lecture notes in bioinformatics)*. https://doi.org/10.1007/978-3-030-22750-0_15.

Palazón-Bru, A., Folgado-De La Rosa, D. M., Cortés-Castell, E., López-Cascales, M. T., & Gil-Guillén, V. F. (2017). Sample size calculation to externally validate scoring systems based on logistic regression models. *PLoS One, 12*(5), 1–11. https://doi.org/10.1371/journal.pone.0176726.

Pattichis, C. S., & Schizas, C. N. (1996). Genetics-based machine learning for the assessment of certain neuromuscular disorders. *IEEE Transactions on Neural Networks, 7*(2), 427–439.

Peduzzi, P., Concato, J., Feinstein, A. R., & Holford, T. R. (1995). Importance of events per independent variable in proportional hazards regression analysis II. Accuracy and precision of regression estimates. *Journal of Clinical Epidemiology, 48*(12), 1503–1510.

Peduzzi, P., Concato, J., Kemper, E., Holford, T. R., & Feinstein, A. R. (1996). A simulation study of the number of events per variable in logistic regression analysis. *Journal of Clinical Epidemiology, 49*(12), 1373–1379.

Podgorelec, V., Kokol, P., Stiglic, B., & Rozman, I. (2002). Decision trees: An overview and their use in medicine. *Journal of Medical Systems*. https://doi.org/10.1023/A:1016409317640.

Quinlan, J. R. (1983). Learning efficient classification procedures and their application to chess end games. *Machine Learning*. https://doi.org/10.1007/978-3-662-12405-5_15.

Ribeiro, M. T., Singh, S., & Guestrin, C. (2016). "Why should I trust you?" Explaining the predictions of any classifier. In *Proceedings of the 22nd ACM SIGKDD international conference on knowledge discovery and data mining, Augus* (pp. 1135–1144).

Schmidt, A. F., & Finan, C. (2018). Linear regression and the normality assumption. *Journal of Clinical Epidemiology*. https://doi.org/10.1016/j.jclinepi.2017.12.006.

Sharma, A., & Paliwal, K. K. (2015). Linear discriminant analysis for the small sample size problem: An overview. *International Journal of Machine Learning and Cybernetics, 6*(3), 443–454.

Srivastava, N., Hinton, G., Krizhevsky, A., Sutskever, I., & Salakhutdinov, R. (2014). Dropout: A simple way to prevent neural networks from overfitting. *Journal of Machine Learning Research, 15*, 1929–1958.

Sun, S., & Zhou, J. (2014). A review of adaptive feature extraction and classification methods for EEG-based brain-computer interfaces. In *2014 International joint conference on neural networks (IJCNN)* (pp. 1746–1753).

Tharwat, A. (2018). Classification assessment methods. *Applied Computing and Informatics*. https://doi.org/10.1016/j.aci.2018.08.003.

Thrall, J. H., Li, X., Li, Q., Cruz, C., Do, S., Dreyer, K., et al. (2018). Artificial intelligence and machine learning in radiology: Opportunities, challenges, pitfalls, and criteria for success. *Journal of the American College of Radiology, 15*(3), 504–508.

**34** Big data in psychiatry and neurology

Tibshirani, R. (1996). Regression shrinkage and selection via the lasso. *Journal of the Royal Statistical Society: Series B: Methodological, 58*(1), 267–288. http://www.jstor.org/stable/2346178.

Valverde-Albacete, F. J., & Peláez-Moreno, C. (2014). 100% classification accuracy considered harmful: The normalized information transfer factor explains the accuracy paradox. *PLoS One*. https://doi.org/10.1371/journal.pone.0084217.

Van Asch, V. (2013). *Macro-and micro-averaged evaluation measures*. Belgium: CLiPS.

Van Rijsbergen, C. J. (1979). *Information retrieval* (2nd ed.). Butterworths.

van Smeden, M., de Groot, J. A. H., Moons, K. G. M., Collins, G. S., Altman, D. G., Eijkemans, M. J. C., et al. (2016). No rationale for 1 variable per 10 events criterion for binary logistic regression analysis. *BMC Medical Research Methodology, 16*(1), 163.

Vandekerckhove, J., Matzke, D., & Wagenmakers, E.-J. (2015). Model comparison and the principle of parsimony. In *Oxford handbook of computational and mathematical psychology* (pp. 300–319). Oxford University Press.

Vandewiele, G., Dehaene, I., Kovács, G., Sterckx, L., Janssens, O., Ongenae, F., et al. (2020). *Overly optimistic prediction results on imbalanced data: Flaws and benefits of applying over-sampling*. http://arxiv.org/abs/2001.06296.

Vogel, A. P., Fletcher, J., & Maruff, P. (2010). Acoustic analysis of the effects of sustained wakefulness on speech. *The Journal of the Acoustical Society of America, 128*(6), 3747–3756. https://doi.org/10.1121/1.3506349.

Vogel, A. P., Magee, M., Torres-Vega, R., Medrano-Montero, J., Cyngler, M. P., Kruse, M., et al. (2020). Features of speech and swallowing dysfunction in pre-ataxic spinocerebellar ataxia type 2. *Neurology, 95*, e194–e205.

Vogel, A. P., Wardrop, M. I., Folker, J. E., Synofzik, M., Corben, L. A., Delatycki, M. B., et al. (2017). Voice in Friedreich Ataxia. *Journal of Voice, 31*(2), 243.e9–243.e19. https://doi.org/10.1016/j.jvoice.2016.04.015.

Webb, G. I., Boughton, J. R., & Wang, Z. (2005). Not so naive Bayes: Aggregating one-dependence estimators. *Machine Learning*. https://doi.org/10.1007/s10994-005-4258-6.

Wickham, H. (2014). Tidy data. *Journal of Statistical Software, 59*(10), 1–23.

Wickham, H., & Grolemund, G. (2016). *R for data science: Import, tidy, transform, visualize, and model data*. O'Reilly Media, Inc.

Wolpert, D. H. (1996). The lack of a priori distinctions between learning algorithms. *Neural Computation, 8*(7), 1341–1390.

Xu, H., Caramanis, C., & Sanghavi, S. (2012). Robust PCA via outlier pursuit. *IEEE Transactions on Information Theory*. https://doi.org/10.1109/TIT.2011.2173156.

Yang, K., Tu, J., & Chen, T. (2019). Homoscedasticity: An overlooked critical assumption for linear regression. *General Psychiatry*. https://doi.org/10.1136/gpsych-2019-100148.

Zellner, A., Keuzenkamp, H. A., & McAleer, M. (2001). *Simplicity, inference and modelling: Keeping it sophisticatedly simple*. Cambridge University Press.

Zhang, H. (2005). Exploring conditions for the optimality of naïve Bayes. *International Journal of Pattern Recognition and Artificial Intelligence, 19*(2), 183–198. https://doi.org/10.1142/S0218001405003983.

Zhang, Z. (2016). Variable selection with stepwise and best subset approaches. *Annals of Translational Medicine, 4*(7), 1–6. https://doi.org/10.21037/atm.2016.03.35.

Zheng, A., & Casari, A. (2018). *Feature engineering for machine learning: Principles and techniques for data scientists*. O'Reilly Media, Inc.

Chapter 2

# Big data in personalized healthcare

## Lidong Wang[a] and Cheryl Alexander[b]

[a]Institute for Systems Engineering Research, Mississippi State University, Starkville, MS, United States, [b]Institute for IT innovation and Smart Health, Vicksburg, MS, United States

## 1 Introduction

Telepsychiatry uses telecommunication in remote treating or assessing psychiatric patients. It reduces patient travel time, shortens waiting periods for appointments, and improves access to healthcare for patients who live outside an urban setting (Chung-Do et al., 2012). It includes computer-based Internet tools, land and cellular telephone lines, videoconferencing, and in-home telehealth communication systems that combine phones with other devices. This provides opportunities to augment mental health services (Kasckow et al., 2014). There is much overlap between psychiatry and neurology. Psychiatry deals with mental disorders (related to aspects of psychology, the human brain, etc.) while neurology studies nervous system disorders and neurological practice depends considerably on neuroscience.

The Human Brain Project was launched which is a ten-year European Flagship to reconstruct the human brain's multiscale organization. The project initially contained platforms such as Neuro-informatics, Neuromorphic Computing, etc. (Amunts et al., 2016). The connectome (the entity of neural connections of the human brain) and brainnetome help explain the brain with diseases or normal functions. Graph measures may be used as biomarkers for neurological or psychiatric diseases (Kopetzky & Butz-Ostendorf, 2018). The China Brain Project with the "one body, two wings" scheme was launched based on brain-inspired intelligence and brain science. Neural theories of brain cognition are the "one body"; the research and development of novel technology for computational brain intelligence and methods for brain diseases diagnosis and treatment are the "two wings" (Poo et al., 2016). The two big long-term projects involve frontier research in neuroscience, neurological diseases, and relevant big data. Both neurological and psychiatric sectors (especially telepsychiatric sectors) generate big data.

Big Data in Psychiatry and Neurology. https://doi.org/10.1016/B978-0-12-822884-5.00017-9
Copyright © 2021 Elsevier Inc. All rights reserved.

36    Big data in psychiatry and neurology

Personalized healthcare and precision healthcare based on big data have become significant topics in health and medical applications. Former President Obama of the USA announced the Precision Medicine Initiative in 2015 to expand access to personalized healthcare information and collect data from a diverse national cohort (Lee, Hamideh, & Nebeker, 2019). The European Commission awarded extra 21 million Euros in 2019 to boost the digital exchange of healthcare data within the European Union (EU) to support personalized healthcare (Green, Carusi, & Hoeyer, 2019). Big data of the human brain can be captured at molecular and neuronal circuitry levels using effective neuroimaging technology. Big data help cure diseases and assists in mental health (Zhong et al., 2015). Brain imaging based on computer tomography (CT) and magnetic resonance imaging (MRI) helps discover the biomarkers of brain pathology that are key for the correct diagnosis and phenotyping of different neurological diseases (Wheater et al., 2019).

The purpose of this chapter is to present big data, its progress, challenges, and applications in general healthcare and personalized healthcare. The following is the arrangement of the rest of the chapter: Section 2 presents the characteristics, methods, and software platforms of big data; Section 3 introduces big data in general healthcare; Section 4 deals with big data and Big Data analytics in personalized healthcare; and the final section is the conclusion.

## 2    Characteristics, methods, and software platforms of big data

Big data has been defined as "datasets whose size is beyond the ability of typical database software tools to capture, store, manage, and analyze." (Minelli, Chambers, & Dhiraj, 2013). Characteristics of big data can be described as "7 Vs," which is described in Table 1 (Altintas & Amarnath Gupta, 2016; Bellini, Di Claudio, Nesi, & Rauch, 2013; Demchenko, Grosso, De Laat, &

**TABLE 1** Characteristics of big data.

| Characteristics | Description |
| --- | --- |
| Volume | Huge amounts of data |
| Velocity | Data generation rate exceeds those of traditional systems |
| Variety | Heterogeneity of data types and various formats |
| Variability | Data change (dynamic) during the processing and lifecycle |
| Valence | Data items are related to each other and connected |
| Veracity | Possible problems of data in veracity (truthfulness, accuracy, and reliability) |
| Value | Added value brought from collected data |

Membrey, 2013; Jagadish et al., 2014; Lavignon et al., 2013; O'Leary, 2013). Table 2 (Ellaway, Pusic, Galbraith, & Cameron, 2014) presents a comparison between the big data method and traditional research methods.

In Table 2, data cleansing is an important step to improve data quality for further data processing. Data cleansing is employed to detect unreasonable, inaccurate, or incomplete data items, and remove or impute the data items for data quality improvement (Chen, Mao, & Liu, 2014). Sometimes, missing data and impurity in large data cause problems. It is critical to create effective big data cleansing methods to enhance data quality so that effective and correct decisions can be made (Fang, Pouyanfar, Yang, Chen, & Iyengar, 2016).

**TABLE 2** The difference between the big data method and traditional research methods.

| Aspects | The big data method | Traditional methods |
| --- | --- | --- |
| Methodology | Reacts to what data are available | Governs what data are essential |
| Data capture instruments | Opportunity mining of preexisting and dynamical data accumulation | Purpose-built data gathering instruments |
| Data acquisition cost for each subject | Low | High |
| Data capture intrusiveness | Low | High |
| Data cleansing | Permitting uncertainty in data quality | Scrupulous confirmation of data quality |
| Sample size | Variable, very large | Variable, small or not very large |
| Sample features | Loosely defined, relying on markers (in the data) for specifying population features | Tightly defined |
| Analytics | New methods and tools needed to parse and report on extremely large data | Prevailing desktop approaches and tools |
| Replicability | Easier to access and analyze data again; experiment may be open-ended | Possibly problematic in sustaining or recreating study resources and contexts |
| Temporality | Dynamic dashboards | Static ultimate reports |

## 38 Big data in psychiatry and neurology

**TABLE 3** Categorized big data technologies according to processing methods.

| Processing methods | Technologies |
| --- | --- |
| Stream processing | Spark, Apache Strom, S4, Splunk, SQLstream |
| Batch processing | Hadoop (HDFS, Hive, HBase), MapReduce, Pig, Sqoop, Apache Mahout |
| Hybrid processing | Apache Flink, Lambdoop, SummingBird |

Data processing can be divided into three main approaches: stream processing, batch processing, and micro-batch processing. Stream processing handles massive data that are continuously generated while micro-batch handles streaming data as a sequence of smaller data blocks (Wang, Xu, Fujita, & Liu, 2016). Table 3 (Lopez, Lobato, & Duarte, 2016; Wang et al., 2016) lists technologies of big data according to processing methods. One of the most used technologies for stream analytics is Apache Storm; Spark is also regarded as a micro-batch processing method (Curry, Kikiras, & Freitas, 2012; Lopez et al., 2016; Oseku-Afful, 2016).

There are the following features for streaming data and relevant data processing: streaming data are constantly changing (dynamic), have a huge volume, only allow one or a small number of scans, and flow in and out with a stable order; data processing needs a fast response. Multilevel and multidimensional online data mining and data analytics should be implemented on streaming data as well (Han, Pei, & Kamber, 2011).

As for real-time streaming data and the stream processing of big data, real-time systems needed to be the ones with low latency in order for that computing could be completed very quickly and a near real-time response can be given (Bhattacharya & Mitra, 2013). Data need to be separated in partitions to be processed in parallel because of a large volume of the data. Fail recovery, fault tolerance, and high availability are vital for the systems of stream processing. A platform needs to have minimal communication overhead between distributed processes during the data transmission. The application of real-time monitoring also needs the distributed stream processing. A major approach to analyzing big data in a distributed manner is the use of MapReduce with Hadoop open sources. But a platform based on this method is not perfect, even sometimes unsuitable in real-time stream processing (Lopez et al., 2016).

Traditional machine learning (ML) algorithms are not appropriate for the classification of big data because: (1) an ML model that is trained on a certain labeled dataset or data domain is possibly not appropriate for other datasets or data domains; (2) an ML model is generally trained using some class types while there may be many various class types in big data (possible growing and dynamic); and (3) an ML model is created based on only a single learning task and it is possibly not appropriate for multiple learning tasks or the knowledge transfer requirement of big data (Suthaharan, 2014); and (4) the memory constraint has been one of major challenges. Even though it is assumed that training samples data are in the main memory during the execution of algorithms, big data do not fit into this situation (Hido, Tokui, & Oda, 2013).

Deep learning is also called deep machine learning, deep structured learning, or hierarchical learning that is based on a multiple layered model. Its algorithms are very effective in the analysis of data with a huge amount and high variety. Deep learning is an efficient method for Big Data analytics where raw data are mostly uncategorized or unlabeled (Najafabadi et al., 2015). Heterogeneous computing and deep learning working with Big Data analytics boost success; high-performance computing (HPC) and deep learning working with Big Data analytics enhance computation intelligence and success (Wu, 2014).

Apache Spark is an open source and a unified engine for the distributed processing of data. It consists of libraries at a high level that enable support of graph processing (GraphX), streaming data (Spark Streaming), ML (MLlib), and SQL queries (Spark SQL). Apache Storm has been developed to offer a real-time framework for the stream processing of data. It offers the capability with good built-in-fault-tolerance for Big Data analytics (Dash, Shakyawar, Sharma, & Kaushik, 2019).

Quantum computing and algorithms can boost Big Data analytics exponentially in computation speed. Furthermore, quantum computation can considerably save the computation power that is needed during the analysis of big data. Although quantum computation remains in its early stage with substantial challenges, it has begun to be implemented in healthcare data. There have been some quantum products for healthcare applications such as quantum microscopes and quantum sensors (Dash et al., 2019).

## 3 Big data in the healthcare area

Healthcare data are summarized in Table 4 (Chen, Chiang, & Storey, 2012; Miller, 2012; Priyanka & Kulennavar, 2014; Shrestha, 2014; Terry, 2013; Yang, Njoku, & Mackenzie, 2014). Some of the healthcare data can be big data. For example, clinical data and notes can be the mixed data with various formats (relational tables—structured; images—unstructured; and notes/texts— unstructured). Electronic health records (EHR) and electronic medical records (EMR) can be structured or unstructured. Sensors and smart devices can generate stream data (big data).

A smartphone can capture the activity data (such as diet and exercise monitoring data) from a patient outside the hospital and record the statistics of activities. In addition, data can also be collected from patients while they stay in the hospital. All captured data can be offloaded to a healthcare big data cloud via

**40** Big data in psychiatry and neurology

**TABLE 4** Healthcare data types.

| Types | Description or examples |
|---|---|
| Clinical data and notes | • Structured data (e.g. structured EHR/EMR, laboratory data)<br>• Unstructured data (e.g. unstructured EHR/EMR, diagnostic testing reports, radiological images, and X-ray images)<br>• Semistructured data (such as copy-paste from other structure sources) |
| Sensor- or device-generated data | • Data generated from various sensors or monitoring devices (e.g., medical monitoring or home monitoring) |
| Genomic data | • Gene expression<br>• Genotyping<br>• DNA sequence |
| Administrative and transaction data | • Insurance claims and relevant financial records<br>• Billing records, scheduling information |
| Social media and Web data | Health plan websites, Facebook, Twitter feeds, blogs, etc. |
| Publications, reference materials, and products (text-based data) | • Publications such as conference papers and journal articles<br>• Health reference materials and practice guidelines<br>• Health products such as drug information |

5G network (the 5th generation of wireless communications technologies facilitating cellular data network). Progresses and developments in key technologies enable the development and implementation of an advanced monitoring system and its application. These technologies include the Internet of Things (IoT), wireless networking technologies (e.g., 5G network), medical Big Data analytics, artificial intelligence (AI) such as deep learning, and wearable computers that are computer-powered devices worn on the body (Chen et al., 2018).

Health data of patients can be collected using wearable devices (installed with exceedingly small sensors) that enable measurement of physiological data. The sensors are united with communication equipment that enables transmission of data to a cloud server or central station. The function of sensors is significant in collecting big data (Clim, Zota, & Tinica, 2019).

IoT may contain billions of devices that are able to communicate and sense. Streaming data coming from the devices are a challenge of traditional methods in data management and contribute to the paradigm of big data (Zaslavsky, Perera, & Georgakopoulos, 2013). IoT often generates big data with variety,

Big data in personalized healthcare **Chapter | 2 41**

heterogeneity, noise, unstructured formats, and high redundancy. Table 5 lists some IoT-enabled healthcare systems (Jagadeeswari, Subramaniyaswamy, Logesh, & Vijayakumar, 2018).

Matplotlib, Scikit-Learn, and Python Pandas are big data tools in healthcare (Rao & Clarke, 2016). A pipeline has been tested in medical image data utilizing Big Data Spark/PySpark tools. The in-memory data processing and the storage results with different data structures are major features of the pipeline. In-memory processing handles data completely in computer memory (e.g., in random access memory), which avoids reading/writing data from and to relatively slow media (e.g., disk drives) (Sarraf & Ostadhashem, 2016). Table 6 (Srinivasan, 2014) presents some major healthcare uses of big data. Table 7 (Poldrack & Gorgolewski, 2014) presents an example of big data-related technologies in handling neuroimaging data.

**TABLE 5** Sensors, advantages, and disadvantages of some IOT-based healthcare systems.

| Systems | Sensors | Advantages | Disadvantages |
| --- | --- | --- | --- |
| Smart health | Mosquito sensors and wearable IoT devices | Feedback in real time | Requirement for security implementation |
| Social network analysis | RFID tags, GPS sensors, climate detector sensors, and wearable IoT sensors | Quick alert generation, high bandwidth efficiency | Much power consumption using various sensors |
| Technology-enabled care (TEC) | Oxygen level measurement, inertial sensors, biopotential signal processing, and blood pressure sensors | Harvesting ambient energy cuts down the power gap and meets the requirements of power. | Using gel in electrodes causes skin irritation or discomfort for users. IoT sensors consume much power |
| OneM2M-based IoT system | *Personal healthcare* sensors and devices | Provides effective healthcare services | Uses Byzantine fault-tolerant algorithm instead of a normal fault-tolerant algorithm |
| Intelligent Real-Time IoT Based System | Wearable sensors and the bedside patient monitor | Requires a doctor to spend more time on decision-making with accurate observation | It is yet to be adopted and its accuracy is still untested |

42 Big data in psychiatry and neurology

**TABLE 6** Major big data use cases in healthcare.

| Use cases in healthcare | Description |
| --- | --- |
| Health monitoring and intervention | Monitor vital sign changes, raise alerts, and help proactive intervention at home and at the bedside |
| Population health management | Make targeted decisions to improve care and outcomes on the population of chronically ill patients |
| Consumer insight and engagement | Create a customer-focused view for *personalized marketing* and engagement strategies |
| Biomedical insights, search and discovery | Facilitate data sharing to support research, clinical trials, and the development of new products |
| Analytics for care management and transitions | Help care transitions through identifying high-risk patients and informing an alternative care plan |
| Translational research | Discover the genetic basis for diseases to help clinicians provide *personalized medicine* |

**TABLE 7** Some technologies for neuroimaging big data.

| Aspects | Description |
| --- | --- |
| Databases | Sharing huge neuroimaging data demands a strong data management system to manage the data and offer researchers with the data. Some systems have been developed, e.g., LORIS, XNAT, and COINS |
| Cloud computing | It is progressively possible to utilize the data stored in a cloud storage system. A main disadvantage of cloud solutions is possibly expensive |
| Bandwidth | A large bandwidth is required to provide large data |
| High-performance computing (HPC) | Access to a high-performance computing system is needed to analyze large MRI (magnetic resonance imaging) data |
| Pipelines | Some pipeline tools were developed to process large neuroimaging data efficiently. There have been multiple pipelining solutions for neuroimaging analysis |

## 4   Big data and big data analytics in personalized healthcare

Precision health refers to personalized healthcare and the use of big datasets that integrate -omics (an area of study in biology ending in -omics, e.g., metabolomics, proteomics, and genomics) data (such as protein, metabolite, genomic sequence, and microbiome information) with clinical information and health outcomes for optimal disease diagnosis, treatment, and prevention that are specific to an individual patient. Effective realization of precision healthcare depends on interprofessional collaboration with statisticians, bioinformatics, genomics medicine, data scientists, and a wide range of multidisciplinary clinical content experts. Precision health is complicated and involves healthcare big data (regarding a patient's personal -omics information). Therefore a challenge of precision healthcare is to create effective methods and IT systems that enable the combination and analysis of big data for patients and healthcare providers (Fu et al., 2020). Personalized healthcare is data driven. With the increased use of EHR/EMR, Big Data technologies help provide proactive and personalized care for patients (Chawla & Davis, 2013).

IoT is one of the sources of clinical data and IoT devices generate continuous streaming data during the monitoring process of a patient's health. IoT often generate big data. Big Data analytics of medical and healthcare systems facilitates personalized medicine (Dash et al., 2019). Wearable IoT sensors, fog computing, mobile computing, and cloud computing help monitor and diagnose patients in real time. IoT plays a significant role in personalized healthcare (Jagadeeswari et al., 2018).

The cost management for the cure of a chronic disease is an important factor in personal treatment and personalized healthcare. A treatment for an individual relies on factors such as behavioral, environmental, social factors, etc. An ideal approach to personalized healthcare is based on Big Data analytics and relevant analytics results or achieved new findings. Big Data analytics helps provide an early diagnosis of a disease and an improvement of a medical treatment (Clim et al., 2019).

Personalized medicine and precision medicine involve the tailoring of medical treatment to features of an individual patient and data-intensive strategies that explain and stratify diseases and patient groups based on a growing number of biomolecular factors. Improving precision in the context of personalized medicine is a strategy for decreased uncertainty of disease diagnosis through molecular biomarkers (Green et al., 2019; Lee et al., 2019). Precision medicine is more research oriented while personalized medicine is more clinical practice oriented (Zhang, 2015).

Personalized medicine aims to find a treatment that is the best for an individual patient and a substantial improvement in the stratification and timing of healthcare using biological information and biomarkers at the level of molecular disease pathways, proteomics, genetics, and metabolomics (Lee et al., 2019; Zhang, 2015). In many aspects, the progress of personalized medicine is driven by the intersection of whole exome sequencing (WES) and Big Data analytics. Personalized medicine, as the tailoring of clinical interventions, is typically pharmacological based on a patient's ability to respond favorably. High costs and limitations in terms of technology have remained main barriers in a wide -omics-based implementation of personalized medicine (Suwinski et al., 2019).

**44** Big data in psychiatry and neurology

A big data use has evolved from the following application: using big data to streamline and personalize medical treatment based on a patient's current data and previous data, maybe captured with sensors or collected through social media (Groves, Kayyali, Knott, & Kuiken, 2016). Big data technologies provide a strong platform for investigating huge amounts of data from various sensors, EHR/EMR, insurance claims, pharmacy refill records, social websites or media, etc. to find cause, evaluate, predict, and decide the best evidence-based treatment schedule or plan (personalized medicine) (Groves et al., 2016).

Big Data analytics have potential in healthcare. At a preclinical stage, it helps facilitate research and speed up the moving of research from a bench to the bedside. At a clinical stage, it enables accurately explaining social determinants of health and clarifying individual phenotypic nuances. It provides the quickest way toward personalized medicine through which health management is personalized or individualized (Cahan, Hernandez-Boussard, Thadaney-Israni, & Rubin, 2019). Linkages derived from big data help prompt an update of patient triage, clinical guidelines, and diagnostic assistance to provide personalized and precise treatment (Yang et al., 2014). Big medical data mining, evidence-based medicine, natural language processing (NLP) in social media for healthcare, etc. are promising areas of big data and personalized medicine (Peek, Combi, Marin, & Bellazzi, 2015).

Precision medicine takes into consideration individual differences in patient genes, microbiomes, family history, environments, and lifestyles to make diagnostic and therapeutic plans precisely tailored to individuals (Zhang, 2015). Precision medicine involves paving ways to collect, translate, and regulate growing amounts of healthcare data (Lee et al., 2019). It deals with data with a range from data collection and management (e.g., data storage, data sharing, and privacy) to data analytics (data integration, data mining, visualization, etc.). There have been the following applications of Big Data analytics in precision medicine (Wu et al., 2017):

- Systems biology modeling using -omics data: obtain results and insights regarding a complicated molecular system.
- General analysis of biomedical big data: many EHR/EMR and -omics are high-dimensional data; feature extraction is often implemented for dimensional reduction and saved computation time.
- -omics data preprocessing: computationally intensive (Big Data analytics helps).
- Biomarker identification using -omics data: most -omics biomarkers are identified through studying the statistically significant difference among groups of samples.
- EHR data preprocessing: e.g., imputing missing values.
- EHR data mining: temporal data mining and static endpoint prediction are used to mine actionable knowledge from EHR data.

Pharmacogenomics is becoming a crucial factor in the concept of personalized medicine (Dopazo, 2014). The selection of drugs and their doses rely on the medical protocol of a healthcare provider and the experience of a medical team. Therefore results are inherently various and frequently suboptimal. Big data technologies have the potential to support decision-making; however, their applications are limited by some challenges. A major challenge is regarding the availability of big data (sufficiently big) with good quality and representative information. A phenotypic personalized medicine (PPM) platform has offered an alternative to personalized medicine that is driven by -omics, systems biology, and big data. It aims to choose the best combination therapy that includes the drug(s) as well as the dose(s) through finding the relationship between a phenotypic response and input stimuli based only on a patient's data: medical log and readouts (Blasiak, Khong, & Kee, 2020).

The use of three-dimensional (3D) printing as a novel manufacturing method for drug products has gained much interest from professionals in healthcare sectors. The 3D printing has emerged as a precise manufacturing technology for personalized and small drug doses for pediatric patients; it has been regarded to be appropriate for on-demand manufacturing. A pharmaceutical company can provide a compounding pharmacy with raw materials of the drug(s) and a guidance on 3D printing; the pharmacy can compound personalized medicines for patients with various needs according to the prescription of a physician (Rautamo et al., 2020).

There is contradiction between big data for population health and the push (from healthcare providers) to practice personalized medicine, which has led to slow applications of big data in healthcare (Sacristán & Dilla, 2015). Sampling bias (some patient cohorts are missing from inputs) yields nonrepresentative algorithmic outputs. There are disparities in mobile sensors, smartphones, and the uses of other devices; therefore the pipelines of healthcare big data often lack demographic diversity. Patients who may benefit most from optimal medical services are the disabled, elderly, poor, or those in rural settings; however, they are the least possible to use the platforms or tools of big data and Big Data analytics. Algorithms used on uncorrected and biased datasets bring out large false negatives or false positives errors. Awareness of data deficiencies, data cleansing, structures for the data inclusiveness, and the mechanisms of data correction facilitate the fulfillment for the potential of big data and Big Data analytics for personalized medicine (Cahan et al., 2019).

Data security is an important issue related to transmitting patients' medical information monitored through IoT devices ( Jagadeeswari et al., 2018). Data integration and data sharing are two major challenges of big data and Big Data analytics in healthcare. The privacy of big data in distributed systems is a major concern of patients due to the data accessed from various locations. Data privacy and security of patients have become primary cybersecurity issues in healthcare. Keeping the data confidential while still conducting analysis accurately is a challenge. There are more challenges for small healthcare providers

(Clim et al., 2019). Data sharing among personal health advisors, patients, and friends enables the provider to bring up more valuable data for an accurate disease diagnosis and analytics. But how they share data in a highly cost-effective manner is also a challenge. (Chen et al., 2018).

## 5 Conclusion

IoT often generates big data with heterogeneity, noise, redundancy, and semi-structured or unstructured characteristics. It plays a significant role in personalized healthcare. Big data technologies enable integration of a patient' relevant data to achieve a 360-degree view of the patient and complete the analysis or prediction of outcomes. Deep learning is a useful method for Big Data analytics and enables the provider to analyze large amounts of unlabeled or uncategorized data. Big data have been used in biomedical research and new products development, population health management, health monitoring and intervention, personalized medicine, etc.

Precision health involves big data about patients' personal -omics information. Personalized healthcare is data driven and big data facilitate personalized patient care. Pharmacogenomics is an important factor of personalized medicine. The 3D printing has the potential to manufacture pharmaceutical products for small and personalized doses and facilitates personalized healthcare for patients with various needs. The challenges of big data and Big Data analytics in general healthcare and personalized healthcare lie in the availability of quality data with representative information, data integration and data sharing, data processing and analytics in real time, data privacy and security of patients, etc.

## Acknowledgment

Authors thank Technology & Healthcare Solutions, USA for great support.

## References

Altintas, I., & Amarnath Gupta, A. (2016). *Introduction to big data, the big data specialization series.* San Diego, USA: University of California.

Amunts, K., Ebell, C., Muller, J., Telefont, M., Knoll, A., & Lippert, T. (2016). The human brain project: Creating a European research infrastructure to decode the human brain. *Neuron, 92*(3), 574–581.

Bellini, P., Di Claudio, M., Nesi, P., & Rauch, N. (2013). Tassonomy and review of big data solutions navigation. In R. Akerkar (Ed.), *Big data computing* (pp. 57–101). Chapman and Hall/CRC. Chapter 2.

Bhattacharya, D., & Mitra, M. (2013). *Analytics on big fast data using real time stream data processing architecture* (p. 34). EMC Corporation.

Blasiak, A., Khong, J., & Kee, T. (2020). CURATE. AI: Optimizing personalized medicine with artificial intelligence. *SLAS Technology: Translating Life Sciences Innovation, 25*(2), 95–105. https://doi.org/10.1177/2472630319890316.

Cahan, E. M., Hernandez-Boussard, T., Thadaney-Israni, S., & Rubin, D. L. (2019). Putting the data before the algorithm in big data addressing personalized healthcare. *NPJ Digital Medicine*, *2*(1), 1–6.

Chawla, N. V., & Davis, D. A. (2013). Bringing big data to personalized healthcare: A patient-centered framework. *Journal of General Internal Medicine*, *28*(3), 660–665.

Chen, H., Chiang, R. H., & Storey, V. C. (2012). Business intelligence and analytics: From big data to big impact. *MIS Quarterly*, *36*, 1165–1188.

Chen, M., Mao, S., & Liu, Y. (2014). Big data: A survey. *Mobile Networks and Applications*, *19*(2), 171–209.

Chen, M., Yang, J., Zhou, J., Hao, Y., Zhang, J., & Youn, C. H. (2018). 5G-smart diabetes: Toward personalized diabetes diagnosis with healthcare big data clouds. *IEEE Communications Magazine*, *56*(4), 16–23.

Chung-Do, J., Helm, S., Fukuda, M., Alicata, D., Nishimura, S., & Else, I. (2012). Rural mental health: Implications for telepsychiatry in clinical service, workforce development, and organizational capacity. *Telemedicine and e-Health*, *18*(3), 244–246.

Clim, A., Zota, R. D., & Tinica, G. (2019). Big data in home healthcare: A new frontier in personalized medicine. Medical emergency services and prediction of hypertension risks. *International Journal of Healthcare Management*, *12*(3), 241–249.

Curry, E., Kikiras, P., & Freitas, A. (2012). *Big data technical working groups* (pp. 1–167). White Paper, BIG Consortium.

Dash, S., Shakyawar, S. K., Sharma, M., & Kaushik, S. (2019). Big data in healthcare: Management, analysis and future prospects. *Journal of Big Data*, *6*(1), 54.

Demchenko, Y., Grosso, P., De Laat, C., & Membrey, P. (2013, May). Addressing big data issues in scientific data infrastructure. In *2013 International conference on collaboration technologies and systems (CTS)* (pp. 48–55). San Diego, CA, USA: IEEE. May 20–24.

Dopazo, J. (2014). Genomics and transcriptomics in drug discovery. *Drug Discovery Today*, *19*(2), 126–132.

Ellaway, R. H., Pusic, M. V., Galbraith, R. M., & Cameron, T. (2014). Developing the role of big data and analytics in health professional education. *Medical Teacher*, *36*(3), 216–222.

Fang, R., Pouyanfar, S., Yang, Y., Chen, S. C., & Iyengar, S. S. (2016). Computational health informatics in the big data age: A survey. *ACM Computing Surveys*, *49*(1), 1–36.

Fu, M. R., Kurnat-Thoma, E., Starkweather, A., Henderson, W. A., Cashion, A. K., Williams, J. K., et al. (2020). Precision health: A nursing perspective. *International Journal of Nursing Sciences*, *7*, 5–12.

Green, S., Carusi, A., & Hoeyer, K. (2019). Plastic diagnostics: The remaking of disease and evidence in personalized medicine. *Social Science & Medicine*, 112318. In press.

Groves, P., Kayyali, B., Knott, D., & Kuiken, S. V. (2016). *The 'big data' revolution in healthcare: Accelerating value and innovation, technical report.* McKinsey & Company.

Han, J., Pei, J., & Kamber, M. (2011). *Data mining: Concepts and techniques.* Elsevier.

Hido, S., Tokui, S., & Oda, S. (2013, December). Jubatus: An open source platform for distributed online machine learning. In *NIPS 2013 workshop on big learning, Lake Tahoe.*

Jagadeeswari, V., Subramaniyaswamy, V., Logesh, R., & Vijayakumar, V. (2018). A study on medical internet of things and big data in personalized healthcare system. *Health Information Science and Systems*, *6*(1), 14.

Jagadish, H. V., Gehrke, J., Labrinidis, A., Papakonstantinou, Y., Patel, J. M., Ramakrishnan, R., et al. (2014). Big data and its technical challenges. *Communications of the ACM*, *57*(7), 86–94.

Kasckow, J., Felmet, K., Appelt, C., Thompson, R., Rotondi, A., & Haas, G. (2014). Telepsychiatry in the assessment and treatment of schizophrenia. *Clinical Schizophrenia & Related Psychoses*, *8*(1), 21–27A.

**48** Big data in psychiatry and neurology

Kopetzky, S. J., & Butz-Ostendorf, M. (2018). From matrices to knowledge: Using semantic networks to annotate the connectome. *Frontiers in Neuroanatomy, 12*, 111.

Lavignon, J. F., Lecomber, D., Phillips, I., Subirada, F., Bodin, F., Gonnord, J., et al. (2013). *European technology platform for high performance computing: ETP4HPC strategic research agenda achieving HPC leadership in Europe, technical report, ETP4HPC.*

Lee, J., Hamideh, D., & Nebeker, C. (2019). Qualifying and quantifying the precision medicine rhetoric. *BMC Genomics, 20*(1), 868.

Lopez, M. A., Lobato, A. G. P., & Duarte, O. C. M. (2016, December). A performance comparison of open-source stream processing platforms. In *2016 IEEE global communications conference (GLOBECOM)* (pp. 1–6). IEEE.

Miller, K. (2012). Big data analytics in biomedical research. *Biomedical Computation Review, 2*, 14–21.

Minelli, M., Chambers, M., & Dhiraj, A. (2013). *Big data, big analytics: Emerging business intelligence and analytic trends for today's businesses. Vol. 578.* John Wiley & Sons.

Najafabadi, M. M., Villanustre, F., Khoshgoftaar, T. M., Seliya, N., Wald, R., & Muharemagic, E. (2015). Deep learning applications and challenges in big data analytics. *Journal of Big Data, 2* (1), 1.

O'Leary, D. E. (2013). 'Big data', the 'internet of things' and the 'internet of signs. *Intelligent Systems in Accounting, Finance and Management, 20*(1), 53–65.

Oseku-Afful, T. (2016). The use of big data analytics to protect critical information infrastructures from cyber-attacks. *Dissertation.* Sweden: Department of Computer Science, Electrical and Space Engineering, Luleå University of Technology.

Peek, N., Combi, C., Marin, R., & Bellazzi, R. (2015). Thirty years of artificial intelligence in medicine (AIME) conferences: A review of research themes. *Artificial Intelligence in Medicine, 65* (1), 61–73.

Poldrack, R. A., & Gorgolewski, K. J. (2014). Making big data open: Data sharing in neuroimaging. *Nature Neuroscience, 17*(11), 1510.

Poo, M. M., Du, J. L., Ip, N. Y., Xiong, Z. Q., Xu, B., & Tan, T. (2016). China brain project: Basic neuroscience, brain diseases, and brain-inspired computing. *Neuron, 92*(3), 591–596.

Priyanka, K., & Kulennavar, N. (2014). A survey on big data analytics in health care. *International Journal of Computer Science and Information Technologies, 5*(4), 5865–5868.

Rao, A. R., & Clarke, D. (2016, October). A fully integrated open-source toolkit for mining healthcare big-data: Architecture and applications. In *2016 IEEE international conference on healthcare informatics (ICHI)* (pp. 255–261). IEEE.

Rautamo, M., Kvarnström, K., Sivén, M., Airaksinen, M., Lahdenne, P., & Sandler, N. (2020). Benefits and prerequisites associated with the adoption of oral 3D-printed medicines for pediatric patients: A focus group study among healthcare professionals. *Pharmaceutics, 12*(3), 229.

Sacristán, J. A., & Dilla, T. (2015). No big data without small data: Learning health care systems begin and end with the individual patient. *Journal of Evaluation in Clinical Practice, 21*(6), 1014.

Sarraf, S., & Ostadhashem, M. (2016, December). Big data application in functional magnetic resonance imaging using apache spark. In *2016 Future technologies conference (FTC)* (pp. 281–284). IEEE.

Shrestha, R. B. (2014). Big data and cloud computing. *Applied Radiology, 43*(3), 32.

Srinivasan, S. (2014). *Big data and analytics in healthcare overview: Fueling the journey toward better outcomes, workshop.* University of Florida. September.

Suthaharan, S. (2014). Big data classification: Problems and challenges in network intrusion prediction with machine learning. *ACM SIGMETRICS Performance Evaluation Review, 41*(4), 70–73.

Suwinski, P., Ong, C., Ling, M. H., Poh, Y. M., Khan, A. M., & Ong, H. S. (2019). Advancing personalized medicine through the application of whole exome sequencing and big data analytics. *Frontiers in Genetics, 10*, 49.

Terry, N. P. (2013). Protecting patient privacy in the age of big data. *SSRN Electronic Journal, 81* (2), 385–415.

Wang, H., Xu, Z., Fujita, H., & Liu, S. (2016). Towards felicitous decision making: An overview on challenges and trends of big data. *Information Sciences, 367*, 747–765.

Wheater, E., Mair, G., Sudlow, C., Alex, B., Grover, C., & Whiteley, W. (2019). A validated natural language processing algorithm for brain imaging phenotypes from radiology reports in UK electronic health records. *BMC Medical Informatics and Decision Making, 19*(1), 184.

Wu, R. (2014). *Deep learning meets heterogeneous computing, workshop.* Baidu Inc.

Wu, P. Y., Cheng, C. W., Kaddi, C. D., Venugopalan, J., Hoffman, R., & Wang, M. D. (2017). Omic and electronic health record big data analytics for precision medicine. *IEEE Transactions on Biomedical Engineering, 64*(2), 263–273.

Yang, S., Njoku, M., & Mackenzie, C. F. (2014). 'Big data' approaches to trauma outcome prediction and autonomous resuscitation. *British Journal of Hospital Medicine, 75*(11), 637–641.

Zaslavsky, A., Perera, C., & Georgakopoulos, D. (2013). *Sensing as a service and big data.* arXiv Preprint. arXiv:1301.0159.

Zhang, X. (2015). Precision medicine, personalized medicine, omics and big data: Concepts and relationships. *Journal of Pharmacogenomics & Pharmacoproteomics, 6*. e144. 9.

Zhong, N., Yau, S. S., Ma, J., Shimojo, S., Just, M., Hu, B., et al. (2015). Brain informatics-based big data and the wisdom web of things. *IEEE Intelligent Systems, 30*(5), 2–7.

Chapter 3

# Longitudinal data analysis: The multiple indicators growth curve model approach

**Thierno M.O. Diallo[a,b] and Ahmed A. Moustafa[c,d]**

[a]*School of Social Science, Western Sydney University, Sydney, NSW, Australia,* [b]*Statistiques & M.N., Sherbrooke, QC, Canada,* [c]*Department of Psychiatry, Wroclaw Medical University, Wroclaw, Poland,* [d]*School of Psychology & Marcs Institute for Brain and Behaviour, Western Sydney University, Sydney, NSW, Australia*

## 1 Introduction

Longitudinal data are frequently collected in psychiatric and neuroscience research. Such data allow researchers to study how cognitive and neural processes change during development and how individuals' characteristics interact to shed light on cognitive, emotional, and behavioral changes. Longitudinal data helps better model multivariate brain–behavior relations in general. For example, Johnson (2011) suggests that the development of essential brain regions such as the frontal lobes is a necessary requirement for gaining psychological competences. Longitudinal data can be used to understand how changes in neural regions are linked to changes in functions associated with these regions. Similarly, longitudinal data can provide tools to test developmental theory (e.g., van den Bos & Eppinger, 2016) by investigating the development of brain regions linked to cognitive control compared to regions linked with mediating emotional responses (Kievit et al., 2018).

One statistical method used to handle longitudinal data is latent curve model (LCM) in the Structural Equation Modeling (SEM) framework (e.g., McArdle & Anderson, 1990; McArdle & Epstein, 1987; Meredith & Tisak, 1990; Rogosa, 1995) or hierarchical linear modeling, random effect modeling, or mixed linear models in the multilevel framework (Bryk & Raudenbush, 1992; Laird & Ware, 1982; Raudenbush & Bryk, 2002). Latent curve models examine changes in an outcome over time by explicitly modeling growth and individual differences in growth over time. This statistical methodology has been widely used in psychiatric and neuroscience research over the last two decades. For instance, Braams, van Duijvenvoorde, Peper, and Crone (2015) used the mixed linear

---

**Big Data in Psychiatry and Neurology. https://doi.org/10.1016/B978-0-12-822884-5.00012-X**
Copyright © 2021 Elsevier Inc. All rights reserved.

51

procedure to provide a better understanding in adolescent risk-taking through the investigation of neural responses to rewards, pubertal development, and risk-taking behavior. The longitudinal analyses showed a quadratic function of growth for nucleus accumbens activity to rewards (peaking in adolescence), as well as for laboratory risk-taking. Nucleus accumbens activity change was found to be related to change in testosterone and self-reported reward-sensitivity, confirming that an adolescent peak in nucleus accumbens activity. Similarly, Ordaz, Foran, Velanova, and Luna (2013) applied the latent curve methodology to study brain regions within circuits known to support motor response control, executive control, and error processing (i.e., aspects of inhibitory control). Interestingly, their findings showed different pattern of results with respect to the development of regions within each circuit over time. In particular, the results showed a hierarchical pattern of maturation of brain activation. No mean change was found for activation in motor response control regions whereas activation in certain executive control regions decreased with age until adolescence. On the other hand, error-processing activation in the dorsal anterior cingulate cortex was associated with continued increases into adulthood. Caselli et al. (2012) used the latent curve models to investigate the possible contribution of TOMM40 (translocase of the outer mitochondrial membrane pore subunit) to Alzheimer's disease onset, by comparing the effects of TOMM40 and APOE (apolipoprotein E) genotype on preclinical longitudinal memory decline. The results showed a quadratic growth for both TOMM40 and APOE. The results also show that both TOMM40 and APOE significantly influenced age-related memory performance, but they appear to do so independently of each other.

Recently, however, big data methods have been used in psychiatric and neuroscience research (Moustafa et al., 2018). The worldwide increase in digital connectivity; the migration of activities to the internet; and the decline in the cost of data collection, storage, and processing, combined with advances in data analytics, are leading to the generation and use of large volumes of data. In this connected world, people, objects, and connections are producing data that provide new insights into people's lives, behaviors, health, needs, and aspirations. Importantly, performing meaningful analysis on these large volumes of data provides interesting opportunities for hypothesis generation and testing, which will enhance clinical practice.

Researchers are often interested in the growth of a latent construct measured by a scale containing multiple indicators (or items) (e.g., the Alzheimer's disease assessment scale consists of 11 items). However, a review of the literature shows that the majority of work in psychiatric and neuroscience research involve latent curve models that focus on fitting univariate LCM (also known as first-order LCM or traditional LCM) to composites of the scales' items. The most commonly used composite scores to represent the construct of interest is the mean of scales' items and the sum of the item scores with lesser extent. To assess the Alzheimer's disease for instance, cognitive measures are sometimes combined through a mean or a sum to create a single composite score for each individual. The univariate LCM can be applied on the composite score to study disease changes over time. However, assessing changes using a single composite score prevents one from taking full advantage of longitudinal data richness and potential. Indeed, a single composite score at each measurement occasion prevents one from investigating measurement invariance, variance partitioning into time-specific and item-residual, and from identifying complex item-residual covariance patterns (Bishop, Geiser, & Cole, 2015; Grilli & Varriale, 2014; Isiordia & Ferrer, 2018; Leite, 2007; Sayer & Cumsille, 2001; Wu, Liu, Gadermann, & Zumbo, 2010). For instance, using the Alzheimer's disease composite score prevents from studying the scale itself before looking into individual changes in Alzheimer condition over time.

Multiple indicators latent curve model (MILCM, McArdle, 1988; Tisak & Meredith, 1990) can be used to address the analytical limitations of univariate latent curve models. MILCM (e.g., Cho, 2019; Cho & Lee, 2019; Guerra, Bassi, & Dias, 2020), sometimes referred to as second-order growth curve model, curve-of-factors model, or latent variable growth curve model, is an extension of the LCM that includes multiple indicators of a latent variable at each measurement occasion. This statistical technique has many advantages including modeling the relationship between the indicators and its underlying latent construct at each measurement occasion as well as assessing the common growth of the multiple indicators simultaneously. Despite its numerous statistical advantages over univariate LCM, MILCM is underutilized in psychiatric and neuroscience research. As a result, the advantage of using MILCM is unknown to numerous psychiatric and neuroscience researchers.

The purpose of the present chapter is to promote the use of MILCM for investigating individuals' development over time in the context of big data in psychiatric and neuroscience research. We first review data reduction techniques that can be used in the context of big data before performing meaningful analysis, including multiple indicators growth curve models. Next, we provide a review of longitudinal measurement invariance, a necessary condition for evaluating changes in the scale measuring the construct before investigating longitudinal changes in the construct (Sayer & Cumsille, 2001). We then describe MILCM and the advantages of using this methodology. Finally, we provide steps for using MILCM.

## 2 Multivariate dimension reduction techniques: Principal component analysis and factor analysis

Large and dynamic data sets that capture diverse characteristics about millions of people are collected at an incredible rate today. Ninety percent of the world's data were generated in the last 3–5 years with the rapid advancement in mobile internet, networking technology, and other digital platforms. These data have the ability to inform important research questions in psychiatric and

**54** Big data in psychiatry and neurology

neuroscience research. The availability of these large amounts of data is not without some challenges when it comes to processing the information in a meaningful way. One of the challenges is due to size. Using data reduction techniques can help reduce the large amount of redundancy and noise contained in the data, and improve computation time and prediction accuracy in drawing inference. Moreover, LCM and MILCM are used on items that measure only one construct. Hence, before using MILCM, one has to test the unidimensionality of the items through multivariate dimension reduction techniques. In this section, we will present principal component analysis and factor analysis, two multivariate techniques that are used for dimension reduction when dealing with large amounts of data. This section will also compare the two methods with respect to using MILCM.

## 2.1 Principal component analysis

Principal component analysis, PCA, was first introduced by Karl Pearson more than a century ago, and has been widely used in many fields of research ever since. PCA is a dimension reduction technique used to reduce the number of observed variables to a smaller number of components. These components, also known as principal components, describe the essence of the data by accounting for most of the variance of the observed variables. The components are used to aggregate the set of observed variables into a synthetic index.

For example, let's assume that $Y_1, Y_2, ..., Y_p$ are the indicators measuring AD Cognitive Assessment Battery. Let us also assume that $Y_1, Y_2, ..., Y_p$ are continuous indicators. PCA can be used to summarize the variation of the indicators by a smaller number of variables (components here). The principal components will be defined as a linear combination (or a weighted sums) of the indicators $X_1, X_2, ..., X_p$, where $X_1, X_2, ..., X_p$ are the standardization (i.e., zero mean and standard deviation of one) of the indicators $Y_1, Y_2, ..., Y_p$.

The first component

$$Z_1 = a_{11}X_1 + a_{12}X_2 + \cdots + a_{1p}X_p \tag{1}$$

is the linear combination that explains as much variance in $X_1, X_2, ..., X_p$. The coefficients $a_{11}, a_{12}, ..., a_{1p}$ are the weight associated with the indicators $X_1, X_2, ..., X_p$, respectively.

The second component

$$Z_2 = a_{21}X_1 + a_{22}X_2 + \cdots + a_{2p}X_p \tag{2}$$

is the linear combination that explains as much variance in $X_1, X_2, ..., X_p$ that is not extracted by the first component. The coefficients $a_{21}, a_{22}, ..., a_{2p}$ are the weight associated with the indicators $X_1, X_2, ..., X_p$, respectively. The second component $Z_2$ is uncorrelated with the first component $Z_1$. The number of components extracted is equal to the number of observed indicators being analyzed.

Here $Z_1, Z_2, ..., Z_p$ uncorrelated principal components are therefore extracted. For dimension reduction purposes however, $q$ components with $q$ smaller than $p$ are retained.

PCA is used when there is greater emphasis on data reduction and less on interpretation as the components $Z_1, Z_2, ..., Z_p$ are uninterpretable. Furthermore, the individuals' scores on each principal component can be computed exactly.

Fig. 1 represents the PCA defined by Eqs. (1) and (2). Following the conventions, circles depict latent variables and rectangles observed variables; arrows connecting latent and/or observed variables depict direct effects.

## 2.2 Factor analysis

Factor Analysis, FA, was introduced by Charles Spearman in 1904 for one-factor model and by Louis Thurstone in 1947 for more than one-factor model. FA is a group of statistical techniques employed to examine patterns of relationships underlying a large number of observed variables (Fabrigar, Wegener, MacCallum, & Strahan, 1999). FA assumes that there are underlying constructs (i.e., a variable not measured directly), also called common factors or latent variables, that are affecting the observed variables (or indicators of the factors). The factors are obtained through the analysis of the covariance matrix and they are the weighted average of the observed variables. FA also assumes that each indicator is measured with a degree of measurement error and the factors may be correlated or uncorrelated. Importantly, FA analyzes the common variance of the indicators. FA includes exploratory factor analysis (EFA) and confirmatory factor analysis (CFA) ( Jennrich & Bentler, 2011).

EFA is a variable reduction technique that seeks to identify the number of underlying factors from which the observed variables are generated. In an EFA a minimum number of common factors are used to explain the correlation matrix of the observed variables. The common factors are then used to represent the data. Scores on each factor (also called factor scores) are estimated for each individual in the sample.

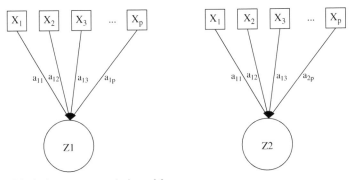

**FIG. 1** Principal component analysis model.

**56** Big data in psychiatry and neurology

CFA, on the other hand, tests a hypothesis about relationships (or factor structure) among observed variables. This is conducted through the imposition of constraints on the EFA model (Jennrich & Bentler, 2011). In other words, CFA is used to verify hypotheses or to confirm a theory (Brown, 2015). CFA is also used to test the equivalence of factor structures across several groups.

Let's recall our example with AD Cognitive Assessment Battery. $Y_1, Y_2, ..., Y_p$ are the indicators measuring AD Cognitive Assessment Battery. $X_1, X_2, ..., X_p$ are the standardization of the indicators $Y_1, Y_2, ..., Y_p$. In an FA framework, $Y_1, Y_2, ..., Y_p$ are considered to arise as a set of regressions on a common factor $(F)$ or on $k$ common factors $(F_1, F_2, ..., F_k)$. In the case of the one-factor model, each indicator $X_1, X_2, ..., X_p$ can be expressed as a regression on the common factor $(F)$ with an error term $(e_i)$, where the common factor accounts for the correlation among $X_1, X_2, ..., X_p$. The one-factor model can be expressed as the following:

$$X_1 = \nu_1 + \lambda_1 F + e_1$$
$$X_2 = \nu_2 + \lambda_2 F + e_2$$
$$\vdots$$
$$X_p = \nu_p + \lambda_p F + e_p$$

$$(3)$$

where $F$ is the common factor; $\nu_1, ..., \nu_p$ are the intercepts in the regression of the factor $F$ on the indicator $X_1, X_2, ..., X_p$; $e_1, ..., e_p$ are the error terms associated with each indicator $X_1, X_2, ..., X_p$; $\lambda_1, \lambda_2, ..., \lambda_p$ are the factor loadings associated with each indicator. The factor loadings are the correlation between the indicators $(X_1, X_2, ..., X_p)$ and the factor $F$.

With $k$ common factors $(F_1, F_2, ..., F_k)$, the common factor model can be written as:

$$X_1 = \nu_1 + \lambda_{11} F_1 + ... + \lambda_{1k} F_k + e_1$$
$$X_2 = \nu_2 + \lambda_{21} F_1 + ... + \lambda_{2k} F_k + e_2$$
$$\vdots$$
$$X_p = \nu_p + \lambda_{p1} F_1 + ... + \lambda_{pk} F_k + e_p$$

$$(4)$$

where $\nu_1, ..., \nu_p$ are the intercepts in the regression of the factors $F_1, F_2, ..., F_k$ on the indicator $X_1, X_2, ..., X_p$; $e_1, ..., e_p$ are the error terms associated with each indicator $X_1, X_2, ..., X_p$; $\lambda_{11}, ..., \lambda_{kp}$ are the factor loadings associated with each factor.

Figs. 2 and 3 represent the EFA defined by one and two factors, respectively. Following usual conventions, circles depict latent variables and rectangles observed variables; arrows connecting latent and/or observed variables depict direct effects and lines with arrows at both extremes represent correlations among latent variables or among items.

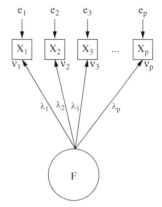

**FIG. 2** Factor analysis with one-factor model.

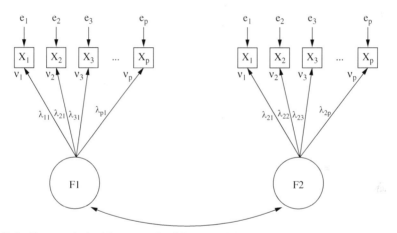

**FIG. 3** Factor analysis with two correlated factors model.

## 2.3 Factor analysis and principal component analysis for multiple indicator growth curve models

PCA and EFA are both variable reduction techniques that can produce similar results when the communities (square of the factor loadings) are close to 1. In other cases, PCA and EFA produce different results. Indeed, PCA and EFA have different aims. PCA explains variations within the data through the maximization of the variance. EFA, on the other hand, explains correlations among the observed variables. PCA can be seen as an estimator for EFA. However, PCA is a biased estimator for EFA since PCA fixes the residual variances of the indicators at zero which is unrealistic in most cases (Fabrigar et al., 1999). Moreover, EFA can provide how much of the variance in each item is

**58** Big data in psychiatry and neurology

accounted for in the model. Therefore EFA is more suited for investigating construct unidimensionality before conducting an MILCM. For discussion on this issue, readers are referred to Hattie (1985).

## 3 Longitudinal measurement invariance

Neurological constructs such as attention, temporal-sequential ordering, spatial ordering, memory, language, neuromotor functions, social cognition, and higher order cognition are not observed directly and as a result these constructs cannot be measured directly as well. Items that are considered to be a manifestation of each latent construct are used to capture the constructs. In other words, the items, which are observed directly, are used to measure each construct. In longitudinal studies, repeated observations on the same indicators (or items) are used to study individual changes in construct over time. To examine the change on the construct of interest, the repeatedly measured indicators need to measure the same construct in the same metric across measurement occasions (Liu et al., 2017). However, measuring the same indicators of a construct over time does not ensure that the same construct is defined over time. For instance, as respondents become older, the respondents' understanding of the different item might evolve. Hence, the construct being measured once is different from the one measured from the one measured prior. In other words, individuals' perceptions or interpretations of the items may change over time resulting in a change in the factor structure rather than to changes on the construct (Horn & McArdle, 1992). As a result, the examination of the relations between the observed indicators and the latent constructs over time is warranted to establish that the same construct is measured across measurement occasions (Meredith, 1964, 1993).

Longitudinal measurement invariance can be used to ensure the equal definition of latent construct over time. For instance, suppose that the indicators $Y_1$, $Y_2$, ..., $Y_p$ are used to measure AD Cognitive Assessment Battery across three measurement occasions:

$$Y_{ijt} = \nu_{jt} + \lambda_{jt}F_{it} + e_{ijt} \tag{5}$$

where $Y_{ijt}$ is item $j(j = 1, .., p)$ at measurement occasion $t(t = 1, 2, 3)$ for individual $i$; $\nu_{tj}$ is the intercept for item $j$ at measurement occasion $t$; $F_{it}$ is the latent factor score for individual $i$ at measurement occasion $t$; $\lambda_{tj}$ is the factor loading connecting item $j$ for individual $i$ at measurement occasion $t$ to the factor; $e_{ijt}$ is the error term associated with item $j$ for individual $i$ at measurement occasion $t$. In Eq. (5) the vector of error terms is usually assumed to be multivariate, normally distributed with a zero mean vector and a variance-covariance matrix $\Theta$.

To test whether the same construct is measured across measurement occasions, a stepwise procedure imposing a series of increasingly constrained models can be followed (Vandenberg & Lance, 2000). The first model, called configural invariance, tests whether the same factor structure is observed across measurement occasions. Thus configural invariance establishes whether the latent construct is characterized by the same items at each measurement occasion. Next, a test of weak factorial invariance is performed. Weak factorial invariance assesses whether the strength of the relationship between the factor and the items is the same over time. This model tests equality of the factor loadings over time. In the previous example, the following test is conducted to establish weak invariance: $\lambda_{11} = \lambda_{12} = \lambda_{13}$, $\lambda_{21} = \lambda_{22} = \lambda_{23}$, $\lambda_{31} = \lambda_{32} = \lambda_{33}$.

A test of strong factorial invariance, on the other hand, assesses whether the means of the construct are compared over time. This model imposes additional constraints on the weak invariance model, namely, equality of the intercepts. Thus, in the example, the following tests are conducted for assessing strong factorial invariance:

$$\lambda_{11} = \lambda_{12} = \lambda_{13}, \lambda_{21} = \lambda_{22} = \lambda_{23}, \lambda_{31} = \lambda_{32} = \lambda_{33}$$

and $\nu_{11} = \nu_{12} = \nu_{13}, \nu_{21} = \nu_{22} = \nu_{23}, \nu_{31} = \nu_{32} = \nu_{33}$. Lastly, a test of strict factorial invariance evaluates whether all elements of the factor model are comparable over time. The strict factorial invariance model imposes additional constraint on the strong factorial model, namely, equality constraints of the residual variances (see Fig. 4).

Fig. 4 represents a longitudinal measurement invariance of a CFA model over three measurement occasions. Following usual conventions, circles depict latent variables and rectangles observed variables; arrows connecting latent and/or observed variables depict direct effects and lines with arrows at both extremes represent correlations among latent variables or among items. All the variables in the frame vary across units, thus the index $i$ is omitted.

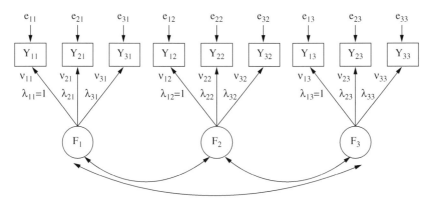

**FIG. 4** Longitudinal measurement invariance.

## 4 Multiple indicators growth curve model

LCM is an approach to growth curve modeling in the SEM framework (e.g., McArdle & Anderson, 1990; McArdle & Epstein, 1987; Meredith & Tisak, 1990; Rogosa, 1995). This statistical methodology is based on the direct estimation of average patterns of intraindividual growth observable in the total sample, as well as interindividual variation around these average trends. A single trajectory that varies continuously across individuals is used to explain the correlations between repeated measures. LCM can be seen as confirmatory factor models (CFA) where change is captured by latent constructs and a residual term (Meredith & Tisak, 1990; Muthén & Curran, 1997). These models are equivalent to hierarchical linear modeling, random effect modeling, or mixed linear models in the multilevel framework (Bryk & Raudenbush, 1992; Laird & Ware, 1982; Raudenbush & Bryk, 2002). The latent trajectory analysis is presented in detail elsewhere (e.g., Diallo & Morin, 2015; Diallo, Morin, & Parker, 2014; Diallo & Lu, 2016; McArdle, 1988; McArdle & Epstein, 1987; Meredith & Tisak, 1984, 1990; Muthén & Curran, 1997; Willett & Sayer, 1994).

Traditional LCM, also called univariate LCM or first-order LCM, is based on a single observed indicator. Here a composite score (e.g., items mean and item total score) is used to represent the construct of interest at each occasion. And LCM is fit to the composite score to evaluate change in the construct (see Fig. 5). The univariate LCM has been extensively utilized in psychiatric and neuroscience research (e.g., Paranjpe et al., 2019; Suzuki et al., 2020; Williams et al., 2019). This method has the advantage of being simple and quick to run. However, investigating change with the univariate LCM limits the full evaluation of longitudinal data as measurement error and individual residuals are indiscernible (e.g., Wänström, 2009).

MILCM on the other hand, an extension of the LCM, overcome the limitation of traditional LCM. MILCM accounts for the relationship between the observed indicators (or items) and the latent construct at each measurement

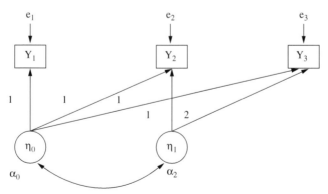

**FIG. 5** Univariate LCM with linear function.

occasion and models the common growth of the indicators. The benefits of using the MILCM approach over the univariate LCM for analyzing change are well documented. Isiordia and Ferrer (2018) provide a summary of the methodological advantages of the MILCM. Specifically, some of the strengths of the MILCM are that they allow for the investigation of longitudinal measurement invariance, they have the ability to partition residual variance into time-specific and item-residual, and they permit the identification of item-residual covariance patterns.

## 4.1 The MILCM equations

MILCM can be formulated as a SEM that consists of a measurement model and a structural model. The measurement model describes the relationship between the indicators and the latent variables over time. The structural model characterizes the development of the latent variable over time.

### 4.1.1 Measurement model

The measurement model studies the relationship between the observed indicators and the latent construct at each time point. It allows for the investigation of longitudinal measurement invariance over time which is a necessary condition for investigating change in a construct. The measurement model assesses whether the mean of the construct is comparable over time. The latent factors at each measurement occasion in the measurement model are often referred to as first-order factors.

Let's recall now our example with AD Cognitive Assessment Battery. $Y_1, Y_2, \ldots, Y_p$ are the indicators measuring AD Cognitive Assessment Battery. The measurement model is defined in Eq. (5). Strong factorial invariance is, however, deemed sufficient for investigating individual changes in a construct over time (Meredith, 1964, 1993). However, partial measurement invariance has also been considered in the literature (Muthén & Muthén, 2010). Partial measurement allows some parameters to be invariant over time while others are allowed to vary over measurement occasions.

### 4.1.2 Structural model

The structural model addresses the growth of the latent variable. In the structural model the estimated average patterns of intraindividual growth are observable in the total sample, as well as interindividual variation around these average trends. A functional form of growth that captures how individuals grow on the construct is specified. Different functional forms of growth can be utilized including linear and nonlinear functions. Polynomial trends (e.g., quadratic, cubic) can easily be specified by extending the linear function of growth with additional latent variables to capture the nonlinear component of change. To represent quadratic trends for instance, an additional latent

**62** Big data in psychiatry and neurology

variable is added to the linear model to capture the nonlinearity present in the data. Another type of function that can be used is the piecewise growth function. Piecewise functions are often used to model transition. In a piecewise linear function for instance, nonlinearity is modeled by including two interrelated linear slopes reflecting the growth trajectory before and after the transition point. For instance, piecewise functions are frequently used in experimental or clinical studies where turning points can be specified as the beginning, or end, of the treatment (Diallo & Morin, 2015). More complex nonlinear functions of growth, such as exponential and logistic can also be used (e.g., Blozis, 2007; Browne & du Toit, 1991; Grimm, Ram, & Hamagami, 2011; Ram & Grimm, 2007).

For the structural model we assume a linear growth. The unconditional growth model, which describes the intraindividual variability, can be specified as the following:

$$F_{it} = \eta_{0i} + b_t \eta_{1i} + \varepsilon_{it} \tag{6}$$

where $F_{it}$ is the latent factor score for individual $i$ at measurement occasion $t(t = 1, 2, 3)$; $\eta_{0i}$ is the intercept growth factor for individual $i$; $\eta_{1i}$ is the slope growth factor for individual $i$; $b_t$ is the time score at measurement occasion $t$; and $\varepsilon_{it}$ is the time-specific residual for individual $i$. The errors $\varepsilon_{it}$ are assumed to be normally distributed with zero mean and variance-covariance matrix $\Psi$.

Growth is represented by imposing constraints on the time score $b_t$ often specified to reflect the passage of time. In Eq. (6) growth is represented by a latent intercept $\eta_{0i}$ and a linear slope $\eta_{1i}$ specified as influencing the repeated measures through fixed loadings that reflect the passage of time. If the time score is coded 0, 1, 2 to reflect the passage of equally spaced time points, then the intercept, $\eta_{0i}$, represents the initial status. The linear component $\eta_{1i}$ represents the instantaneous rate of change at the initial assessment.

Since the latent variables $\eta_{0i}$ and $\eta_{1i}$ are random variables, they can be described by the average level observed in the sample $(\mu_{\eta_0}, \mu_{\eta_1})$, plus a variance component reflecting interindividual deviations from the growth factor means $(\zeta_{\mu_{\eta_0}}, \zeta_{\mu_{\eta_1}})$:

$$\begin{aligned} \eta_{0i} &= \mu_{\eta_0} + \zeta_{\mu_{\eta_0}} \\ \eta_{1i} &= \mu_{\eta_1} + \zeta_{\mu_{\eta_1}} \end{aligned} \tag{7}$$

The vector of random effects $(\zeta_{\mu_{\eta_0}}, \zeta_{\mu_{\eta_1}})$ is assumed to be bivariate normally distributed with a zero mean vector and variance-covariance matrix $\Phi$. The variance-covariance matrix $\Phi$ contains the interindividual variance of the intercept factor, the interindividual variance on the slope factor, and the covariance between the intercept and slope factors.

As can be seen from Eqs. (5)–(7), three types of errors are associated with MILCMs. First, measurement errors $e_{ijt}$ which quantify the error associated with each indicator in measuring the latent factor $F$ (i.e, AD Cognitive

Assessment Battery). The second type of error $\varepsilon_{it}$ characterizes deviations of the latent construct $F_{it}$ from the straight line. Finally, $(\zeta_{\mu_{\eta 0}}, \zeta_{\mu_{\eta 1}})$ represents deviations of an individual's growth from the average population growth. These three sets of errors are assumed to be independent in the MILCM.

The conditional growth model, which describes the interindividual variability, introduces predictors into the model. This model can be specified as the following:

$$\begin{aligned} \eta_{0i} &= \alpha_0 + \gamma_{01} X_{1i} + \gamma_{02} X_{2i} + \ldots + \gamma_{0p} X_{pi} + \varsigma_{0i} \\ \eta_{1i} &= \alpha_1 + \gamma_{11} X_{1i} + \gamma_{12} X_{2i} + \ldots + \gamma_{1p} X_{pi} + \varsigma_{1i} \end{aligned} \quad (8)$$

where $X_{1i}, X_{2i}, \ldots, X_{pi}$ are $p$ time-invariant predictors; $\gamma_{01}, \ldots, \gamma_{1p}$ are sets of regression coefficients associated with the predictors; $\varsigma_{0i}$ and $\varsigma_{1i}$ are residuals for $\eta_{0i}$ and $\eta_{1i}$, respectively.

Fig. 6 represents an MILCM defined by Eqs. (6) and (7) over three measurement occasions. Following usual conventions, circles depict latent variables and rectangles observed variables; arrows connecting latent and/or observed variables depict direct effects and lines with arrows at both extremes represent correlations among latent variables or among items. All the variables in the frame vary across units, thus the index $i$ is omitted.

## 4.2 Specification details

MILCM in the SEM framework is a latent variable modeling technique. Because latent variables are unobservable by definition, they have no scale

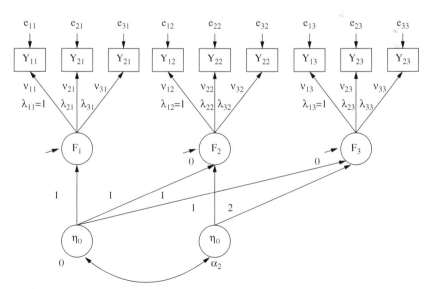

**FIG. 6** MILCM with linear function.

**64**  Big data in psychiatry and neurology

or units of measurement. In order to represent a latent variable, a scale must be defined. Moreover, for the MILCM model parameters to be estimated, it is necessary that the measurement and the structural models are identified. In other words, for the estimation to take place, it is indispensable that at least one algebraic solution is possible expressing the model parameters as a function of the observed mean and variances covariances. In this section we discuss the identification, the scaling, and covariance structure in MILCM.

### 4.2.1 Identification and scaling

The most common method used to identify and scale the CFA models in the measurement model at each measurement occasion (i.e., the first-order factors) is the marker variable method (Bollen & Curran, 2006). The same referent indicator is defined for the latent variable $F$ at each measurement occasion. Specifically, the intercept and the factor loading of the same indicator are fixed at zero and one, respectively, at each measurement occasion. The remaining intercepts and factor loadings are freely estimated. The structural model is identified by fixing the mean of the intercept growth factor mean at zero. In this method, the mean and variance of the growth factors ($\eta_{0i}$ and $\eta_{1i}$) will be determined by the mean of the referent indicator variable. This is because the scaling of the factor mean of the latent variable $F$ at each measurement occasion is based on the mean of the referent indicator (Newsom, 2015). Other identification and scaling methods can also be used. Interested readers are referred to Newsom (2015).

### 4.2.2 Covariance structure

Different variance-covariance structures can be specified and tested in both the measurement and the structural models. For instance, homogeneity of the residual variances associated with the first-order factors at each measurement occasion can be tested as well as correlated first-order indicators residual variances. More complex covariance structure can also be tested (see for instance Grilli & Varriale).

## 5  Steps in fitting an MILCM

The specification of an MILCM entails different choices involving both the measurement and the structural models. Indeed, specifying an MILCM necessitates both a mean and covariance structure present in both the measurement and the structural models. In this section we are going to provide some steps that can be followed when fitting an MILCM. It is important to note that theory has to be used when selecting the right model. In general, we recommend using theory and other knowledge such as the data collection process along with the steps outlined in this section.

The first step in fitting an MILCM is to perform preliminary analysis. During this step, descriptive statistics as well as bivariate statistics are performed. Another analysis that can be carried out is an EFA to investigate the unidimensionality of the latent construct. In the case of multidimensional constructs, one MILCM has to be run for each construct, either separately or combined, in a bivariate MILCM fashion. For an example of bivariate MILCM, readers are referred to Isiordia and Ferrer (2018).

The second step in fitting an MILCM is to investigate full and partial longitudinal measurement invariance in the measurement model. This step aims to study the first-order factors and to ensure an equal definition of latent constructs over time. The CFA models in the measurement model can be scaled and identified using the marker variable method presented previously.

The CFA models can be evaluated using goodness of fit. Following common practice in SEM, goodness of fit can be evaluated using a variety of fit indices: the root mean square error of approximate (RMSEA; Browne & Cudeck, 1992) and its 90% confidence interval (90% CI; MacCallum, Browne, & Sugawara, 1996), comparative fit index (CFI, Bentler, 1990), and Tucker–Lewis index (TLI; Bentler & Bonett, 1980; Tucker & Lewis, 1973). CFI and TLI values greater than or equal to 0.95 and RMSEA up to 0.10 (Browne & Cudeck, 1992; MacCallum et al., 1996) are considered to be minimally sufficient criteria for acceptable model fit. For nested models, the Chi square difference test is used. However, measurement equivalence can be also evaluated using practical fit indices (i.e., TLI and RMSEA). Evidence of measurement invariance is then supported when the difference in the TLI and the RMSEA between nested models is smaller than 0.05 (Cheung & Rensvold, 2002; Meade, Johnson, & Braddy, 2006).

Once the full (or partial) longitudinal measurement invariance has been established, the next step is to specify the structural part of the MILCM. The first step here is to characterize the mean structure of the MILCM. Different functional forms of growth can be considered afterward, for instance, fixing the mean of the intercept growth factor for identification purposes. The linear function is the simplest functional form of growth to consider. Then, nonlinear functional form of growth can be investigated. A class of nonlinear function where some of the time coefficients are estimated rather than fixed can be fit to the data as well. These models are called latent basis models or free time score (Bollen & Curran, 2006; McArdle, 1988). Another class of nonlinear growth trajectory that can be studied is the polynomial functional form of growth and the piecewise function. This class is characterized by the fact that there is a linear relationship between the dependent variable (repeated measurements which are defined by the first-order factors) and the parameters associated with the trajectory. Among this family, the quadratic and linear piecewise are the most common. The adequacy of this function can be assessed with the goodness fit as well as the normalized results.

**66** Big data in psychiatry and neurology

The next step consists of investigating different error structures that fit the data well. For instance, one can consider testing the hypothesis of homogenous variance of the first-order factors across time points. Indicators' residual correlations can also be studied. More generally, different specifications of variance-covariance structure of measurement errors can be examined. For details, interested readers can refer to Grilli and Varriale (2014) who provide some guidelines on the specifications of variance–covariance structure of measurement errors in MILCM.

## References

Bentler, P. M. (1990). Comparative fit indexes in structural models. *Psychological Bulletin, 107*(2), 238.

Bentler, P. M., & Bonett, D. G. (1980). Significance tests and goodness of fit in the analysis of covariance structures. *Psychological Bulletin, 88*(3), 588.

Bishop, J., Geiser, C., & Cole, D. A. (2015). Modeling latent growth with multiple indicators: A comparison of three approaches. *Psychological Methods, 20*, 43–62.

Blozis, S. A. (2007). On fitting nonlinear latent curve models to multiple variables measured longitudinally. *Structural Equation Modeling, 14*, 179–201.

Bollen, K. A., & Curran, P. J. (2006). Latent curve models. In *A structural equation perspective*. New York: Wiley.

Braams, B. R., van Duijvenvoorde, A. C. K., Peper, J. S., & Crone, E. A. (2015). Longitudinal changes in adolescent risk-taking: A comprehensive study of neural responses to rewards, pubertal development, and risk-taking behavior. *Journal of Neuroscience, 35*(18), 7226–7238.

Brown, T. A. (2015). *Confirmatory factor analysis for applied research* (2nd ed.). New York: NY Guilford Publications.

Browne, M. W., & Cudeck, R. (1992). Alternative ways of assessing model fit. *Sociological Methods & Research, 21*(2), 230–258.

Browne, M. W., & du Toit, S. H. C. (1991). Models for learning data. In L. Collins, & J. L. Horn (Eds.), *Best methods for the analysis of change* (pp. 47–68). Washington, DC: APA.

Bryk, A. S., & Raudenbush, S. W. (1992). *Hierarchical linear models: Applications and data analysis methods*. Newbury Park, CA: Sage.

Caselli, R. J., Dueck, A. C., Huentelman, M. J., Lutz, M. W., Saunders, A. M., Reiman, E. M., et al. (2012). Longitudinal modeling of cognitive aging and the TOMM40 effect. *Alzheimer's & Dementia, 8*, 490–495.

Cheung, G. W., & Rensvold, R. B. (2002). Evaluating goodness-of-fit indexes for testing measurement invariance. *Structural Equation Modeling, 9*(2), 233–255.

Cho, S. (2019). Self-control, risky lifestyles, and bullying victimization among Korean youth: Estimating a second-order latent growth model. *Journal of Child and Family Studies, 28*, 2131–2144.

Cho, S., & Lee, J. R. (2019). Joint growth trajectories of bullying perpetration and victimization among Korean adolescents: Estimating a second-order growth mixture model-factor-of-curves with low self-control and opportunity correlates. *Crime & Delinquency, 66*, 1–42.

Diallo, T. M. O., & Lu, H. (2016). Consequences of misspecifying across-cluster time-specific residuals in multilevel latent growth curve models. *Structural Equation Modeling, 24*, 359–382.

Diallo, T. M. O., & Morin, A. J. S. (2015). Power of latent growth curve models to detect piecewise linear trajectories. *Structural Equation Modeling, 22*, 449–460.

Diallo, T. M. O., Morin, A. J. S., & Parker, P. D. (2014). Statistical power of latent growth curve models to detect quadratic growth. *Behavior Research Methods, 46*, 357–371.

Fabrigar, L. R., Wegener, D. T., MacCallum, R. C., & Strahan, E. J. (1999). Evaluating the use of exploratory factor analysis in psychological research. *Psychological Methods*, *4*, 272–299.

Grilli, L., & Varriale, R. (2014). Specifying measurement error correlations in latent growth curve models with multiple indicators. *Methodology*, *10*, 117–125.

Grimm, K. J., Ram, N., & Hamagami, F. (2011). Nonlinear growth curves in developmental research. *Child Development*, *82*, 1357–1371.

Guerra, M., Bassi, F., & Dias, J. G. (2020). A multiple-indicator latent growth mixture model to track courses with low-quality teaching. *Social Indicators Research*, *147*, 361–381.

Hattie, J. (1985). Methodology review: Assessing unidimensionality of tests and items. *Applied Psychological Measurement*, *9*(2), 139–164.

Horn, J. L., & McArdle, J. J. (1992). A practical and theoretical guide to measurement invariance in aging research. *Experimental Aging Research*, *18*, 117–144.

Isiordia, M., & Ferrer, E. (2018). Curve of factors model: A latent growth modeling approach for educational research. *Educational and Psychological Measurement*, *78*(2), 203–231.

Jennrich, R. I., & Bentler, P. M. (2011). Exploratory bi-factor analysis. *Psychometrika*, *76*, 537–549.

Johnson, M. H. (2011). Interactive specialization: A domain-general framework for human functional brain development? *Developmental Cognitive Neuroscience*, *1*, 7–21.

Kievit, R. A., Brandmaier, A. M., Ziegler, G., van Harmelen, A. L., de Mooij, S. M. M., Moutoussis, M., et al. (2018). Developmental cognitive neuroscience using latent change score models: A tutorial and applications. *Developmental Cognitive Neuroscience*, *33*, 99–117.

Laird, N. M., & Ware, J. H. (1982). Random-effects models for longitudinal data. *Biometrics*, *38*, 963–974.

Leite, W. L. (2007). A comparison of latent growth models for constructs measured by multiple items. *Structural Equation Modeling*, *14*, 581–610.

Liu, Y., Millsap, R. E., West, S. G., Tein, J.-Y., Tanaka, R., & Grimm, K. J. (2017). Testing measurement invariance in longitudinal data with ordered-categorical measures. *Psychological Methods*, *22*(3), 486–506.

MacCallum, R. C., Browne, M. W., & Sugawara, H. M. (1996). Power analysis and determination of sample size for covariance structure modeling. *Psychological Methods*, *1*(2), 130.

McArdle, J. J. (1988). Dynamic but structural equation modeling of repeated measures data. In J. R. Nesselroade, & R. B. Cattell (Eds.), *Vol. 2*. *The handbook of multivariate experimental psychology* (pp. 561–614). New York: Plenum Press.

McArdle, J. J., & Anderson, E. (1990). Latent variable growth models for research on aging. In J. E. Birren, & K. W. Schaie (Eds.), *The handbook of the psychology of aging* (pp. 21–43). New York: Plenum.

McArdle, J. J., & Epstein, D. B. (1987). Latent growth curves within developmental structural equation models. *Child Development*, *58*, 110–133.

Meade, A. W., Johnson, E. C., & Braddy, P. W. (2006). The utility of alternative fit indices in tests of measurement invariance. In *Paper presented at the annual academy of management conference*. GA: Atlanta.

Meredith, W. (1964). Notes on factorial invariance. *Psychometrika*, *29*, 177–186.

Meredith, W. (1993). Measurement invariance, factor analysis, and factorial invariance. *Psychometrika*, *58*, 525–543.

Meredith, W., & Tisak, J. (1984). "Tuckerizing" curves. In *Paper presented at the annual meeting of the psychometric society, Santa Barbara, CA*.

Meredith, W., & Tisak, J. (1990). Latent curve analysis. *Psychometrika*, *55*, 107–122.

Moustafa, A. A., Diallo, T., Amoroso, N., Zaki, N., Hassan, M., & Alashwal, H. (2018). Applying big data methods to understanding human behavior and health. *Frontiers in Computational Neuroscience*, *12*, 84.

## 68 Big data in psychiatry and neurology

Muthén, B. O., & Curran, P. J. (1997). General longitudinal modeling of individual differences in experimental designs: A latent variable framework for analysis and power estimation. *Psychological Methods*, *2*, 371–402.

Muthén, B. O., & Muthén, L. (2010). *Mplus short course. Topic 4. Advanced growth modeling, missing data analysis, and survival analysis.* Retrieved from http://www.statmodel.com/course_materials.shtml.

Newsom, J. T. (2015). *Chapter 2 in longitudinal structural equation modeling: a comprehensive introduction.* New York: Routledge.

Ordaz, S. J., Foran, W., Velanova, K., & Luna, B. (2013). Longitudinal growth curves of brain function underlying inhibitory control through adolescence. *Journal of Neuroscience*, *33*(46), 18109–18124.

Paranjpe, M. D., Chen, X., Liu, M., Paranjpe, I., Leal, J. P., Wang, R., et al. (2019). The effect of ApoE ε4 on longitudinal brain region-specific glucose metabolism in patients with mild cognitive impairment: A FDG-PET study. *NeuroImage: Clinical*, *22*. Art. No. 101795.

Ram, N., & Grimm, K. J. (2007). Using simple and complex growth models to articulate developmental change: Matching theory to method. *International Journal of Behavioral Development*, *31*, 303–316.

Raudenbush, S. W., & Bryk, A. S. (2002). *Hierarchical linear models: Applications and data analysis methods.* Thousand Oaks, CA: Sage.

Rogosa, D. R. (1995). Myths and methods: "Myths about longitudinal research," plus supplemental questions. In J. M. Gottman (Ed.), *The analysis of change* (pp. 3–65). Hillsdale, NJ: Erlbaum.

Sayer, A. G., & Cumsille, P. E. (2001). Second–order latent growth models. In L. M. Collins, & A. G. Sayer (Eds.), *New methods for the analysis of change* (pp. 179–200). Washington, DC: American Psychological Association.

Suzuki, K., Hirakawa, A., Ihara, R., Iwata, A., Ishii, K., Ikeuchi, T., et al. (2020). Effect of apolipoprotein E ε4 allele on the progression of cognitive decline in the early stage of Alzheimer's disease. *Alzheimer's & Dementia: Translational Research & Clinical Interventions*, *6*, 1.

Tisak, J., & Meredith, W. (1990). Descriptive and associative developmental models. In A. von Eye (Ed.), *Vol. 2. Statistical methods in developmental research* (pp. 387–406). San Diego, CA: Academic Press.

Tucker, L. R., & Lewis, C. (1973). A reliability coefficient for maximum likelihood factor analysis. *Psychometrika*, *38*(1), 1–10.

Vandenberg, R. J., & Lance, C. E. (2000). A review and synthesis of the measurement invariance literature: Suggestions, practices, and recommendations for organizational research. *Organizational Research Methods*, *3*, 4–70. https://doi.org/10.1177/109442810031002.

van den Bos, W., & Eppinger, B. (2016). Developing developmental cognitive neuroscience: From agenda setting to hypothesis testing. *Developmental Cognitive Neuroscience*, *17*, 138–144.

Wänström, L. (2009). Sample sizes for two-group second-order latent growth curve models. *Multivariate Behavioral Research*, *44*, 588–619.

Willett, J. B., & Sayer, A. G. (1994). Using covariance structure analysis to detect correlates and predictors of individual change over time. *Psychological Bulletin*, *116*, 363–381.

Williams, O. A., An, Y., Armstrong, N. M., Shafer, A. T., Helphery, J., Kitner-Triolo, M., et al. (2019). Apolipoprotein E epsilon4 allele effects on longitudinal cognitive trajectories are sex and age dependent. *Alzheimer's & Dementia*, *15*(12), 1558–1567.

Wu, A. D., Liu, Y., Gadermann, A. M., & Zumbo, B. D. (2010). Multiple-indicator multilevel growth model: A solution to multiple methodological challenges in longitudinal studies. *Social Indicators Research*, *97*, 123–142.

Chapter 4

# Challenges and solutions for big data in personalized healthcare

## Tim Hulsen
*Department of Professional Health Solutions & Services, Philips Research, Eindhoven, The Netherlands*

## 1 Introduction

The 21st century holds two promises: "big data" and "personalized healthcare." These two concepts seem to be opposites at a first glance: big data is about analyzing, systematically extracting information from, or otherwise dealing with datasets that are too large or complex to be dealt with by traditional data-processing application software (Hong et al., 2018), and personalized healthcare is about finding the right treatment for a specific patient. However, big data can help in achieving personalized healthcare. First of all, it is important to realize that "big data" can mean "long data" (or "tall data") as well as "wide data." "Long data" means that the data contain a large number of patients, whereas "wide data" means that the data contain a large number of data items (presumed that each row equals one patient and each column equals one data item). Especially wide data is of interest for personalized healthcare, since it means that we have a larger number of data items for the subject we are investigating, as well as for patients that are similar to him/her, which will increase the chance that we successfully can compare both datasets. Data is becoming wider because more measurements are being done through both medical devices and wearables (see Fig. 2 of Estrada-Galiñanes & Wac, 2019). For machine learning, the dataset also needs to have enough features to create a good statistical model (although having more features does not always mean better models). Long data can be useful as well, because it enables us to create patient cohorts from a large dataset that share characteristics with the patient we are investigating. For machine learning algorithms, the dataset needs to be long enough to be split up into a training set, a validation set, and a test set. Nonlinear algorithms particularly need long data (Brownlee, 2017). If the dataset is much wider than it is long, there is a risk of overfitting (Domingos, 2012).

Big Data in Psychiatry and Neurology. https://doi.org/10.1016/B978-0-12-822884-5.00016-7
Copyright © 2021 Elsevier Inc. All rights reserved.

For personalized healthcare, the use of big data and artificial intelligence (AI) holds a big promise. In September 2019, the company Deep Genomics announced that it had used AI to identify a treatment for a genetic disorder called Wilson's disease, a rare disorder that prevents sufferers from metabolizing trace amounts of copper found in food (Alsever, 2020). They fed a big dataset related to more than 200,000 gene mutations into so-called training algorithms, teaching their computers to find connections between the gene mutations and the faulty proteins they encode, which in turn seem to drive certain human diseases. In the case of Wilson's disease, the algorithms found just such a connection with very high speed. They had deciphered precisely how a mutation known as Met645Arg leads to a crucial defect in an essential copper-metabolizing protein. Machine learning and deep learning algorithms are also used more and more to predict disease outcomes, finding correlations that would probably never have been found by human analysis.

Physicians also get more comfortable with using AI in their daily job. According to the Philips Future Health Index 2019 (Philips, 2019) this holds particularly for workflow applications such as staffing and patient scheduling (64%), and to a lesser degree for clinical applications, including flagging anomalies (59%), diagnosis (47%), and recommending treatment plans (47%). Philips also sees six areas in which AI is changing the future of healthcare (Philips, 2020a): (1) Precision diagnosis: from workflow optimization to clinical decision support; (2) Cancer care: improving accuracy of early detection for precision diagnosis; (3) Computational pathology: providing the pathologist a helping hand; (4) Acute care: spotting early signs of deterioration for timely intervention; (5) Image-guided therapy: helping the clinician focus on the procedure and the patient; (6) Access to care: could AI be the great healthcare equalizer? Especially the last area is an interesting one: currently, there are many differences in the quality of healthcare, depending on the region and care setting. Next-generation technologies such as AI, machine learning, and blockchain offer us the possibility to change this by democratizing healthcare and providing access to care in any setting.

The advancement of big data also means that new methods dedicated to improving data collection, storage, cleaning, processing, and interpretation for medical research continue to be developed. This does not only affect the 'hot' area of data science, but also data management, data stewardship, and data governance. Exploiting new tools and methods to extract meaning from big data has the potential to drive real change in clinical practice, and combining this novel data-driven research with the classical hypothesis-driven research will have a large impact on personalized healthcare (Hulsen et al., 2019). However, significant challenges remain. In this book chapter we discuss the challenges (and possible solutions) posed to biomedical research by our increasing ability to collect, store, and analyze large datasets. Important challenges include: (1) the need for standardization of data content, format, and clinical definitions, adhering to the FAIR guiding principles (Wilkinson et al., 2016); (2) the need

## Challenges and solutions for big data Chapter | 4 **71**

for collaborative networks with sharing of both data and expertise, for example through a federated approach; (3) stricter privacy and ethics regulations, in particular the GDPR in the European Union; and (4) a need to reconsider how and when analytic methodology (data science) is taught to medical researchers. Overcoming these challenges will help to make a success of the use of big data in medical and translational research (Hulsen, 2019a).

## 2 Standardization

### 2.1 Interoperability and reusability

One of the main issues when using big data for personalized healthcare is that data cannot be integrated with other data, limiting its (re)usability: the data stands on its own and cannot be read properly by applications or workflows for analysis, storage, and processing. Data integration is usually performed by a data manager or data steward (Hartter, Ryan, Mackenzie, Parker, & Strasser, 2013), a function that has been gaining importance over the past years, but until recently have had to do their job without having a clear set of rules to guide them. In 2016 however, the FAIR Guiding Principles for scientific data management and stewardship (Alsever, 2020) were published. FAIR stands for the four foundational principles—Findability, Accessibility, Interoperability, and Reusability—that serve to guide data producers and publishers as they navigate around the obstacles around data management and stewardship. The publication offers guidelines around these four principles. Interoperability and reusability are both part of the FAIR Guiding Principles and are closely connected to standardization: using a broadly applicable language for knowledge representation (ontologies, I1), vocabularies (I2), and domain-relevant community standards (R1.3). In the following paragraphs, several ontologies, vocabularies, and standards will be discussed. They have been split up in four sections: (1) standards for clinical data, (2) standards for -omics data, (3) standards for imaging data, and (4) standards for biosample data.

In February 2019, the Heads of Medicine Agencies (HMA) and the European Medicines Agency (EMA) jointly created a Big Data Task Force, and published a summary report (HMA-EMA Joint Big Data Taskforce, 2019), describing in detail what big data is, and what their view is on data standardization, data quality, data sharing and access, data linkage and integration, and data analytics. In January 2020, the Big Data Task Force presented 10 Priority Recommendations (HMA-EMA Joint Big Data Taskforce, 2020) meant to evolve its approach to data use and evidence generation, in order to make best use of big data to support innovation and public health. In the Netherlands, the recently announced collaboration Standards Developing Organizations Netherlands (SDO-NL) (Nictiz, 2020) aims to optimize healthcare by a better exchange of information for patients and professionals, based on existing standards such as HL7 and SNOMED. The Health Research Infrastructure

## 72  Big data in psychiatry and neurology

(Health-RI) (Health-RI, 2020) is a Dutch initiative to facilitate and stimulate an integrated health data infrastructure accessible for researchers, citizens, and care providers. An important part of this is standardization, which is why Health-RI is also very much involved in initiatives around Open Science and the implementation of the FAIR Guiding Principles.

## 2.2  Standards for clinical data

The term "clinical data" is used here as a placeholder for (usually textual or numerical) data that is collected through an EHR, an eCRF, patient registry, questionnaires, lab results, etc. This type of data exists for a long time, and thus many ontologies, vocabularies, and standards to describe clinical data exist.

### 2.2.1  SNOMED CT

The Systematized Nomenclature of Medicine Clinical Terms (SNOMED CT) (Donnelly, 2006) is a systematically organized computer processable collection of medical terms providing codes, terms, synonyms, and definitions used in clinical documentation and reporting. It was started in 1965 as the Systematized Nomenclature of Pathology (SNOP) and was further developed into a logic-based healthcare terminology. SNOMED CT was created in 1999 (and released in 2002) by the merger, expansion, and restructuring of two large-scale termi-nologies: SNOMED Reference Terminology (SNOMED RT) and the Clinical Terms Version 3 (CTV3). The July 31, 2019 release of the SNOMED CT International Edition included 350,830 current concepts (SNOMED, 2019). SNOMED CT covers all kinds of clinical findings such as diagnosis, signs, and symptoms. It includes tens of thousands of surgical, therapeutic, and diagnostic procedures. It includes observables (e.g., heart rate) and also includes concepts representing body structures, organisms, substances, pharma-ceutical products, physical objects, physical forces, specimens, and many other types of information that may need to be recorded in or around the health record. The core component types in SNOMED CT are (1) concepts, (2) descriptions, and (3) relationships. Every concept represents a unique clinical meaning, which is referenced using a unique, numeric, and machine-readable SNOMED CT identifier. Two types of description are used to represent every concept: Fully Specified Name (FSN) and Synonym. The FSN represents a unique, unambiguous description of a concept's meaning. A synonym represents a term that can be used to display or select a concept. A concept may have several synonyms. A relationship represents an association between two concepts. Relationships are used to logically define the meaning of a concept in a way that can be processed by a computer. The relationship type (or attribute) is used to represent the meaning of the association between the source and destination concepts.

### 2.2.2 LOINC

Logical Observation Identifiers Names and Codes (LOINC) (Forrey et al., 1996) was developed to provide a definitive standard for identifying clinical information in electronic reports. The LOINC database provides a set of universal names and ID codes for identifying laboratory and clinical test results in the context of existing observation report messages. One of the main goals of LOINC is to facilitate the exchange and pooling of results for clinical care, outcomes management, and research. LOINC codes are intended to identify the test result or clinical observation. Other fields in the message can transmit the identity of the source laboratory and special details about the sample. The Regenstrief LOINC Mapping Assistant (RELMA) (Regenstrief Institute, 2020) can help users to map their local terms or lab tests to universal LOINC codes.

### 2.2.3 CDISC ODM

The Clinical Data Interchange Standards Consortium (CDISC) Operational Data Model (ODM) (CDISC, 2020) was introduced in 1999 and designed to facilitate the regulatory-compliant acquisition, archive, and interchange of metadata and data for clinical research studies. CDISC ODM is a vendor-neutral, platform-independent format for interchange and archive of clinical study data. The model includes the clinical data along with its associated metadata, administrative data, reference data, and audit information. It is an XML schema that provides number of constructs for modeling electronic Case Report Forms (eCRFs). CDISC ODM is also used in sending forms data from a clinical trial system to an electronic health record (EHR) system.

### 2.2.4 NCI Thesaurus

National Cancer Institute Thesaurus (NCIt) (Fragoso, de Coronado, Haber, Hartel, & Wright, 2004) provides reference terminology for many National Cancer Institute (NCI) and other systems. It covers vocabulary for clinical care, translational and basic research, and public information and administrative activities. NCIt currently features over 100,000 textual definitions and over 400,000 cross-links between concepts, and is updated frequently by a team of subject matter experts.

### 2.2.5 HL7 FHIR

Health Level Seven International (HL7) provides a number of standards for electronic health information. The Fast Healthcare Interoperability Resources (FHIR) (Bender & Sartipi, 2013) is a standard describing data formats and elements and an application programming interface (API) for exchanging electronic health records. FHIR aims to define the key entities involved in healthcare information exchange as resources. Each resource is a distinct identifiable entity. Example resources include "patient," "device," and "document."

The current version (4.0.1) contains 145 resources in 5 main categories: "Foundation," "Base," "Clinical," "Financial," and "Specialized."

### 2.2.6 Project-specific codebooks

In spite of the abovementioned standards, in research projects it is not always feasible to map all data items to either one of them. Sometimes researchers try to make discoveries by studying new types of data, which could mean that there are data items that are not included in any standard yet. In this kind of projects, it is essential that there is a very precisely defined codebook (Hulsen, 2019a). Within several prostate cancer projects the author has worked on, such as Movember GAP3 (Hulsen et al., 2016), ERSPC (Hulsen et al., 2019), and RE-IMAGINE (Moss et al., 2020), this codebook was created with much detail through close collaboration with the clinical experts.

## 2.3 Standards for -omics data

The "omics" are various disciplines in biology whose names end with -omics, such as genomics, transcriptomics, and proteomics. In spite of the dynamics of these disciplines and the resulting wide range of data sources and data types, some standards in the form of ontologies have been created.

### 2.3.1 Gene Ontology

The Gene Ontology (GO) resource (Ashburner et al., 2000; The Gene Ontology Consortium, 2019) provides a computational representation of our current scientific knowledge about the functions of proteins and noncoding RNA molecules produced by genes from many different organisms, including humans. It is the world's largest source of information on the functions of genes. This knowledge is both human-readable and machine-readable, and is a foundation for computational analysis of large-scale molecular biology and genetics experiments in biomedical research. GO describes our knowledge of the biological domain with respect to three aspects (The Gene Ontology Consortium, 2020): (1) Molecular Function: molecular-level activities performed by gene products. Molecular function terms describe activities that occur at the molecular level. GO molecular function terms represent activities rather than the entities that perform the actions, and do not specify where, when, or in what context the action takes place. (2) Cellular Component: the locations relative to cellular structures in which a gene product performs a function, either cellular compartments, or stable macromolecular complexes of which they are parts. (3) Biological Process: the larger processes, or 'biological programs' accomplished by multiple molecular activities.

## 2.3.2 Human Phenotype Ontology

Deep phenotyping has been defined as the precise and comprehensive analysis of phenotypic abnormalities in which the individual components of the phenotype are observed and described. The three components of the Human Phenotype Ontology (HPO) (Kohler et al., 2017) project are the phenotype vocabulary, disease-phenotype annotations, and the algorithms that operate on these. These components are being used for computational deep phenotyping and precision medicine as well as integration of clinical data into translational research.

## 2.4 Standards for imaging data

"Imaging" in healthcare usually refers to medical imaging such as magnetic resonance imaging (MRI), UltraSound (US) imaging, Positron Emission Tomography—Computed Tomography (PET/CT) imaging, or X-ray imaging, but it can also refer to other imaging types such as histopathological imaging.

### 2.4.1 DICOM

The main standard for storing medical images is Digital Imaging and Communications (DICOM), which was created by ACR-NEMA in 1985 and published in 1992 (Bidgood & Horii, 1992). DICOM files have metadata embedded into the file itself. A DICOM data object consists of a number of attributes ('DICOM tags'), including items such as name, ID, etc., and also one special attribute containing the image pixel data (DICOM—Digital Imaging and Communications in Medicine, 2020). Currently, there are over 4000 DICOM tags. A full list can be found at the website of the DICOM library (DICOM Library, 2020). DICOM supports many modalities, such as CT, MR, PET, and X-ray.

### 2.4.2 RadLex

RadLex is a lexicon of terms relevant to diagnostic and interventional radiology (Langlotz, 2006). This vocabulary has been developed by the Radiological Society of North America (RSNA) to provide uniform terminology for clinical practice, research, and education in medical imaging. To meet the day-to-day terminological requirements of radiologists, it imports vocabularies imported from external ontologies. Among its clinical applications, RadLex has been used to encode the results of radiologic procedures and to search the content of radiology reports.

### 2.4.3 Quantitative Imaging Biomarker Ontology

The Quantitative Imaging Biomarker Ontology (QIBO) (Buckler et al., 2013) is the main ontology for imaging biomarkers. It consists of 488 terms spanning the following upper classes: Experimental subject, biological intervention, imaging

76    Big data in psychiatry and neurology

agent, imaging instrument, image postprocessing algorithm, biological target, indicated biology, and biomarker application.

## 2.5    Standards for biosample data

Standardization of biosamples allows for data generated from one individual/ cohort to be used in other related studies and removes the need to generate the same biological data over and over again (if permitted by privacy and data sharing laws).

### 2.5.1    OBIB

The Ontology for BIoBanking (OBIB) (Brochhausen et al., 2016) is an ontology built for annotation and modeling of biobank repository and biobanking administration. OBIB is based on a subset of the Ontology for Biomedical Investigation (OBI), has the Basic Formal Ontology (BFO) as its upper ontology, and is developed following OBO Foundry principles. The first version of OBIB resulted from the merging of two existing biobank-related ontologies, OMIA-BIS, and biobank ontology.

### 2.5.2    MIABIS

The Minimum Information About BIobank data Sharing (MIABIS) (Norlin et al., 2012) was developed in 2012 by the Biobanking and BioMolecular Resources Research Infrastructure of Sweden (BBMRI.se). The wide acceptance of the first version of MIABIS encouraged evolving it to a more structured and descriptive standard. In 2013 a working group was formed under the Biobanking and BioMolecular Resources Research Infrastructure (BBMRI-ERIC), aiming to continue the development of MIABIS through a multicountry governance process. MIABIS 2.0 Core was published in 2016 (Merino-Martinez et al., 2016) and contains 22 attributes describing Biobanks, Sample Collections, and Studies according to a modular structure that makes it easier to adhere to and to extend the standard. This integration standard will make a great contribution to the discovery and exploitation of biobank resources and lead to a wider and more efficient use of valuable bioresources, thereby speeding up the research on human diseases.

### 2.5.3    BioSCOOP

The Biobank Sample Communication Protocol (BioSCOOP) ( Jarczak et al., 2019) is a new approach for the transfer of information between biobanks. It has the form of a well-documented JSON API which describes an organized data format for a list of attributes describing the donor with particular emphasis on the phenotype, anthropological measurements, medical data, and sample material.

### 2.5.4 OBI

The Ontology for Biomedical Investigations (OBI) (Bandrowski et al., 2016) is a generic biomedical ontology that provides terms with precisely defined meanings to describe all aspects of how investigations in the biological and medical domains are conducted. OBI reuses ontologies that provide a representation of biomedical knowledge from the Open Biological and Biomedical Ontologies (OBO) project and adds the ability to describe how this knowledge was derived. OBI covers all phases of the investigation process, such as planning, execution, and reporting. It represents information and material entities that participate in these processes, as well as roles and functions. OBI is being used in a wide range of projects covering genomics, multiomics, immunology, and catalogs of services. The current release (of 2019-11-20) contains 3479 classes.

## 3 Data sharing and integration

### 3.1 Data ownership

Medical institutes appear to be the owners of patient data, because the data reside on computers within their walls. However, the data actually is the property of the patient and the access and use of that data outside of the institution requires patient consent (Allen, Adams, & Flack, 2019). This inhibits the rapid exploitation of the large volume of retrospective data available in clinical records: while retrospective hypothesis-driven research can be undertaken on specific, anonymized data as with any research, once the study has ended the data should be destroyed (Hulsen, Jamuar, et al., 2019). This is, of course, a big loss for data scientists, for whom these retrospective datasets could be a "pot of gold," especially if they are annotated well. Therefore, instead of relying on retrospective data, it could be a better option to start new studies with an informed consent for hypothesis-driven research, and perhaps even with a 'blanket consent' for the future reuse of their data. The discussion about data ownership and data sharing has been brought to the fore by the start of the General Data Protection Regulation (GDPR) of the European Union in 2018, initiating an international debate on data sharing in health (Knoppers & Thorogood, 2017).

Kalkman, Mostert, Gerlinger, van Delden, and van Thiel (2019) performed a systematic review of literature and ethical guidelines for principles and norms pertaining to data sharing for international health research. They observed an abundance of principles and norms with considerable convergence at the aggregate level of four overarching themes: societal benefits and value; distribution of risks, benefits, and burdens; respect for individuals and groups; and public trust and engagement. However, at the level of principles and norms they identified substantial variation in the phrasing and level of detail, the number and content of norms considered necessary to protect a principle, and the contextual approaches in which principles and norms are used.

## 3.2 Support for data sharing

The idea that data should be shared as much as possible to enable scientific progress is not new, but is gaining momentum, mainly because of the increasing power of big data analyses, machine learning, deep learning, etc. Datasets are increasingly shared through the study website, a public repository or elsewhere, as can be seen in reviews such as for prostate cancer data (Hulsen, 2019b). Publishers also increasingly expect from researchers that they share their data together with the publication, and funding programs such as Horizon 2020 obligate researchers to share data in a public repository after the study has finished. The EU also stresses in their action plan (Auffray et al., 2016) around big data in health research that the sharing of data from patient registries could have large benefits for healthcare. However, there are still many challenges around data sharing, at the ethical/legal, cultural, financial, and/or technical levels (Figueiredo, 2017).

Data sharing initiatives are opportunities to increase the speed of knowledge discovery and scientific progress. The reuse of research data has the potential to bring new views from multiple analyses of the same dataset. However, data sharing also poses challenges to the scientific community. Figueiredo (2017) reviews the impact that data sharing has in science and society and presents guidelines to improve the efficient sharing of research data.

### 3.2.1 Open Science

The Science in Transition movement (Dijstelbloem, Huisman, Miedema, & Mijnhardt, 2013) states that "science has become a self-referential system where quality is measured mostly in bibliometric parameters and where societal relevance is undervalued," emphasizing that researchers tend to care mostly about publications and not so much how the data can benefit society. Science in Transition also points to the fact that more than 70% of researchers have tried and failed to reproduce another scientist's experiments, and more than half have failed to reproduce their own experiments (Baker, 2016). A possible solution here is 'Open Science': the practice of science in a sustainable manner which gives others the opportunity to work with, contribute to, and make use of the scientific process. This allows nonscientific users to influence the scientific research world with questions and ideas and help gather research data (Nationaal Platform Open Science, n.d.).

### 3.2.2 Sharing with industry

Healthcare Business and Technology wrote about "How data sharing could change the entire healthcare industry" (Ketchum, 2018). It discusses the partnership announced by Apple in 2018 with 13 major healthcare systems, which will allow Apple to download patients' electronic health data onto its devices (with patient permission). This type of data sharing could transform the U.S.

healthcare industry by empowering patients and improving clinical care. It could even streamline care processes: hospital staff would spend less time on making data available to patients. Patients could be helped faster, by having their questions answered by artificial intelligence algorithms using their data.

The Mayo Clinic Platform (Cohen, 2020) is a new cloud-based clinical data analytics platform, housing deidentified patient data, which external partners can link up to via application programming interfaces, as well as establishing standard templates for compliance and legal agreements. Mayo Clinic Platform's first partner on the clinical data analytics platform is Nference, a software startup that Mayo is an investor in. Nference develops analytics, machine learning, and natural language processing tools to aid data scientists. Mayo Clinic hopes to work with pharmaceutical companies to commercialize new therapies. Mayo itself wouldn't commercialize those therapies, though the system could receive royalties from insights generated on the platform, which could then be reinvested into Mayo's clinical practice, research, and education work.

## 3.3 Data sharing initiatives

There are many initiatives around the world supporting the sharing of medical data (Hulsen, 2020). Some of them enable data sharing through a central database, and others give patients the possibility to share their EHR data. A relatively new development in healthcare is data federation, which offers a way to share data without sending the data to central database.

### 3.3.1 GIFT-Cloud

GIFT-Cloud (Doel et al., 2017) is a data sharing and collaboration platform for medical imaging research. It was built to support GIFT-Surg, an international research collaboration that is developing novel imaging methods for fetal surgery, but GIFT-Cloud also has general applicability to other areas of imaging research. It simplifies the transfer of imaging data from clinical to research institutions, facilitating the development and validation of medical research software, and the sharing of results back to the clinical partners. GIFT-Cloud supports collaboration between multiple healthcare and research institutions while satisfying the demands of patient confidentiality, data security, and data ownership.

### 3.3.2 Personalized Consent Flow

The Personalized Consent Flow (Rake et al., 2017) is a new consent model that allows people to control their personally collected health data and determine to what extent they want to share these for scientific research purposes. The proposed consent flow, which aims to be "personalized, transparent, and very simple," is characterized by three main features (Fig. 1 of Rake et al., 2017).

**80** Big data in psychiatry and neurology

First, users are asked general questions about sharing data. If they want to share data for scientific research, they may opt for a single-study consent or a multi-study consent. Users are able to decide which data will be shared for specific studies and with whom. Second, users can choose to share existing data that they have collected passively, to share prospectively, collect data, or both. For prospective studies, researchers can invite specific users to collect selected data during a specific time period. Users can also be notified about future studies by signing up for the research program. Third, expiration dates are connected to each consent choice, which ensures that a user reconsiders his decision. A default expiration date of one year will be assigned, but users may also select personal expiration dates, such as an expiration date connected to the duration of the study. Obviously, users can quit sharing data at any time. During all steps, users are informed about implications of consent options.

### 3.3.3  Sync for Science

In the US, a collaboration among EHR vendors, the National Institutes of Health (NIH), the Office of the National Coordinator for Health IT (ONC), and Harvard Medical School's Department of Biomedical Informatics was started, named Sync for Science (S4S) (Sync for Science, 2017). The vision of S4S is to allow individuals to access their health data and share these data with researchers to support studies that generate insights into human health and disease (Mandel, 2018). Because different EHRs collect and store health data differently, S4S has focused on promoting both authorization and standardization to make it possible for EHR systems to release high-quality data that researchers can readily consume. The first study to use S4S technology was the All of Us Research Program (The All of Us Research Program Investigators, 2019), which started in 2018 and expected to continue until at least 2028. S4S is broader than any single study: the goal is to ensure that any research study has a consistent, transparent, inexpensive, high-quality way to request EHR data donations. People are in charge of their own data: they can decide by themselves where and how to share their data, e.g., to drive scientific understanding of conditions and causes they care about.

### 3.3.4  DistibutedLearning.ai

The growing complexity of cancer diagnosis and treatment requires datasets that are larger than currently available in a single hospital or even in cancer registries. However, sharing patient data is difficult due to patient privacy and data protection needs. Privacy preserving distributed learning technology has the potential to overcome these limitations. The general idea behind distributed learning is that sites share a (statistical) model and model parameters instead of sharing sensitive data: each site runs computations on a local data store that generate these aggregated statistics. In this setting, organizations can collaborate by exchanging aggregated data/statistics while keeping the underlying data safely on site and undisclosed. DistributedLearning.ai (Integraal Kankercentrum Nederland, 2020) provides a way to use distributed learning technology, using the open source software Vantage.

### 3.3.5 Personal Health Train

The Personal Health Train (PHT) (Dutch Techcentre for Life Sciences, 2015; van Soest et al., 2018) aims to connect distributed health data and create value by increasing the use of existing health data for citizens, healthcare, and scientific research. The key concept in the PHT is to bring algorithms to the data where they happen to be, rather than bringing all data to a central place (federalization instead of centralization). The PHT is designed to give controlled access to heterogeneous data sources, while ensuring privacy protection and maximum engagement of individual patients and citizens. As a prerequisite, health data is made FAIR (Findable, Accessible, Interoperable and Reusable) (Fragoso et al., 2004). Stations containing FAIR data may be controlled by individuals, (general) physicians, biobanks, hospitals, and public or private data repositories. The PHT was applied recently to a project with 20,000+ lung cancer patients (Deist et al., 2020).

### 3.3.6 DataSHIELD

Research in modern biomedicine and social science requires sample sizes so large that they can often only be achieved through a pooled coanalysis of data from several studies. But the pooling of information from individuals in a central database that may be queried by researchers raises important ethico-legal questions and can be controversial. DataSHIELD (Gaye et al., 2014) provides a novel technological solution that can circumvent some of the most basic challenges in facilitating the access of researchers and other healthcare professionals to individual-level data. It facilitates research in settings where sharing the data itself is not possible (due to government restrictions, intellectual property issues, or data size). DataSHIELD has been used by the Healthy Obese Project and the Environmental Core Project of the Biobank Standardization and Harmonization for Research Excellence in the European Union (BioSHaRE-EU Kaye et al., 2016) for the federated analysis of 10 datasets across 8 European countries.

## 3.4 Data integration

Data that is shared over multiple partners usually needs to be integrated into a central database to enable analysis of the aggregated data. A number of options exist here, from open-source solutions to closed-course solutions.

## 82 Big data in psychiatry and neurology

### 3.4.1 tranSMART

The Movember Global Action Plan 3 (GAP3) database (Bruinsma et al., 2017; Hulsen et al., 2016) is an example of how data from (currently) 28 partners worldwide can be collected and integrated into a central database, using open-source software. This database currently contains data from more than 21,000 prostate cancer patients, undergoing active surveillance (Fig. 1). The clinical and (derived) imaging data has been gathered using a common data model (or "codebook"), specifically designed to answer the research questions defined by the principal investigators at the start of the GAP3 project. This data can be browsed through the open-source, web-based tranSMART platform (Scheufele et al., 2014), which is only accessible for a selected group of users. tranSMART supports some statistical analyses, such as correlation analysis, logistic regression, and survival analysis. Genomics data can be analyzed in tranSMART through built-in analysis methods such as group tests and heatmaps. The Movember GAP3 tranSMART instance is connected to R-studio to enable the statisticians to execute their own R scripts on the database. The Movember GAP3 infrastructure was (for the larger part) reused for the European Randomized Study of Screening for Prostate Cancer (ERSPC) study (Hulsen, Van der Linden, et al., 2019). This database contains data from almost 270,000 patients from 9 European institutes who were screened for prostate cancer.

### 3.4.2 i2b2

Informatics for Integrating Biology and the Bedside (i2b2) (Murphy et al., 2010) is an open-source clinical data warehousing and analytics research platform used at over 250 locations worldwide. i2b2 enables sharing, integration, standardization, and analysis of heterogenous data from healthcare and research. i2b2 and tranSMART are both part of the i2b2 tranSMART Foundation (since 2017).

### 3.4.3 Clinical Data Lake (CDL)

For the RE-IMAGINE project (Moss et al., 2020), a central database will be built using the HealthSuite Digital Platform (HSDP) Philips Clinical Data Lake (CDL) (Philips, 2018). CDL is a service for clinical research data management, which will be available as a micro-service: a service that can be instantiated by HSDP users for their data collection needs, with access for as few or as many people as the context demands. CDL supports standards such as FHIR, DICOM, and NIFTI, and contains pipelines for deidentification, data ingestion, data curation, annotation, data querying, and consent management.

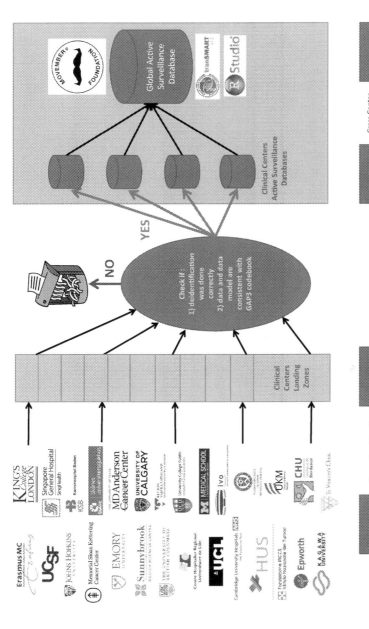

**FIG. 1** The IT infrastructure for the Movember GAP3 infrastructure integrates data from 28 institutes.

### 3.4.4 cBioPortal

The open source cBioPortal for Cancer Genomics (Gao et al., 2013) provides visualization, analysis, and download of large-scale cancer genomics datasets, as well as integration with clinical data. A public instance of cBioPortal (https://www.cbioportal.org) is hosted and maintained by Memorial Sloan Kettering Cancer Center. It provides access to data by The Cancer Genome Atlas as well as many carefully curated published datasets. The cBioPortal software can be used to create local instances that provide access to private data.

## 4 Privacy and ethics

### 4.1 Stricter regulations

When using big data to achieve personalized healthcare, one of the biggest hurdles that needs to be taken is the stricter privacy and ethics regulations that have been taken into effect around the world. The most prominent example is the General Data Protection Regulation (GDPR) (The European Parliament and the Council of the European Union, 2016) of the European Union, which is in effect since May 25, 2018. The GDPR applies if the data controller (the organization that collects data), data processor (the organization that processes data on behalf of the data controller), or data subject is based in the EU. For science, this means that all studies performed by European institutes/companies and/or on European citizens will be subject to the GDPR, with the exception of data that is fully anonymized (Intersoft Consulting, 2018c). The GDPR sets out seven key principles: lawfulness, fairness, and transparency; purpose limitation; data minimization; accuracy; storage limitation; integrity and confidentiality (security); and accountability. The GDPR puts some constraints on data sharing, e.g., if a data controller wants to share data with another data controller, he/she needs to have an appropriate contract in place, particularly if that other data controller is located outside the EU (Data Protection Network, 2018). If a data controller wants to share data with a third party, and that third party is a processor, then a Data Processor Agreement (DPA) needs to be made. Apart from this DPA, the informed consent that the patient signs before participating in a study needs to state clearly for what purposes their data will be used (Intersoft Consulting, 2018a). Penalties for noncompliance can be significant: up to €20 million or 4% of annual turnover. Because health data are "sensitive," potential discrimination has been addressed in legislation and a more proportionate approach is applied to balance privacy rights against the potential benefits of data sharing in science (Knoppers & Thorogood, 2017). In fact, processing of biological or health data is prohibited unless the subject gives "explicit consent," processing is necessary for the purposes of provision of services or management of health system, or processing is necessary for reasons of public interest in the area of public health. Besides GDPR, there are new privacy laws elsewhere in the world as well, such as the California Consumer Privacy

Act (CCPA) (State of California, 2018), which is stricter than the existing Health Insurance Portability and Accountability Act (HIPAA) of the USA (Rothstein & Tovino, 2019). The CCPA gives California residents the right to: (1) know what personal information is being collected about them; (2) know whether their personal information is sold or disclosed and to whom; (3) say no to the sale of personal information; (4) access their personal information; (5) equal service and price, even if they exercise their privacy rights. The CCPA is similar to the GDPR, but the CCPA also grants consumers the right to sue if someone gets unauthorized access to their personal data, if they can show that the breach occurred because of a lack of "reasonable security procedures" such as encryption or redacting identifying data.

## 4.2 Explicit consent

The question of "explicit consent" of patients for their healthcare data to be used for research purposes provoked intense debate already during negotiations for GDPR, but ultimately the advocacy groups of greater privacy defeated the research groups which lobbied that restricting access to billions of terabytes of data would hold back research in Europe. The research groups feared that GDPR would make healthcare innovation more laborious in Europe as lots of people would not give their consent. It would also add another layer of staff to take consent, making it more expensive. The stricter data protection laws in Europe compared to the US and China would enable the latter countries to move ahead in producing innovative healthcare technology (Smith, 2018). Within the GDPR, the data subject can withdraw the consent at any time, after which the data controller needs to remove all his/her personal data (the "right to be forgotten") (Intersoft Consulting, 2018b). Moreover, informed consents might contain so much and so complicated information (with > 20 different opt-in or opt-out checkboxes, using scientific terms) that patients are getting overwhelmed. Because of all issues around data sharing, scientists might consider (whenever possible) to share only aggregated data which cannot be traced back to individual data subjects, or to raise the abstraction level, sharing insights instead of data (Staten, 2018).

## 4.3 Privacy and ethics in industry

In the industry (health-tech, pharma, and others), GDPR and related laws have a very large effect. Infringement of the GDPR by a company can lead to huge fines: up to €20 million, or 4% of annual global turnover (whichever is higher). Therefore companies take privacy even more serious than before. For example, Philips has just published the Philips Data Principles (Philips, 2020b) and the Philips AI Principles (Philips, 2020c). The Philips Data Principles (Table 1) focus on security, privacy, and benefit to customers, patients and society as a

**86** Big data in psychiatry and neurology

**TABLE 1** The Philips Data Principles.

| | | |
|---|---|---|
| | Security | We ensure the security of all data entrusted to us. We operate under global security policies that guide our activities to protect against vulnerabilities and manage any incidents |
| | Privacy | We handle all personal data with integrity in compliance with all applicable privacy regulations of the countries in which we operate. We adhere to the Philips Code of Conduct—our binding corporate rules—that governs data transfers and processing within our company. When our business partners process personal data on our behalf, we ensure that they comply with our security and privacy requirements |
| | Beneficial | We aim to create innovative solutions that benefit our customers, patients, and society as a whole. We use your personal data in line with your reasonable expectations |

**TABLE 2** The Philips AI Principles.

| | | |
|---|---|---|
| | Well-being | We design our solutions to benefit the health and well-being of individuals and to contribute to the sustainable development of society |
| | Oversight | We design AI-enabled solutions to augment and empower people, with appropriate human supervision |
| | Robustness | We develop AI-enabled solutions that are intended to do no harm, with appropriate protection against deliberate or inadvertent misuse |
| | Fairness | We develop and validate solutions using data that is representative of the target group for the intended use, and we aim to avoid bias or discrimination |
| | Transparency | We disclose which functions and features of our offerings are AI-enabled, the validation process, and the responsibility for ultimate decision-taking |

whole. The Philips AI Principles (Table 2) are about well-being, oversight, robustness, fairness, and transparency. Other companies working in the medical domain have similar principles and guidelines in place, to make sure that the use of big data and data science does not interfere with security and privacy regulations.

## 5 Teaching data science

### 5.1 Need for more training

From the previous paragraphs it is clear that big data, if standardized and shared in a correct manner, can open up the road to interesting developments and discoveries in the area of personalized healthcare. However, increasing availability of data has been matched by a shortage of those with the skills to analyze and interpret those data: data analysts and data scientists. Data volume has increased faster than predicted and, although the current shortage of data analysts such as bioinformaticians was foreseen already in the twentieth century (MacLean & Miles, 1999), corrective measures are still required to encourage skilled data analysts to work on (bio)medical problems. Including prior knowledge of relevant domains demonstrably improves the performance of models built on big data (Camacho, Collins, Powers, Costello, & Collins, 2018; Libbrecht & Noble, 2015), suggesting that the perfect data analyst should not only be trained in data science but also in biomedicine. However, teaching data scientists about biomedicine is only half of the puzzle. Healthcare professionals should also learn the basics of data science, so that they have a better understanding of the data science, and will be able to discuss results with data scientists.

### 5.2 Training data science to medical students

Artificial intelligence driven by machine learning or deep learning algorithms is a branch in computer science that is rapidly gaining popularity within the healthcare sector. Recent regulatory approvals of AI-driven companion diagnostics and other products are glimmers of a future in which these tools could play a key role by defining the way medicine will be practiced. Educating the next generation of medical professionals with the right machine learning techniques will enable them to become part of this emerging data science revolution (Kolachalama & Garg, 2018). Healthcare practitioners use insights from digital medical devices, software, and computer-based tools every day. These tools are dependent on AI more and more. For example, an AI-enabled magnetic resonance imaging (MRI) machine can help a radiologist identify and diagnose a tumor. A clinical decision support (CDS) tool can predict the disease outcome of a patient using AI. How these devices make such decisions and how accurate these decisions are is in the heart of medical care, as doctors use these insights into make countless decisions every day (Saqr & Tedre, 2019). Furthermore, patients are increasingly using their own AI-based devices to monitor their own health and activity through heart rate monitors and other wearables. Therefore it is crucial that healthcare practitioners understand at least the basics of AI. To make this possible, medical curricula should embrace the fundamentals of big data, modern inferential techniques, and principles of computational thinking. To grow familiar with big data, medical students will need to learn to see and interpret the world through computational and data-driven lenses. By understanding the principles involved in computational sciences, healthcare

88 Big data in psychiatry and neurology

practitioners will have improved problem-solving skills well suited for the latest technology, will better understand the insights generated by the devices they use, will design better research, and, most importantly, will deliver a better service to patients. The students will not need to learn all ins and outs of data science, but have a general understanding of what data-driven research is, how artificial intelligence, machine learning, and deep learning work, and how data-driven research and hypothesis-driven research can complement each other (Hulsen, Jamuar, et al., 2019).

## 5.3 Available courses in Clinical Data Science

Studies in the field of translational research usually collect an abundance of data types: clinical data (demographics, death/survival data, questionnaires, etc.), imaging data (MR, UltraSound, PET, CT, and derived values), biosample data (values from blood, urine, etc.), molecular data (genomics, proteomics, etc.), digital pathology data, data from wearables (blood pressure, heart rate, insulin level, etc.), and much more. To combine and integrate these data types, the scientist needs to understand both informatics (data science, data management, and data curation) and the specific disease area. As there are very many disease areas which all require their own expertise, we will focus here on the informatics side: data integration in translational research. Although this field is relatively new, there are a number of online and offline trainings available. As for the online trainings, Coursera offers a course on "Big Data Integration and Processing" (UC San Diego, 2020), as well as a specialization in "Clinical Data Science" (University of Colorado, 2020). The i2b2 tranSMART Foundation, which is developing an open-source/open-data community around the i2b2, tranSMART, and OpenBEL translational research platforms, has an extensive training program available as well (i2b2 tranSMART Foundation, 2020). As for the offline trainings, ELIXIR offers a number of workshops and courses around data management and data analysis (ELIXIR, 2020). The European Bioinformatics Institute (EBI) has created a 4-day course specific for big data analysis in systems biology (European Bioinformatics Institute, 2020).

## 6 Discussion

The challenges described in this chapter are selected on basis of the current situation in the fields of big data and personalized healthcare. If we look into the future, different challenges might lie ahead. Concerning big data and data science, the Internet of Things (IoT) (Islam, Kwak, Kabir, Hossain, & Kwak, 2015) will mean that data from different systems (such as wearables) can be shared and integrated more easily. The emergence of even faster internet connections (e.g., 5G) (Li, 2019), combined with steadily increasing computing power, will speed up the importance of the use of artificial intelligence and deep learning.

For personalized healthcare there are interesting new developments on the horizon as well, such as gene editing and genetic engineering. Over the past 20 years, only 22 gene therapy drugs have been approved (Ma, Wang, Xu, He, & Wei, 2019), but thanks to gene editing techniques such as CRISPR-Cas9 (Cong et al., 2013) and prime editing (Anzalone et al., 2019), this number will likely grow significantly in the coming decades. The continued integration of data science and personalized healthcare will pose a challenge for privacy and ethics as well; how are we going to make sure that these new techniques are not misused?

Another interesting development in healthcare is medical crowdsourcing (Dissanayake, Nerur, Singh, & Lee, 2019), which offers hope to patients who suffer from complex health conditions or rare diseases that are difficult to diagnose. Medical crowdsourcing platforms such as CrowdMed (CrowdMed, 2020) empower patients to make use of the "wisdom of the crowd" by providing access to a vast pool of diverse medical knowledge. Greater participation in crowdsourcing increases the likelihood of finding the right solution, but it might also lead to increased "noise," which makes the identification of the most likely solution from a large pool of recommendations difficult. The challenge for medical crowdsourcing platforms is to increase participation of both patients and solution providers, while simultaneously increasing the efficacy and accuracy of solutions.

The Digital Twin (Bruynseels, Santoni de Sio, & van den Hoven, 2018) is another concept that will continue to develop in the (near) future. Digital Twins stand for a specific engineering paradigm, where individual physical artifacts are paired with digital models that dynamically reflect the status of those artifacts. These artifacts could be machines, organs, or even complete persons. When applied to persons, a Digital Twins can be seen as an "in silico" representation of an individual that dynamically reflects molecular status, physiological status, and life style over time.

The fear in popular culture is that AI will take over the world: "Algorithms evolve, push us aside and render us obsolete" (Note, n.d.), but perhaps humanity should not be too concerned about AI. Within healthcare, AI systems will not replace human clinicians on a large scale, but rather will augment their efforts to care for patients (Davenport & Kalakota, 2019). Over time, these clinicians may move toward tasks that draw on uniquely human skills like empathy and creativity. The only ones who might lose their jobs over time are those that refuse to work alongside AI. Finally, giving tasks away to AI will give clinicians more time to have face-to-face contact with patients, which is something that is often lacking in the current situation.

## Competing interest statement

Dr. Hulsen is employed by Philips Research.

# References

Allen, J., Adams, C., & Flack, F. (2019). The role of data custodians in establishing and maintaining social licence for health research. *Bioethics*, *33*, 502–510.

Alsever, J. (2020). *Medicine by machine: Is A.I. the cure for the world's ailing drug industry?* From https://fortune-com.cdn.ampproject.org/c/s/fortune.com/longform/ai-artificial-intelligence-medicine-healthcare-pharmaceutical-industry/amp/.

Anzalone, A. V., Randolph, P. B., Davis, J. R., Sousa, A. A., Koblan, L. W., Levy, J. M., et al. (2019). Search-and-replace genome editing without double-strand breaks or donor DNA. *Nature*, *576*, 149–157.

Ashburner, M., Ball, C. A., Blake, J. A., Botstein, D., Butler, H., Cherry, J. M., et al. (2000). Gene ontology: Tool for the unification of biology. The Gene Ontology Consortium. *Nature Genetics*, *25*, 25–29.

Auffray, C., Balling, R., Barroso, I., Bencze, L., Benson, M., Bergeron, J., et al. (2016). Making sense of big data in health research: Towards an EU action plan. *Genome Medicine*, *8*, 71.

Baker, M. (2016). 1,500 scientists lift the lid on reproducibility. *Nature*, *533*, 452–454.

Bandrowski, A., Brinkman, R., Brochhausen, M., Brush, M. H., Bug, B., Chibucos, M. C., et al. (2016). The ontology for biomedical investigations. *PLoS One*, *11*, e0154556.

Bender, D., & Sartipi, K. (2013). HL7 FHIR: An agile and RESTful approach to healthcare information exchange. In *Proceedings of the 26th IEEE international symposium on computer-based medical systems, IEEE* (pp. 326–331).

Bidgood, W. D., Jr., & Horii, S. C. (1992). Introduction to the ACR-NEMA DICOM standard. *Radiographics*, *12*, 345–355.

Brochhausen, M., Zheng, J., Birtwell, D., Williams, H., Masci, A. M., Ellis, H. J., et al. (2016). OBIB-a novel ontology for biobanking. *Journal of Biomedical Semantics*, *7*, 23.

Brownlee, J. (2017). *How much training data is required for machine learning?* From https://machinelearningmastery.com/much-training-data-required-machine-learning/.

Bruinsma, S. M., Nieboer, D., Hulsen, T., Zhang, L., Kirk-Burnnand, R., Gledhill, S., et al. (2017). International AS registry: The Movember Foundation's global action plan prostate cancer active surveillance initiative. In L. Klotz (Ed.), *Active surveillance for localized prostate cancer* (pp. 135–147). Cham: Humana Press.

Bruynseels, K., Santoni de Sio, F., & van den Hoven, J. (2018). Digital twins in health care: Ethical implications of an emerging engineering paradigm. *Frontiers in Genetics*, *9*, 31.

Buckler, A. J., Liu, T. T., Savig, E., Suzek, B. E., Rubin, D. L., & Paik, D. (2013). Quantitative imaging biomarker ontology (QIBO) for knowledge representation of biomedical imaging biomarkers. *Journal of Digital Imaging*, *26*, 630–641.

Camacho, D. M., Collins, K. M., Powers, R. K., Costello, J. C., & Collins, J. J. (2018). Next-generation machine learning for biological networks. *Cell*, *173*, 1581–1592.

CDISC. (2020). *CDISC ODM-XML*. CDISC. https://www.cdisc.org/standards/data-exchange/odm.

Cohen, J. K. (2020). *Mayo Clinic's new data-sharing initiative launches first project*. From https://www.modernhealthcare.com/information-technology/mayo-clinics-new-data-sharing-initiative-launches-first-project.

Cong, L., Ran, F. A., Cox, D., Lin, S., Barretto, R., Habib, N., et al. (2013). Multiplex genome engineering using CRISPR/Cas systems. *Science*, *339*, 819–823.

CrowdMed. (2020). *CrowdMed*. From https://www.crowdmed.com/.

Data Protection Network. (2018). *GDPR and data processing agreements*. From https://www.dpnetwork.org.uk/gdpr-data-processing-agreements/.

DICOM—Digital Imaging and Communications in Medicine. (2020). *DICOM key concepts*. From https://www.dicomstandard.org/concepts/.

Davenport, T., & Kalakota, R. (2019). The potential for artificial intelligence in healthcare. *Future Healthcare Journal*, *6*, 94–98.

Deist, T. M., Dankers, F. J. W. M., Ojha, P., Marshall, M. S., Janssen, T., Faivre-Finn, C., et al. (2020). Distributed learning on 20 000+ lung cancer patients—The Personal Health Train. *Radiotherapy and Oncology*, *144*, 189–200.

DICOM Library. (2020). *DICOM tags*. From https://www.dicomlibrary.com/dicom/dicom-tags/.

Dijstelbloem, H., Huisman, F., Miedema, F., & Mijnhardt, W. (2013). *Why science does not work as it should and what to do about it*. From http://www.scienceintransition.nl/app/uploads/2013/10/Science-in-Transition-Position-Paper-final.pdf.

Dissanayake, I., Nerur, S., Singh, R., & Lee, Y. (2019). Medical crowdsourcing: Harnessing the "wisdom of the crowd" to solve medical mysteries. *Journal of the Association for Information Systems*, *20*, 4.

Doel, T., Shakir, D. I., Pratt, R., Aertsen, M., Moggridge, J., Bellon, E., et al. (2017). GIFT-Cloud: A data sharing and collaboration platform for medical imaging research. *Computer Methods and Programs in Biomedicine*, *139*, 181–190.

Domingos, P. M. (2012). A few useful things to know about machine learning. *Communications of the ACM*, *55*, 78–87.

Donnelly, K. (2006). SNOMED-CT: The advanced terminology and coding system for eHealth. *Studies in Health Technology and Informatics*, *121*, 279.

Dutch Techcentre for Life Sciences. (2015). *Personal health train*. From https://www.dtls.nl/fair-data/personal-health-train/.

European Bioinformatics Institute. (2020). *Systems biology: From large datasets to biological insight*. From https://www.ebi.ac.uk/training/events/2020/systems-biology-large-datasets-biological-insight-0.

ELIXIR. (2020). *ELIXIR workshops and courses*. From https://elixir-europe.org/events?f[0]=field_type:15.

Estrada-Galiñanes, V., & Wac, K. (2019). Collecting, exploring and sharing personal data: Why, how and where. *Data Science*, 1–28.

Figueiredo, A. S. (2017). Data sharing: Convert challenges into opportunities. *Frontiers in Public Health*, *5*, 327.

Forrey, A. W., McDonald, C. J., DeMoor, G., Huff, S. M., Leavelle, D., Leland, D., et al. (1996). Logical observation identifier names and codes (LOINC) database: A public use set of codes and names for electronic reporting of clinical laboratory test results. *Clinical Chemistry*, *42*, 81–90.

Fragoso, G., de Coronado, S., Haber, M., Hartel, F., & Wright, L. (2004). Overview and utilization of the NCI thesaurus. *Comparative and Functional Genomics*, *5*, 648–654.

Gao, J., Aksoy, B. A., Dogrusoz, U., Dresdner, G., Gross, B., Sumer, S. O., et al. (2013). Integrative analysis of complex cancer genomics and clinical profiles using the cBioPortal. *Science Signaling*, *6*, pl1.

Gaye, A., Marcon, Y., Isaeva, J., LaFlamme, P., Turner, A., Jones, E. M., et al. (2014). Data-SHIELD: Taking the analysis to the data, not the data to the analysis. *International Journal of Epidemiology*, *43*, 1929–1944.

Hartter, J., Ryan, S. J., Mackenzie, C. A., Parker, J. N., & Strasser, C. A. (2013). Spatially explicit data: Stewardship and ethical challenges in science. *PLoS Biology*, *11*, e1001634.

Health-RI. (2020). *Health-RI—Enabling data driven health*. From https://www.health-ri.nl/.

HMA-EMA Joint Big Data Taskforce. (2019). *HMA-EMA joint big data taskforce—Summary report*. HMA-EMA. https://www.ema.europa.eu/en/documents/minutes/hma/ema-joint-task-force-big-data-summary-report_en.pdf.

## 92 Big data in psychiatry and neurology

HMA-EMA Joint Big Data Taskforce. (2020). *Priority recommendations of the HMA-EMA joint big data task force.* HMA-EMA. https://www.ema.europa.eu/en/documents/other/priority-recommendations-hma-ema-joint-big-data-task-force_en.pdf.

Hong, L., Luo, M., Wang, R., Lu, P., Lu, W., & Lu, L. (2018). Big data in health care: Applications and challenges. *Data and Information Management, 2*, 175.

Hulsen, T. (2019a). The ten commandments of translational research informatics. *Data Science, 2*, 341–352.

Hulsen, T. (2019). An overview of publicly available patient-centered prostate cancer datasets. *Translational Andrology and Urology, 8*(Suppl. 1), S46–S77.

Hulsen, T. (2020). Sharing is caring—Data sharing initiatives in healthcare. *International Journal of Environmental Research and Public Health, 17*(9), 3046.

Hulsen, T., Jamuar, S. S., Moody, A. R., Karnes, J. H., Varga, O., Hedensted, S., et al. (2019). From big data to precision medicine. *Frontiers in Medicine (Lausanne), 6*, 34.

Hulsen, T., Obbink, H., Van Der Linden, W., De Jonge, C., Nieboer, D., Bruinsma, S., et al. (2016). 958 integrating large datasets for the Movember global action plan on active surveillance for low risk prostate cancer. *European Urology Supplements, 15*(3), e958.

Hulsen, T., Van der Linden, W., De Jonge, C., Hugosson, J., Auvinen, A., & Roobol, M. J. (2019). Developing a future-proof database for the European randomized study of screening for prostate cancer (ERSPC). *European Urology Supplements, 18*(1), e1766.

i2b2 tranSMART Foundation. (2020). *The i2b2 tranSMART foundation 2020 training program.* From https://transmartfoundation.org/the-i2b2-transmart-foundation-training-program/.

Intersoft Consulting. (2018c). *Recital 26—Not applicable to anonymous data.* From https://gdpr-info.eu/recitals/no-26/.

Intersoft Consulting. (2018a). *Art. 7 GDPR—Conditions for consent.* Intersoft Consulting. https://gdpr-info.eu/art-7-gdpr/.

Intersoft Consulting. (2018). *Art. 17 GDPR—Right to erasure ("right to be forgotten").* Intersoft Consulting. https://gdpr-info.eu/art-17-gdpr/.

Integraal Kankercentrum Nederland. (2020). *DistributedLearning.ai—Privacy preserving federated learning.* From https://distributedlearning.ai/.

Islam, S. M. R., Kwak, D., Kabir, M. H., Hossain, M., & Kwak, K. (2015). The internet of things for health care: A comprehensive survey. *IEEE Access, 3*, 678–708.

Jarczak, J., Lach, J., Borówka, P., Gałka, M., Baćko, M., Marciniak, B., et al. (2019). BioSCOOP— Biobank sample communication protocol. New approach for the transfer of information between biobanks. *Database (Oxford), 2019*, baz105.

Kalkman, S., Mostert, M., Gerlinger, C., van Delden, J. J. M., & van Thiel, G. J. M. W. (2019). Responsible data sharing in international health research: A systematic review of principles and norms. *BMC Medical Ethics, 20*, 21.

Kaye, J., Briceno Moraia, L., Curren, L., Bell, J., Mitchell, C., Soini, S., et al. (2016). Consent for biobanking: The legal frameworks of countries in the BioSHaRE-EU project. *Biopreservation and Biobanking, 14*, 195–200.

Ketchum, K. (2018). *How data sharing could change the entire healthcare industry.* From http://www.healthcarebusinesstech.com/data-sharing/.

Knoppers, B. M., & Thorogood, A. M. (2017). Ethics and big data in health. *Current Opinion in Systems Biology, 4*, 53–57.

Kohler, S., Vasilevsky, N. A., Engelstad, M., Foster, E., McMurry, J., Ayme, S., et al. (2017). The human phenotype ontology in. *Nucleic Acids Research, 45*(2017), D865–D876.

Kolachalama, V. B., & Garg, P. S. (2018). Machine learning and medical education. *npj Digital Medicine, 1*, 54.

Langlotz, C. P. (2006). RadLex: A new method for indexing online educational materials. *Radiographics, 26*, 1595–1597.

Li, D. (2019). 5G and intelligence medicine-how the next generation of wireless technology will reconstruct healthcare? *Precision Clinical Medicine, 2*, 205–208.

Libbrecht, M. W., & Noble, W. S. (2015). Machine learning applications in genetics and genomics. *Nature Reviews. Genetics, 16*, 321–332.

Ma, C.-C., Wang, Z.-L., Xu, T., He, Z.-Y., & Wei, Y.-Q. (2019). The approved gene therapy drugs worldwide: From 1998 to 2019. *Biotechnology Advances*, 107502.

MacLean, M., & Miles, C. (1999). Swift action needed to close the skills gap in bioinformatics. *Nature, 401*, 10.

Mandel, J. (2018). *Sync for Science: Empowering individuals to participate in health research.* From https://blog.verily.com/2018/03/sync-for-science-empowering-individuals.html.

Merino-Martinez, R., Norlin, L., van Enckevort, D., Anton, G., Schuffenhauer, S., Silander, K., et al. (2016). Toward global biobank integration by implementation of the minimum information about BIobank data sharing (MIABIS 2.0 Core). *Biopreservation and Biobanking, 14*, 298–306.

Moss, C. L., Santaolalla, A., Hulsen, T., Ruwe, B., Coolen, T., Attard, G., et al. (2020). *ReIMAGINE: Development of an innovative digital platform to promote the use of artificial intelligence for accurate risk stratification of prostate cancer, AI and big data in Cancer: From innovation to impact.* Boston, USA: Elsevier. https://app.oxfordabstracts.com/events/1623/program-app/session/11667.

Murphy, S. N., Weber, G., Mendis, M., Gainer, V., Chueh, H. C., Churchill, S., et al. (2010). Serving the enterprise and beyond with informatics for integrating biology and the bedside (i2b2). *Journal of the American Medical Informatics Association: JAMIA, 17*, 124–130.

Muse, Algorithm, Simulation Theory, 2018. https://www.musewiki.org/Algorithm_(song).

Nationaal Platform Open Science. *Open science.* From https://www.openscience.nl/en/open-science.

Nictiz. (2020). *SDO-Nederland.* From https://www.nictiz.nl/samenwerking/sdo-nederland/.

Norlin, L., Fransson, M. N., Eriksson, M., Merino-Martinez, R., Anderberg, M., Kurtovic, S., et al. (2012). A minimum data set for sharing biobank samples, information, and data: MIABIS. *Biopreservation and Biobanking, 10*, 343–348.

Philips. (2018). *Philips HealthSuite Digital Platform.* Philips. https://cf-s3-9b8a2d91-c007-4d01-985a-ef737a876dbd.s3.amazonaws.com/s3fs-public/PDF/hsdp-brochure.pdf.

Philips. (2019). *Philips future health index 2019.* Philips. https://images.philips.com/is/content/PhilipsConsumer/Campaigns/CA20162504_Philips_Newscenter/Philips_Future_Health_Index_2019_report_transforming_healthcare_experiences.pdf.

Philips. (2020a). *Six areas in which AI is changing the future of healthcare.* From https://www.philips.com/a-w/about/news/archive/features/20200107-six-areas-in-which-ai-is-changing-the-future-of-healthcare.html.

Philips. (2020b). *Philips data principles.* From https://www.philips.com/a-w/about/philips-data-principles.html.

Philips. (2020c). *Philips AI principles.* From https://www.philips.com/a-w/about/artificial-intelligence/philips-ai-principles.html.

Rake, E. A., van Gelder, M., Grim, D. C., Heeren, B., Engelen, L., & van de Belt, T. H. (2017). Personalized consent flow in contemporary data sharing for medical research: A viewpoint. *BioMed Research International, 2017*, 7147212.

Regenstrief Institute. (2020). *Regenstrief LOINC mapping assistant (RELMA).* From https://loinc.org/relma/.

# 94  Big data in psychiatry and neurology

Rothstein, M. A., & Tovino, S. A. (2019). California takes the Lead on data privacy law. *Hastings Center Report*, *49*, 4–5.

Saqr, M., & Tedre, M. (2019). Should we teach computational thinking and big data principles to medical students? *International Journal of Health Sciences*, *13*, 1–2.

Scheufele, E., Aronzon, D., Coopersmith, R., McDuffie, M. T., Kapoor, M., Uhrich, C. A., et al. (2014). tranSMART: An open source knowledge management and high content data analytics platform. *AMIA Joint Summits on Translational Science proceedings*, *2014*, 96–101.

Smith, D. W. (2018). *GDPR runs risk of stifling healthcare innovation*. From https://eureka.eu.com/gdpr/gdpr-healthcare/.

SNOMED. (2019). *SNOMED CT 5-Step Briefing*. From https://www.snomed.org/snomed-ct/five-step-briefing.

State of California. (2018). *The California Consumer Privacy Act of 2018*. State of California. https://leginfo.legislature.ca.gov/faces/billTextClient.xhtml?bill_id=201720180AB375.

Staten, J. (2018). *GDPR and the end of reckless data sharing*. From https://www.securityroundtable.org/gdpr-end-reckless-data-sharing/.

Sync for Science. (2017). *Sync for Science—Helping patients sharing EHR data with researchers*. From http://syncfor.science/.

The All of Us Research Program Investigators. (2019). The "All of Us" research program. *New England Journal of Medicine*, *381*, 668–676.

The European Parliament and the Council of the European Union. (2016). Regulation (EU) 2016/679 of the European Parliament and of the council of 27 April 2016 on the protection of natural persons with regard to the processing of personal data and on the free movement of such data, and repealing directive 95/46/EC (general data protection regulation). *Official Journal of the European Union*, *L119*, 1–88. https://eur-lex.europa.eu/legal-content/EN/TXT/PDF/?uri=CELEX:32016R0679.

The Gene Ontology Consortium. (2019). The gene ontology resource: 20 years and still GOing strong. *Nucleic Acids Research*, *47*, D330–D338.

The Gene Ontology Consortium. (2020). *Gene ontology overview*. From http://geneontology.org/docs/ontology-documentation/.

University of Colorado. (2020). *Clinical data science*. From https://www.coursera.org/specializations/clinical-data-science.

UC San Diego. (2020). *Big data integration and processing*. From https://www.coursera.org/learn/big-data-integration-processing.

van Soest, J., Sun, C., Mussmann, O., Puts, M., van den Berg, B., Malic, A., et al. (2018). Using the personal health train for automated and privacy-preserving analytics on vertically partitioned data. *Studies in Health Technology and Informatics*, *247*, 581–585.

Wilkinson, M. D., Dumontier, M., Aalbersberg, I. J., Appleton, G., Axton, M., Baak, A., et al. (2016). The FAIR guiding principles for scientific data management and stewardship. *Science Data*, *3*, 160018.

## Chapter 5

# Data linkages in epidemiology

Sinéad Moylett

*Laboratory of Neuroimmunology, KU Leuven, Leuven, Belgium*

## 1 Introduction

As the number and size of big data resources within healthcare has grown coupled with innovative statistics methods, new sources of knowledge remain largely untapped. By combining state-of-the-art technical and interdisciplinary expertise, researchers have begun to exploit the potential of big healthcare data and generate clinically relevant findings that are likely to have direct practical applications. By linking different sources of local and national routinely- and non-routinely-collected healthcare data that examines all stages from prodromal markers throughout diagnosis and disease course up to final outcomes, the resulting applications span a whole host of areas including predictive modeling and clinical decision support, disease or safety surveillance, public health, and research. Within recent times, a growing number of databases and cohorts can be seen where routinely- and non-routinely-collected information are combined to further advance the scope of health data available for investigation. This includes the use of both structured and unstructured information, such as electronic medical records (EMRs) detailing patient referrals, assessments, hospital stay notes, etc. Thanks to technological advances, the wealth of knowledge contained in EMRs can be combined with other healthcare data to provide information on millions of individuals' prodromal stages, diagnoses, treatments, and outcomes for several years pre- and post-diagnosis. Not only does it advance our ability to learn more about common diseases (e.g., cardiovascular disease (CVD), cancer), but also furthermore opens up new opportunities to investigate rare conditions and those with high misdiagnoses rates.

Large-scale cohort and longitudinal studies have been cornerstones in epidemiology for decades, providing information that can aid us in measuring disease and other aspects of health, identifying the causes of ill-health and possible mechanisms for intervention to improve health (Webb, Bain, & Page, 2020). Such studies have been utilized to address major public health concerns spanning prevention, diagnosis, and treatment, including vaccination, CVD diagnostic practices, and the complex needs of aging populations (Centers for Disease

Big Data in Psychiatry and Neurology. https://doi.org/10.1016/B978-0-12-822884-5.00008-8
Copyright © 2021 Elsevier Inc. All rights reserved.

**96** Big data in psychiatry and neurology

Control and Prevention, 2011; Lee et al., 2018). One of the greatest difficulties with such research is that they are resource heavy. They demand a strong research infrastructure and a large amount of human resource hours over long periods of time. Utilizing existing data collected as part of the healthcare system and innovative web-based and (bio)informatics technologies allows researchers to develop large-scale cohorts with greater depth of health information which are less time consuming and expensive. A great example can be seen in New Zealand where the Integrated Data Infrastructure (IDI) contains person-centered microdata from a range of government agencies, StatsNZ surveys (an official data agency collecting information from people and organizations through censuses and surveys), and nongovernment organizations across eight broad categories: health, education and training, benefits and social services, justice, people and communities, population, income and work, and housing. All the data is located in one national database that researchers can utilize to address complex health, social, and economic issues (StatsNZ, 2018). Without a national infrastructure like the IDI, linking different sources of health information can be difficult. Healthcare systems include an array of disciplines and support services all collecting information in different formats, structures, and databases. This chapter presents a discussion of European and UK investigations that have sought to link routinely- and non-routinely-collected health information, both structured and unstructured data from local and national sources, combined with the use of innovative web-based and (bio)informatic technologies, to improve healthcare for whole populations and specific disease groups.

## 2 Linking local and national routinely-collected data

When creating research databases from routinely-collected healthcare data great attention must be paid to ethical and governance challenges, including information autonomy, data protection, and the risks associated with disclosure (Soni et al., 2020). Healthcare data contains vast amounts of information and there is huge potential for health research; however, patient identifiers (PIs; e.g., name, date of birth, address) are regularly present within such data. Great care must be taken to ensure patient confidentiality is upheld and patient trust in the use of patient data maintained. In line with patient privacy rights and in order to exercise information autonomy, the standard requirement across most jurisdictions when using healthcare data for research purposes is to "consent or anonymise" (Fernandes et al., 2013). Given the huge patient numbers involved in research databases developed from healthcare data seeking informed consent from each individual is time consuming, demanding on resources, and impractical in many instances (e.g., due to their condition, patients may lack the ability to give informed consent). Furthermore, seeking informed consent can lead to biased sampling for a variety of reasons in categories such as mental health, sexual and reproductive health, domestic violence, substance abuse, and genetic information (Soni et al., 2020). For example, due to the stigma within many societies about mental health conditions, patients will refuse to share information for research resulting in low numbers and reduced generalizability (Powell, Fitton, & Fitton, 2006; Yawn, Yawn, Geier, Xia, & Jacobsen, 1998). Many databases opt for anonymized, pseudonymized, or deidentified approaches to ensure information autonomy, data protection, and confidentiality are upheld. An anonymized database removes all PIs, each individual is given an identifier number and this number cannot be linked in any way to the original data. Within a pseudonymized database, all PIs have been removed, each individual is given an identifier, and the original identifier is held in a securely linked table held by a trusted third party (who is separate from the research). On the other hand, a deidentified database is one where all PIs are replaced (with a configuration of lettering or number that would not be present anywhere else in the database, e.g., ZZZZZZZ) or removed completely (Fernandes et al., 2013). Care needs to be given to each nation's specific guidelines stipulating what is considered a PI as guidelines can vary, from specific lists of PI variables to more general descriptions.

One of the largest anonymized databases within the United Kingdom is the Clinical Practice Research Datalink (CPRD; https://www.cprd.com/). CPRD is an ongoing primary care database that includes data from 738 general practices in the United Kingdom, including over 19 million patients (as of September 2018) and data on demographics, symptoms, tests, diagnoses, prescriptions, therapies, health-related behaviors, and referrals to secondary care (Herrett et al., 2015; Wolf et al., 2019). As part of the development of CPRD, the database has grown to include data linkages to datasets from secondary care, disease-specific cohorts (e.g., cancer registries), and national mortality records greatly enhancing the breath of data accessible and allowing the database to be widely used internationally for epidemiology. Given the size of CPRD and its depth of data, it has been used for a variety of health research applications: to assess the recording of particular diseases within primary care and its effect on incidence rates (Tate et al., 2017); the feasibility and quality of code list generation in EMR-based databases (Watson, Nicholson, Hamilton, & Price, 2017); the common definitions used in antidepressant prescribing across the United Kingdom and four other European countries impacting prevalence (Abbing-Karahagopian et al., 2014); and to describe and predict the clinical features and management of numerous diseases including systemic lupus erythematosus (Rees et al., 2017), bowel disease (Stapley et al., 2017), brain tumors (Hamilton & Kernick, 2007), and self-harm (Carr et al., 2016). This clearly demonstrates the wealth of knowledge that can be gained from such databases for an array of public health concerns. Similarly on a national scale within Belgium, the Intego and linked Intego-IMA databases are a pseudonymized collection of healthcare data compiled from multiple sources such as local primary care facilities, laboratories, and insurance. Intego contains a cohort of >440,000 patients between 1994 and 2018, yearly containing $\pm 2\%$ of the total

Flemish population that is representative by age and gender. Intego-IMA is compiled from national insurance data including pharmacy delivered medications and important outcome measures like mortality, hospitalizations, and cardiac interventions. It contains a cohort of >70,000 patients between 2011 and 2015 (Delvaux et al., 2018; Truyers, Goderis, Dewitte, Van den Akker, & Buntinx, 2014). As well as having ethical approval for all purely observational research using the pseudonymized data within Intego, it also has approval for the development of risk prediction algorithms based on real-life longitudinal data.

## 2.1 Development of diagnostic algorithms: Structured data

An essential component of preventative medicine is risk prediction. Clinical risk prediction models guide clinical reasoning, decision-making, and health policy by using multiple predictors to estimate the probability that a certain outcome is present/will occur within a specific time period for an individual and/or among a specific population (Moons et al., 2012). In order to be impactful, these models require large sample sizes to ensure adequate power for the necessary analyses, and databases like CPRD and Intego go a long way in contributing to public health epidemiology by allowing data to be used for predictive modeling as well as information gathering, highlighting the utility of big healthcare data for reducing disease incidence and burden. Among noncommunicable diseases (NCDs), CVD is the leading cause of death, illness, and health burden globally (GBD 2015 DALYs and HALE Collaborators, 2016; GBD 2015 Mortality and Causes of Death Collaborators, 2016). Despite substantial decreases in CVD mortality within Europe over the last four decades, it remains the most common cause of death (Tadayon, Wickramasinghe, & Townsend, 2019; Townsend et al., 2016). Furthermore, recent epidemiological investigations have found a concerning trend that CVD rates are plateauing and no longer declining (Roth et al., 2017). An aging population means that more individuals are living with chronic diseases, such as CVD, along with the associated treatments and effects. As a result, years lost due to disability (YLD) are declining slower than rates of mortality (Global Burden of Disease Study 2013 Collaborators, 2015). Despite an abundance of CVD risk prediction models, e.g., Framingham (Anderson, Odell, Wilson, & Kannel, 1991; Wilson et al., 1998), SCORE (Conroy et al., 2003), and QRISK (Hippisley-Cox, Coupland, & Brindle, 2017), their usefulness is limited due to a number of restrictions: (1) lack of consideration of changing population and CVD rates (Pate, Emsley, Ashcroft, Brown, & van Staa, 2019; Tadayon et al., 2019); (2) limited sample sizes, risk predictors, and data sources—the predominant use of cohort studies rather than real-world healthcare data (Collins & Altman, 2010; Damen et al., 2016); (3) lack of validation (Damen et al., 2016); and (4) the use of rigid statistical methods and tools (Kennedy, Wiitala, Hayward, & Sussman, 2013). Due to restricted statistical methods and the widespread use of cross-sectional data from cohort studies, standard models do not consider the cumulative effects over time of risk factors on outcomes, limiting their ability to guide clinical decisions and health policy.

Some recent investigations have demonstrated an improvement of CVD risk prediction models when multiple real-world measurements of clinical risk factors (e.g., blood pressure, cholesterol) from over time have been included (Mehta et al., 2018; Paige et al., 2017). However progress is slow and to date, only one study has considered the impact of CVD preventative medication and treatment changes over time for CVD risk prediction models using real-world data. In their example using EMR data, Mehta et al. (2018) showed that 5-year CVD risk models that include baseline pharmacotherapy will account for the majority of the effect of treatment during follow-up among people up to 75 years of age. Building from this study, suggestions for future research include considering the role of medication and other treatment changes beyond 5 years, and the use of novel analytical methods and different data sources that would allow the inclusion of temporal dynamics for multiple medications and their dosages. The recent availability of big healthcare datasets that provide real-world clinical information on millions of diagnoses and treatments with several years of follow-up coupled with the use of statistical methods that incorporate longitudinal effects into models (Fieuws & Verbeke, 2006; Verbeke & Molenberghs, 2000) offers an unprecedented opportunity to address the earlier identified research gaps for multiple diseases. A better understanding of the role of treatments and their temporal dynamics within disease outcomes may help clinicians and guideline developers to focus on patients for whom screening and treatments are more likely to have an effect and improve the prevention and treatment of numerous diseases.

## 3 Linking routinely- and non-routinely-collected data

When establishing and creating databases like CPRD and Intego, drawing on the knowledge and skills of multiple disciplines is key. Intego is based within the Department of Public Health and Primary Care in KU Leuven, along with one of the most international-recognized biostatistics groups, the Leuven Biostatistics and Statistical Bioinformatics Centre (L-BioStat). Such interdisciplinarity between public health, medicine, epidemiology, and advanced biostatistics is paramount to the success of such large healthcare research databases. Another example is that of HealthWise Wales (HWW), a government-funded initiative based within the Division of Population Medicine in Cardiff University that seeks to create a population-level cohort to investigate the widest possible range of social, environmental, and biological determinants of health and wellbeing by combining routinely- and non-routinely-collected health information from multiple sources (Hurt et al., 2019). Slightly differently to the other databases mentioned within this chapter, HWW does seek consent from participants. Participants agree to be contacted about research studies and biannual data collections, and furthermore, to give access to their routinely-collected health-care data (Ford et al., 2009; Lyons et al., 2009). HWW is employing a more cost-effective and efficient method of recruitment and follow-up by using a web-based application, designed specifically for the project, which is accessible to participants through the main HWW website (https://www.healthwisewales. gov.wales/). The web-based application allows participants to register for the project, complete the biannual data collection, and receive information about other research projects all in one online location. Recruitment is ongoing with 21,779 participants alive and currently registered. Furthermore since its beginning in, 2015, HWW has facilitated the recruitment of 43,826 participants to 15 different studies. Beyond being a population cohort, HWW acts as a resource for other researchers by: providing information to participants about other research projects, allowing researchers to complete secondary analyses on cohort data via the researcher portal, and allowing researchers to include specific topics within the biannual data collections that can be linked with the routinely-collected healthcare data (Hurt et al., 2019). The platform has been used by researchers to collect data on a wide range of diseases and health practices:

**(1)** on more common disease like respiratory tract infections, skin cancer, and bowel cancer;

**(2)** on prescriptions—assessing acceptability of putting the costs of medicines on dispensing labels and views on the potential for redispensing medicines returned unused to pharmacies;

**(3)** on healthcare behaviors such as seeking general practitioner (GP) consultation for dental problems (Cope, Wood, Francis, & Chestnutt, 2018); and

**(4)** on rarer conditions—investigating the incidence rates of hidradenitis suppurativa within Wales (Ingram, Collins, Atkinson, & Brooks, 2020).

HWW is a rare example of an ongoing and evolving database that links routinely- and non-routinely-collected healthcare data to not only create a population-level longitudinal cohort but also additionally a resource of actively engaged potential participants for other researchers to avail of.

## 4 Linking structured and unstructured routinely-collected data

As previously mentioned, healthcare data regularly contains PIs (e.g., name, date of birth, address). Each database discussed has gone through rigorous ethical and data protection procedures to ensure that patient confidential is maintained, data is protected and stored securely, and only bona fide researchers can utilize the data for investigations that fall within public health interests. EMRs contain a wealth of information, which when combined with structured data, can bring together enough individual information about a disease to draw conclusions that apply to the whole disease population. However, EMRs contain large amounts of patient identifiable and highly sensitive confidential information. Technological advances in (bio)informatics have opened up a range of opportunities where EMRs can be used to provide greater detail about patient presentation, clinical course, treatment, and outcomes that is often lacking in many structured healthcare databases. Databases such as Clinical Record Interactive Search (CRIS; Perera et al., 2016; Stewart et al., 2009) and Clinical Records Anonymization and Text Extraction (CRATE; Cardinal, 2017) within the UK show how deidentification tools and natural language processing (NLP) algorithms can be utilized in the creation of research databases from sensitive EMRs. The combination of these deidentified EMRs with structured health data from both local and national databases allows researchers greater capability to examine in-depth diseases that in the past were hard to access and/or where information was lacking about the disease group.

## 4.1 CRIS and CRATE databases

Developed within the South London and Maudsley NHS Foundation Trust (SLaM) Biomedical Research Centre, CRIS (Perera et al., 2016; Stewart et al., 2009) is one of the first and largest databases of its kind. SLaM is one of the United Kingdom's largest mental health and dementia care providers, serving a geographic catchment of four South London boroughs (Lambeth, Lewisham, Southwark, and Croydon) with a population of over 1.2 million residents. In 2007–2008, the CRIS database was developed with National Institute for Health Research (NIHR) funding to provide researchers access to anonymized copies of SLaM's EMRs within a robust governance framework (Perera et al., 2016; Stewart et al., 2009). CRIS has received ethical approval as an anonymized data resource (Oxford Research Ethics Committee C, reference 08/H0606/7115), utilizing a bespoke deidentification algorithm designed to create dictionaries using PIs entered into dedicated source fields and then identify, match, and mask them (with ZZZZZ) when they appear in medical texts (Fernandes et al., 2013). By placing the deidentification algorithm within a carefully designed security model, the CRIS database removes all PIs and allows researchers access to a rich variety of information from routine clinical notes. As well as combining local structured and unstructured health data, the CRIS database has established a number of data linkages with a national dataset of all acute (and mental health) hospitalizations, including episode dates and diagnoses (Hospital Episode Statistics, 2020; Sinha, Peach, Poloniecki, Thompson, & Holt, 2013). Combining detailed local health data with national hospital statistics, the CRIS database can provide long-term in-depth information about a patient from their first admission within the healthcare system to their final outcomes.

With its wealth of health data, the CRIS database within SLaM has been used across a wide range of research projects. In an investigation of neuroleptic malignant syndrome (NMS, a rare but potentially fatal complication of

**102** Big data in psychiatry and neurology

antipsychotic treatment), Chang, Harrison, Lee, Taylor, and Stewart (2012) found 485 initial results. Following further examination, 183 were revealed to be suspected NMS cases, of which 43 cases fulfilled at least 1 set of the 6 diagnostic criteria. Among the EMRs, relatively poor agreement was found among published diagnostic criteria for NMS. Despite this, the study found that among those cases that met any diagnostic criteria, certain symptoms were significantly more common when compared to cases that did not meet diagnostic criteria (e.g., pyrexia, extrapyramidal symptoms, altered consciousness, autonomic symptoms, and elevated CK concentrations). As well as examining the detailed symptom presentation of rare conditions, the CRIS database has allowed investigations into long-term outcomes for a range of disorders. The lower rates of standardized mortality ratios (SMRs) and life expectancies at birth for those who suffer from personality disorder compared to the general population have highlighted the significant public health burden of mental illnesses for such individuals (Fok et al., 2012). Additionally in a cohort of 11,567 people with schizophrenia spectrum disorders, it was found that comorbid physical health conditions and analgesic medication prescription were associated with higher risk of falls and fractures (Stubbs et al., 2018). By examining the hospital admission data for the cohort, Stubbs and colleagues (2018) found that 579 (incidence rate 12.79 per 1000 person-years) and 528 (11.65 per 1000 person-years) had at least one reported hospital admission due to a fall or fracture, respectively, and 822 patients had at least either a recorded fall or a fracture during this period, displaying the large amount of healthcare usage among this disease group.

Considering the large public health burden of our growing aging populations, the CRIS database has been used for a number of investigations to access the costs and consequences of some of the largest cost drivers in dementia: hospitalization and medications (Mueller et al., 2018; Mueller et al., 2019). As our aging populations grow, more and more individuals are living longer with multiple comorbidities and their associated treatments. Large-scale investigations into the effects of polypharmacy (the concurrent use of five or more medications) are scarce. Given that polypharmacy and multimorbidity are more common among dementia compared to nondementia patients, the CRIS database containing ample information about hospital use and medications—both at local and national levels—is uniquely placed to answer some of the most pressing questions concerning these interactions and their effects for dementia patients. Utilizing the data from 4668 dementia patients and comparing those using 0–3 medications with those using 4–6 or ≥7 medications, Mueller et al. (2018) displayed that polypharmacy was linked to increased hazards of emergency department attendance (hazard ratio 1.20/1.35), hospitalization (hazard ratio 1.12/1.32), unplanned hospital admission (hazard ratio 1.12/1.25), and death within 2 years (hazard ratio 1.29/1.39) after controlling for potential confounders. Building on these findings in a cohort of 12,148 dementia patients divided into three groups: reference group (0–4 medications), polypharmacy (5–9 medication), and excessive polypharmacy (≥10 medications),

Mueller et al. (2019) found no significant differences in cognitive improvement nor long-term decline (measured by the MMSE) in relation to polypharmacy. The findings of both studies show the differences that can exist for specific disease groups across different measures, and why access to a wide range of clinical measures in databases like CRIS is vital for gaging a complete understanding of disease progression.

Following its development and success at SLaM, CRIS has begun to be installed as a resource in other mental health trusts throughout the United Kingdom. Firstly within Cambridgeshire and Peterborough NHS Foundation Trust (CPFT), and subsequently in the Royal Devon & Exeter NHS Foundation Trust (RDE) and University Hospital Southampton NHS Foundation Trust (UHS) providing the potential for a national data resource on routine care and hitherto unrealized case numbers spanning multiple sites. Following the design of the deidentification process used in the CRIS database within SLaM, CRATE within CPFT was developed to use relational databases for deidentification of PIs, relies on only open-source technologies, includes a web front end for researchers accessing the resulting research database, and additionally a web front end for implementing a specific consent-for-contact model through which patients may be approached about participation in research studies directly or via their clinicians (Cardinal, 2017). CRATE is approved under NHS Research Ethics reference 12/EE/0407 and via Caldicott Guardian and information governance approvals within CPFT. Similarly to CRIS, the wealth of information contained within CRATE through the combination of structured and unstructured routinely-collected information has allowed researchers to examine some of the more complicated dimensions of psychiatric disorders. Two investigations have been completed examining the healthcare usage, medications, and comorbidities that exist within schizophrenia, highlighting even further the complicated course and burden of this disease for affected individuals. By examining all the records of those with a coded International Classification of Disease (ICD-10) diagnosis of schizophrenia between 2005 and 2012 ($n = 1485$), Cardinal, Savulich, Mann, and Fernández-Egea (2015) found a positive correlation between the use of certain psychotropic drugs (sulpiride, mirtazapine, venlafaxine, and clozapine–aripiprazole and clozapine–amisulpride) and reduced admission days. By taking a more retrospective approach, Fernandez-Egea, Walker, Ziauddeen, Cardinal, and Bullmore (2020) displayed how birth weight, family history of diabetes, and age at clozapine initiation all predicted glucose dysregulation onset. Among 190 clozapine-treated schizophrenic patients with 10 years of follow-up, 80% of those with two risk factors (abnormal birth weight and a family history of diabetes) were found to have developed diabetes, compared to 56% with only abnormal birth weight, 40% with only a family history of diabetes, and 20% in those with neither. It's well founded that those taking antipsychotics gain weight (Leucht et al., 2013), and studies such as this by Fernandez-Egea et al. (2020) display how treatments for schizophrenia can interact with diabetic risk factors to make certain subgroups more susceptible to developing diabetes.

## 104 Big data in psychiatry and neurology

### 4.1.1 Natural language processing

As demonstrated previously, there is a huge body of work where structured health information is being utilized to develop diagnostic algorithms for diseases like CVD. A key feature of the development of CRIS within SLaM as a data resource has been in its use of NLP processing for the automated detection of key constructs from text fields in EMRs (Cunningham, Maynard, & Bontcheva, 2011; Cunningham, Tablan, Roberts, & Bontcheva, 2013). Using General Architecture for Text Engineering (GATE; https://gate.ac.uk) software, NLP algorithms take unstructured text as input and, through analyzing both the syntactic locations of words within the text and their underlying semantic meanings, bring back information required by the user. For example in CRIS, NLP algorithms are used to detect parameters which are stated as applying to the patient (rather than a friend or relative) and which are described as present rather than absent for the patient. NLP algorithms developed to date at SLaM include cognitive test scores (e.g., MMSE), social care receipt (e.g., home care, meals on wheels), diagnostic statement text, educational attainment, pharmacotherapy and psychotherapy receipt, smoking and illicit drug use, treatment adherence/compliance, common adverse drug events, and over 60 individual psychotic and affective symptoms. NLP algorithms allow a more automatic method of extracting entities from EMRs, have been extensively validated, and have supported a range of research studies (Sinha et al., 2013; Wu et al., 2013).

In a basic way, NLP algorithms can be used to identify the presence of a particular aspect (e.g., symptom, diagnosis, treatment) within the patient's EMR. In the earlier days of NLP use, Toyabe (2012) examined the differences between when and where inpatient falls were reported. By employing 170 syntactic rules, they found that inpatient falls were recorded most frequently in progress notes (100%), incident reports (65.0%), and image order entries (12.5%). However, the resulting $F$-scores showed that the incident reports and image order entries were more accurate for true falls than progress notes and discharge summaries. In a more recent examination, Patterson et al. (2019) developed and validated a pragmatic, rules-based NLP algorithm that assessed emergency department EMRs from older adults for fall occurrence. Compared to manual extraction by human coders (the "gold standard" of extraction), the algorithm had demonstrated a recall (sensitivity) of 95.8%, specificity of 97.4%, precision of 92.0%, and F1 score of 0.939 for identifying fall events within emergency department EMRs. Previously in order to collect such data, a labor-intensive manual chart review and extraction would need to be completed. NLP algorithms allow the data to be collected with excellent precision and recall in a significantly less demanding process. The data can then be used for a range of purposes such as identify patients for targeted interventions, quality measure development, and epidemiologic surveillance.

In more sophisticated formats, NLP algorithms can search through EMRs to identify and amalgamate features suggesting a particular disease may be present. Teixeira et al. (2017) examined EMRs from 631 individuals in order to develop a phenotyping algorithm that would identify individuals with hypertension and other risk factors suggestive of CVD. By including diagnostic, treatment, assessment, and costing information, it was found that random forests using billing codes, medications, vitals, and concepts had the best performance with a median area under the receiver operator characteristic curve (AUC) of 0.976. One of the key advantages of NLP algorithms is that they capture information from the free text records that is regularly missing from the structured information collected within healthcare settings. This advantage has huge potential for identifying prodromal symptoms to aid with more timely diagnoses, identifying lifestyle-based risk factors that could suggest mechanisms for preventative medicine, and for assisting in the diagnosis and treatment of diseases that carry stigma. Velupillai et al. (2019) developed a simple lexicon- and rule-based NLP algorithm for identifying young individuals exhibiting suicidal risk behavior, with over 80% accuracy at both document and patient level on data from 200 patients ($\sim$5000 documents). Many of the behaviors that lead to NCDs are lifestyle based and similarly to behaviors related to mental health, this information is rarely captured in a structured format within EMRs. However, this information may exist within free text documents conducted during a patient assessment or interview. NLP algorithms can tap into this information in a quick way and give greater insight into how these lifestyle behaviors are impacting disease development. Recent studies have examined the associated lifestyle behaviors for AD, hypertension, and diabetes (Shoenbill et al., 2020; Zhou et al., 2019). For AD, it was found that those with AD compared to cognitively unimpaired individuals were significantly exposed to more potential risk factors ($\chi^2 = 120.31$, $P < 0.001$), such as high fat diets, vitamin D deficiency, physical inactivity, smoking, and sleep disorders (Zhou et al., 2019). Compared to other areas of development within NLP algorithms, the assessment of lifestyle factors is relatively new and further research is required.

A large amount of the research and development conducted to date for NLP algorithms has examined assessments and treatments completed during the clinical course and their outcomes for patients (Fu et al., 2019; Pons, Braun, Hunink, & Kors, 2016; Pruitt, Naidech, Van Ornam, Borczuk, & Thompson, 2019; Yim, Yetisgen, Harris, & Kwan, 2016; Zhang et al., 2019). NLP algorithms have been developed and tested for a suite of treatment outcomes:

- identifying the presence of adverse outcomes (e.g., deep venous thrombosis (DVT), pulmonary embolisms (PE)) postsurgery (Selby, Narain, Russo, Strong, & Stetson, 2018);
- examining the risks associated with different drugs (e.g., vedolizumab, tumor necrosis factor inhibitor (TNFi)) for patient groups (Cai et al., 2018);
- not just the presence of a particular drug, but furthermore the prescribing physicians reasoning for inclusion (Li, Salmasian, Harpaz, Chase, & Friedman, 2011); and

**106** Big data in psychiatry and neurology

○ identification of potential drug use (e.g., problem opioid use) and the associated healthcare costs (Carrell et al., 2015; Masters et al., 2018).

One of the key aims for the future of NLP algorithms is to place the algorithms within the healthcare database where they can process data in real time and assist clinicians in their decision-making. One such example in Spain developed and implemented a clinical decision support system (CDSS) based on NLP and artificial intelligence (AI) techniques to provide recommendations to primary care physicians (Cruz, Canales, Muñoz, Pérez, & Arnott, 2019). By including recommendations across a wide range of diseases (e.g., heart failure, lower back pain, osteoporosis) such as "SSRIs should not be prescribed to patients with chronic lower back pain unless they are also suffering from depression," the CDSS increased adherence to suggested clinical pathways and reduced clinical variability. A number of gaps in the literature need to be addressed before such CDSS can be implemented more widely, such as clinical data standardization, vast majority of NLP algorithms have been developed in the English language, recognition of relationships between entries, and the extraction of temporal data (Kreimeyer et al., 2017; Névéol, Dalianis, Velupillai, Savova, & Zweigenbaum, 2018; Sheikhalishahi et al., 2019). However, the Spanish example highlights the huge potential NLP algorithms have for assisting healthcare professionals in real time to provide more timely and accurate diagnosis and treatment.

## 4.2 Development of diagnostic algorithms: Unstructured data

The Lewy-CRATE/CRIS is an exciting project currently underway that displays how databases like CRIS and CRATE, with their linkages to local and national structured and unstructured routinely-collected information and suite of NLP algorithm development, can be used to address the needs of one of the most misdiagnosed forms of dementia. Conducted by teams in the University of Cambridge and King's College London (KCL) using CRIS and CRATE databases in SLaM and CPFT, Lewy-CRATE/CRIS will identify a cohort of ~1500 dementia with Lewy bodies (DLB) cases and several thousand non-DLB disease dementia controls to allow a detailed examination of their early presentation, diagnosis, clinical course, and outcomes. DLB is the second commonest cause of dementia after Alzheimer's disease (AD; McKeith, 2006), but accurate recognition and early diagnosis remain suboptimal. Current estimates indicate that less than 4% of people with dementia are diagnosed with DLB, less than half the number predicted on the basis of pathological and epidemiological studies. For example, a pathological series indicate Lewy body pathology in up to 20% of dementia cases (McKeith, 2006), while epidemiological studies have shown prevalence rates of 10% (Vann Jones & O'Brien, 2014). There are several factors that likely contribute to the underdiagnosis of DLB, including a lack of awareness by clinicians of some of the diagnostic features, a failure to ask about these during patient assessment, and a lack of appreciation of the many and varied ways in which DLB can present (Gore, Vardy, & O'Brien, 2015). While symptoms like visual hallucinations are usually elicited, other core features such as REM sleep disorder (RBD) and low mood may not be. However, observations may be made in clinical notes that subjects are frequently sleepy or drowsy, or slowed in movements, or have falls or speech which is difficult to follow, which can give important clues as to a possible DLB syndrome (Gore et al., 2015). In addition, there may be other symptoms or clinical factors contained within hospital records which should alert clinicians to a DLB diagnosis, but which are currently not known (Palmqvist, Hansson, Minthon, & Londos, 2009). Improved and early recognition and diagnosis of DLB is an important outcome since accurate recognition is essential to optimize management (McKeith, 2006; Stinton et al., 2015). DLB specific symptoms including sleep disturbance, parkinsonism, fluctuation, autonomic symptoms, and falls are frequently managed suboptimally unless DLB is accurately recognized and diagnosed (McKeith, 2006; Stinton et al., 2015).

In the first part of the project, the deidentified EMRs within the CRIS and CRATE databases will be searched utilizing data mining techniques in order to identify all diagnoses of DLB between 2005 and 2019 in order to create one of the world's largest retrospective DLB cohorts. As mentioned, DLB is difficult to diagnose given the range of symptoms that patients present with. Furthermore, there is no diagnostic F code for DLB in the ICD-10. The ability to search through deidentified EMRs for diagnostic statements related to DLB gives this research project a unique advantage: the high numbers of DLB cases seen in routine mental healthcare and because much of the key information for disorders like DLB will be predominantly contained in text fields, the project stands to create one of the largest unbiased DLB cohorts seen in research. Much of what has been learned to date has been from relatively small cohorts, which are not necessarily representative of all those with DLB as they have "opted in" to studies, often run by tertiary centers with particular expertise in DLB. Despite its frequency, and perhaps related to the problem of underrecognition, we still understand very little about the presentation, natural history, and clinical course of DLB, which hinders greater understanding of the disorder and the search for effective treatments. Sex differences in the prevalence of the disease are still unclear (McKeith, 2006; Price et al., 2017; Vann Jones & O'Brien, 2014), and differences in cognitive decline and mortality have been reported but remain to be confirmed in larger more unselected series (Boström, Jönsson, Minthon, & Londos, 2007; Williams, Xiong, Morris, & Galvin, 2006).

To date, the research teams have already produced results from the CRIS and CRATE databases displaying the greater burden, varying symptom profile, and reduced survival in DLB cases compared with other forms of dementia. In a cohort from CPFT comparing 251 DLB cases with 222 AD cases utilizing local structured and unstructured routinely-collected data, Price et al. (2017) found that survival among the DLB cases was markedly shorter (3.72 years) in comparison to AD (6.95 years). Interestingly, the investigation found that the reduction in survival was independent of many of the expected predictor factors (e.g., age, sex, physical comorbidity, or antipsychotic prescribing). While examining the cohort further, Moylett et al. (2019) found that DLB cases presented to healthcare centers with a wide range of presenting complaints—memory loss (27.1%), hallucinations (25.4%), and low mood (25.1%) being the most common—and varying initial diagnoses (e.g., AD, mild cognitive impairment, depression). Rates of RBD were considerably lower (8.4%) than would be expected of a DLB diagnosis and this is believed to be due to under-reporting and lack of recognition. In a cohort from SLaM comparing 194 DLB cases with 776 AD cases utilizing local and national structured and unstructured data, Mueller et al. (2017) found that DLB cases had higher healthcare usage compared to AD or the general catchment population. In the year following dementia diagnosis, the rates of hospital admissions were higher among the DLB cases (crude incidence rate ratio 1.50; 95% confidence interval: 1.28–1.75) than the AD or catchment population (indirectly standardized hospitalization rate 1.22; 95% confidence interval: 1.06–1.39). Further analysis revealed poorer physical health early in the disease course to be the main indicator for this finding. The project and its findings show how the use of local and national structured and unstructured routinely-collected data allowed a large amount of diagnostic, temporal, and neuropsychological information to be gathered on the whole cohort and enabled an in-depth examination to be completed.

By growing the cohort and establishing more linkages to national data, the Lewy-CRATE/CRIS project will be able to examine more in depth the patterns of early predictors, presentations and symptoms associated with DLB to facilitate earlier diagnosis, and better management and treatments. Ways of better managing DLB are urgently needed, since the mainstay of current treatment consists of the use of modest symptomatic treatments and dementia support (Department of Health, 2009; McKeith, 2006; Stinton et al., 2015). Furthermore by drawing on the large amounts of healthcare data available across both databases and new technological advances, the teams are currently working to develop a bespoke DLB NLP algorithm which will process information from routinely-collected EMRs to identify potential DLB cases. DLB cases and non-DLB disease dementia controls will be split into development/test and validation sets (equal numbers in each group), where the development/test sets will be used to develop the NLP algorithm with iterative performance and development stages, and the validation set will be used as an independent cohort to test the NLP algorithm's accuracy. Both bottom-up and top-down approaches are being taken for the NLP algorithm development. In the bottom-up approach, techniques for extracting lexical statistics including bag-of-words (BOW) model and nested latent Dirichlet allocation (LDA) models (Resnik et al., 2015), along with novel neural models of NLP such as Convolutional Neural Network (CNN; Kim, 2014), are being tested in order to develop an NLP that's informed by the contents of the EMRs. In the top-down approach, the core and suggestive features of DLB according to the latest diagnostic criteria (McKeith et al., 2017) have informed the symptoms that a bespoken DLB NLP algorithm will search for within the EMRs. By combining the current suite of NLP algorithms with CRIS and CRATE with new developmental work, a DLB NLP algorithm will be created that searches for a set of DLB diagnostic symptoms (e.g., hallucinations, falls, parkinsonian features) and determines the probability that the particular case may have DLB. This will be important to determine whether the NLP algorithm may be useful to increase diagnostic rates for DLB by identifying subjects who may fulfill clinical criteria but have not yet been diagnosed clinically.

## 5 Conclusion

By combining the wealth of information contained within two EMR research databases, the Lewy-CRATE/CRIS project will create one of the world's largest retrospective DLB cohorts, combining local and national healthcare databases with deidentified EMRs to learn as much as possible about the disease's presentation, progression, and outcomes. Furthermore, it will create a bespoke DLB NLP algorithm to aid with earlier diagnosis and thus, help with paving the way for future projects within the area of big data and its application in epidemiology. As the presence and advancements in deidentified EMR databases grow, there are possibilities for utilizing NLP algorithms in "real-time" to assist clinicians while they make diagnoses. The NLP algorithm can be placed within databases like CRIS and CRATE so that it searches case records of all newly assessed cases with cognitive impairment (which are uploaded daily to the databases) in order to flag potential DLB cases, and this information will be fed back to the clinicians. This may be of help to all clinicians but is likely to be of particular help to more junior clinicians or those of a nonmedical background who may consider DLB diagnosis less frequently. Highlighting how big databases in healthcare can go beyond collecting information and have real-world practical applications that can assist clinicians to deliver timely and accurate diagnoses. Throughout the chapter, different databases and projects have been discussed that show the varied role big healthcare databases can have by combing sources of structured and unstructured, local and national, routinely- and non-routinely-collected information to illuminate the path of many diseases from prodromal stages through diagnosis and disease course to final outcomes. These projects are informing better diagnosis and treatment of disease by identifying prodromal markers for earlier diagnosis, assessing the healthcare usage among disease groups, making suggestions for personalized preventive strategies and treatment that could be implemented among different disease subgroups to reduce negative outcomes, and creating diagnostic aids for assisting healthcare professionals.

Big databases and their applications created for research from healthcare data are not without their faults. A review in 2016 highlighted some of the main challenges existing around data structure, security, data standardization, storage and transfers, and managerial skills such as data governance (Kruse, Goswamy,

**110** Big data in psychiatry and neurology

Raval, & Marawi, 2016). Ethical and governance guidelines and procedures are continually evolving to incorporate the growing size of big healthcare databases and the technological and statistical advancements seen in recent times. As previously mentioned, data is collected from multiple sources within the healthcare system through different formats and structures. Great care must be taken to ensure data consistency is maintained when combining data. Furthermore, clinical records are not created for the purpose of research and therefore, missing data is a regular occurrence. Advancements in statistical methods for dealing with missing data are going a long way to assist researchers when completing analyses. Calls have been made for greater consideration of big data analytics and validation within clinical practice so that the opportunities and benefits present within big healthcare data and its applications can be fully realized (Lee & Yoon, 2017). Furthermore, it should be noted that the vast majority of these big healthcare databases are located with higher-income countries with advanced IT resources, meaning generalizability of results outside of higher-income countries may be limited.

In order for big data applications in healthcare to be truly impactful, findings must be used to inform guidelines and policy. One of the great advantages of the projects and databases discussed in this chapter is the huge sample sizes included, meaning the results are applicable to large numbers of individuals within the respective nations and disease groups. The utility of big data in healthcare for reducing disease incidence and burden is perfectly in line with the European Commission's "Digital Agenda for Europe" calling for greater and more sustainable use of information and communication technologies to provide better healthcare (European Commission, 2010). Epidemiology is a particular strength of European research, and large real-world healthcare datasets that have recently become available open up a new era with immense possibilities. The Max Planck Institute for Demographic Research (MPIDR; https://www.demogr.mpg.de/en) in Rostock, Germany, is an internationally renowned leader in the use of high-quality population data for identifying and quantifying how major demographic, behavioral, and structural conditions affect population health. Conducting basic research into demographic processes, researchers at the MPIDR analyze the underlying causes of demographic change, describe contemporary demographic trends, produce forecasts for the future direction of demographic processes, highlight the potential consequences facing society, and assist decision-makers across various political and social institutions by providing them with information and expert advice. Since 2009, the MPIDR hosts "Population Europe" (https://population-europe.eu/ )—the collaborative network of Europe's leading demographic research institutes and centers—whose main activity is the dissemination of policy relevant demographic research findings. It is only through such collaborative efforts including researchers, IT technicians, healthcare workers, statisticians, and policy makers that the true potential of big data and its applications within healthcare can be realized.

## References

Abbing-Karahagopian, V., Huerta, C., Souverein, P. C., de Abajo, F., Leufkens, H. G., Slattery, J., et al. (2014). Antidepressant prescribing in five European countries: Application of common definitions to assess the prevalence, clinical observations, and methodological implications. *European Journal of Clinical Pharmacology*, *70*(7), 849–857. https://doi.org/10.1007/s00228-014-1676-z.

Anderson, K. M., Odell, P. M., Wilson, P. W., & Kannel, W. B. (1991). Cardiovascular disease risk profiles. *American Heart Journal*, *121*(1 Pt 2), 293–298. https://doi.org/10.1016/0002-8703(91)90861-b.

Boström, F., Jönsson, L., Minthon, L., & Londos, E. (2007). Patients with Lewy body dementia use more resources than those with Alzheimer's disease. *International Journal of Geriatric Psychiatry*, *22*(8), 7130–7719. https://doi.org/10.1002/gps.1738.

Cai, T., Lin, T. C., Bond, A., Huang, J., Kane-Wanger, G., Cagan, A., et al. (2018). The association between arthralgia and vedolizumab using natural language processing. *Inflammatory Bowel Diseases*, *24*(10), 2242–2246. https://doi.org/10.1093/ibd/izy127.

Cardinal, R. (2017). Clinical records anonymisation and text extraction (CRATE): An open-source software system. *BMC Medical Informatics and Decision Making*, *17*, 50. https://doi.org/10.1186/s12911-017-0437-1.

Cardinal, R. N., Savulich, G., Mann, L. M., & Fernández-Egea, E. (2015). Association between antipsychotic/antidepressant drug treatments and hospital admissions in schizophrenia assessed using a mental health case register. *NPJ Schizophrenia*, *1*, 15035. https://doi.org/10.1038/npjschz.2015.35.

Carr, M. J., Ashcroft, D. M., Kontopantelis, E., While, D., Awenat, Y., Cooper, J., et al. (2016). Clinical management following self-harm in a UK-wide primary care cohort. *Journal of Affective Disorders*, *197*, 182–188. https://doi.org/10.1016/j.jad.2016.03.013.

Carrell, D. S., Cronkite, D., Palmer, R. E., Saunders, K., Gross, D. E., Masters, E. T., … Von Korff, M. (2015). Using natural language processing to identify problem usage of prescription opioids. *International Journal of Medical Informatics*, *84*(12), 1057–1064. https://doi.org/10.1016/j.ijmedinf.2015.09.002.

Centers for Disease Control and Prevention. (2011). Ten great public health achievements. *Morbidity and Mortality Weekly Report*, *60*(19), 619–623. https://www.cdc.gov/mmwr/preview/mmwrhtml/mm6019a5.htm.

Chang, C.-K., Harrison, S., Lee, W., Taylor, D., & Stewart, R. (2012). Ascertaining instances of neuroleptic malignant syndrome in a secondary mental healthcare electronic medical records database: The SLAM BRC case register. *Therapeutic Advances in Psychopharmacology*, *2*(2), 75–83. https://doi.org/10.1177/2045125312438215.

Collins, G. S., & Altman, D. G. (2010). An independent and external validation of QRISK2 cardiovascular disease risk score: A prospective open cohort study. *BMJ*, *340*, c2442. https://doi.org/10.1136/bmj.c2442.

Conroy, R. M., Pyörälä, K., Fitzgerald, A. P., Sans, S., Menotti, A., De Backer, G., et al. (2003). Estimation of ten-year risk of fatal cardiovascular disease in Europe: The SCORE project. *European Heart Journal*, *24*(11), 987–1003. https://doi.org/10.1016/S0195-668X(03)00114-3.

Cope, A. L., Wood, F., Francis, N. A., & Chestnutt, I. G. (2018). Patients' reasons for consulting a GP when experiencing a dental problem: A qualitative study. *British Journal of General Practice*, *68*(677), e877–e883. https://doi.org/10.3399/bjgp18X699749.

Cruz, N. P., Canales, L., Muñoz, J. G., Pérez, B., & Arnott, I. (2019). Improving adherence to clinical pathways through natural language processing on electronic medical records. *Studies in Health Technology and Informatics*, *264*, 561–565. https://doi.org/10.3233/SHTI190285.

# 112 Big data in psychiatry and neurology

Cunningham, H., Maynard, D., & Bontcheva, K. (2011). *Text processing with GATE* (Version 6). GATE.

Cunningham, H., Tablan, V., Roberts, A., & Bontcheva, K. (2013). Getting more out of biomedical documents with GATE's full lifecycle open source text analytics. *PLoS Computational Biology*, *9*(2), e1002854. https://doi.org/10.1371/journal.pcbi.1002854.

Damen, J. A., Hooft, L., Schuit, E., Debray, T. P., Collins, G. S., Tzoulaki, I., et al. (2016). Prediction models for cardiovascular disease risk in the general population: Systematic review. *BMJ*, *353*, i2416. https://doi.org/10.1136/bmj.i2416.

Delvaux, N., Aertgeerts, B., van Bussel, J. C., Goderis, G., Vaes, B., & Vermandere, M. (2018). Health data for research through a nationwide privacy-proof system in Belgium: Design and implementation. *JMIR Medical Informatics*, *6*(4), e11428. https://doi.org/10.2196/11428.

Department of Health. (2009). *Living well with dementia: A national dementia strategy. Vol. 2009*. https://www.gov.uk/government/publications/living-well-with-dementia-a-national-dementia-strategy.

European Commission. (2010). *A digital agenda for Europe*. https://eur-lex.europa.eu/legal-content/en/ALL/?uri=CELEX%3A52010DC0245.

Fernandes, A. C., Cloete, D., Broadbent, M. T., Hayes, R. D., Chang, C. K., Jackson, R. G., et al. (2013). Development and evaluation of a de-identification procedure for a case register sourced from mental health electronic records. *BMC Medical Informatics and Decision Making*, *13*, 71. https://doi.org/10.1186/1472-6947-13-71.

Fernandez-Egea, E., Walker, R., Ziauddeen, H., Cardinal, R. N., & Bullmore, E. T. (2020). Birth weight, family history of diabetes and diabetes onset in schizophrenia. *BMJ Open Diabetes Research & Care*, *8*(1). https://doi.org/10.1136/bmjdrc-2019-001036, e001036.

Fieuws, S., & Verbeke, G. (2006). Pairwise fitting of mixed models for the joint modeling of multivariate longitudinal profiles. *Biometrics*, *62*(2), 424–431. https://doi.org/10.1111/j.1541-0420.2006.00507.x.

Fok, M. L., Hayes, R. D., Chang, C. K., Stewart, R., Callard, F. J., & Moran, P. (2012). Life expectancy at birth and all-cause mortality among people with personality disorder. *Journal of Psychosomatic Research*, *73*(2), 104–107. https://doi.org/10.1016/j.jpsychores.2012.05.001.

Ford, D. V., Jones, K. H., Verplancke, J. P., Lyons, R. A., John, G., Brown, G., et al. (2009). The SAIL databank: Building a national architecture for e-health research and evaluation. *BMC Health Services Research*, *9*, 157. https://doi.org/10.1186/1472-6963-9-157.

Fu, S., Leung, L. Y., Wang, Y., Raulli, A. O., Kallmes, D. F., Kinsman, K. A., et al. (2019). Natural language processing for the identification of silent brain infarcts from neuroimaging reports. *JMIR Medical Informatics*, *7*(2), e12109. https://doi.org/10.2196/12109.

GBD 2015 DALYs and HALE Collaborators. (2016). Global, regional, and national disability-adjusted life-years (DALYs) for 315 diseases and injuries and healthy life expectancy (HALE), 1990-2015: A systematic analysis for the global burden of disease study 2015. *Lancet*, *388* (10053), 1603–1658. https://doi.org/10.1016/S0140-6736(16)31460-X.

GBD 2015 Mortality and Causes of Death Collaborators. (2016). Global, regional, and national life expectancy, all-cause mortality, and cause-specific mortality for 249 causes of death, 1980–2015: A systematic analysis for the Global Burden of Disease Study 2015. *Lancet*, *388* (10053), 1459–1544. https://doi.org/10.1016/S0140-6736(16)31012-1.

Global Burden of Disease Study 2013 Collaborators. (2015). Global, regional, and national incidence, prevalence, and years lived with disability for 301 acute and chronic diseases and injuries in 188 countries, 1990-2013: A systematic analysis for the global burden of disease study 2013. *Lancet*, *386*(9995), 743–800. https://doi.org/10.1016/S0140-6736(15)60692-4.

Data linkages in epidemiology Chapter | 5  **113**

Gore, R. L., Vardy, E. R. L. C., & O'Brien, J. T. (2015). Delirium and dementia with Lewy bodies: Distinct diagnoses or part of the same spectrum? *Journal of Neurology, Neurosurgery & Psychiatry, 86*(1), 50–59. https://doi.org/10.1136/jnnp-2013-306389.

Hamilton, W., & Kernick, D. (2007). Clinical features of primary brain tumours: A case-control study using electronic primary care records. *The British Journal of General Practice, 57* (542), 695–699.

Herrett, E., Gallagher, A. M., Bhaskaran, K., Forbes, H., Mathur, R., van Staa, T., et al. (2015). Data resource profile: Clinical practice research datalink (CPRD). *International Journal of Epidemiology, 44*(3), 827–836. https://doi.org/10.1093/ije/dyv098.

Hippisley-Cox, J., Coupland, C., & Brindle, P. (2017). Development and validation of QRISK3 risk prediction algorithms to estimate future risk of cardiovascular disease: Prospective cohort study. *BMJ, 357*, j2099. https://doi.org/10.1136/bmj.j2099.

Hospital Episode Statistics. (2020). https://digital.nhs.uk/data-and-information/data-tools-and-services/data-services/hospital-episode-statistics.

Hurt, L., Ashfield-Watt, P., Townson, J., Heslop, L., Copeland, L., Atkinson, M. D., et al. (2019). Cohort profile: HealthWise Wales. A research register and population health data platform with linkage to National Health Service data sets in Wales. *BMJ Open, 9*(12). https://doi.org/10.1136/bmjopen-2019-031705, e031705.

Ingram, J. R., Collins, H., Atkinson, M. D., & Brooks, C. J. (2020). Prevalence of hidradenitis suppurativa is one percent of the population of Wales using the Secure Anonymised Information Linkage (SAIL) databank. *British Journal of Dermatology, 183*, 950–952. https://doi.org/10.1111/bjd.19210.

Kennedy, E. H., Wiitala, W. L., Hayward, R. A., & Sussman, J. B. (2013). Improved cardiovascular risk prediction using nonparametric regression and electronic health record data. *Medical Care, 51*(3), 251–258. https://doi.org/10.1097/MLR.0b013e31827da594.

Kim, Y. (2014). *Convolutional neural networks for sentence* classification [Paper presentation]. In *2014 Conference on Empirical Methods in Natural Language Processing (EMNLP), Doha, Qatar* (pp. 1746–1751). https://www.aclweb.org/anthology/D14-1181.pdf.

Kreimeyer, K., Foster, M., Pandey, A., Arya, N., Halford, G., Jones, S. F., et al. (2017). Natural language processing systems for capturing and standardizing unstructured clinical information: A systematic review. *Journal of Biomedical Informatics, 73*, 14–29. https://doi.org/10.1016/j.jbi.2017.07.012.

Kruse, C. S., Goswamy, R., Raval, Y., & Marawi, S. (2016). Challenges and opportunities of big data in health care: A systematic review. *JMIR Medical Informatics, 4*(4), e38. https://doi.org/10.2196/medinform.5359.

Lee, S. J., Larson, E. B., Dublin, S., Walker, R., Marcum, Z., & Barnes, D. (2018). A cohort study of healthcare utilization in older adults with undiagnosed dementia. *Journal of General Internal Medicine, 33*(1), 13–15. https://doi.org/10.1007/s11606-017-4162-3.

Lee, C. H., & Yoon, H. J. (2017). Medical big data: Promise and challenges. *Kidney Research and Clinical Practice, 36*(1), 3–11. https://doi.org/10.23876/j.krcp.2017.36.1.3.

Leucht, S., Cipriani, A., Spineli, L., Mavridis, D., Orey, D., Richter, F., et al. (2013). Comparative efficacy and tolerability of 15 antipsychotic drugs in schizophrenia: A multiple-treatments meta-analysis. *Lancet, 382*(9896), 951–962. https://doi.org/10.1016/S0140-6736(13)60733-3.

Li, Y., Salmasian, H., Harpaz, R., Chase, H., & Friedman, C. (2011). Determining the reasons for medication prescriptions in the EHR using knowledge and natural language processing. *AMIA .. annual symposium proceedings. AMIA symposium, 2011*, 768–776.

## 114 Big data in psychiatry and neurology

Lyons, R. A., Jones, K. H., John, G., Brooks, C. J., Verplancke, J. P., Ford, D. V., et al. (2009). The SAIL databank: Linking multiple health and social care datasets. *BMC Medical Informatics and Decision Making*, *9*, 3. https://doi.org/10.1186/1472-6947-9-3.

Masters, E. T., Ramaprasan, A., Mardekian, J., Palmer, R. E., Gross, D. E., Cronkite, D., et al. (2018). Natural language processing-identified problem opioid use and its associated health care costs. *Journal of Pain & Palliative Care Pharmacotherapy*, *32*(2–3), 106–115. https://doi.org/10.1080/15360288.2018.1488794.

McKeith, I. G. (2006). Diagnosis and management of dementia with Lewy bodies: Third report of the DLB consortium. *Neurology*, *66*(September 2004), 1455. https://doi.org/10.1212/01.wnl.0000224698.67660.45.

McKeith, I. G., Boeve, B. F., Dickson, D. W., Halliday, G., Taylor, J. P., Weintraub, D., et al. (2017). Diagnosis and management of dementia with Lewy bodies: Fourth consensus report of the DLB consortium. *Neurology*, *89*(1), 88–100. https://doi.org/10.1212/WNL.0000000000004058.

Mehta, S., Jackson, R., Wells, S., Harrison, J., Exeter, D. J., & Kerr, A. J. (2018). Cardiovascular medication changes over 5 years in a national data linkage study: Implications for risk prediction models. *Clinical Epidemiology*, *10*, 133–141. https://doi.org/10.2147/CLEP.S138100.

Moons, K. G., Kengne, A. P., Woodward, M., Royston, P., Vergouwe, Y., Altman, D. G., et al. (2012). Risk prediction models: I. Development, internal validation, and assessing the incremental value of a new (bio)marker. *Heart*, *98*(9), 683–690. https://doi.org/10.1136/heartjnl-2011-301246.

Moylett, S., Price, A., Cardinal, R. N., Aarsland, D., Mueller, C., Stewart, R., et al. (2019). Clinical presentation, diagnostic features, and mortality in dementia with Lewy bodies. *Journal of Alzheimer's Disease*, *67*(3), 995–1005. https://doi.org/10.3233/JAD-180877.

Mueller, C., Molokhia, M., Perera, G., Veronese, N., Stubbs, B., Shetty, H., et al. (2018). Polypharmacy in people with dementia: Associations with adverse health outcomes. *Experimental Gerontology*, *106*, 240–245. https://doi.org/10.1016/j.exger.2018.02.011.

Mueller, C., Perera, G., Rajkumar, A. P., Bhattarai, M., Price, A., O'Brien, J. T., et al. (2017). Hospitalization in people with dementia with Lewy bodies: Frequency, duration, and cost implications. *Alzheimer's & Dementia*, *10*, 143–152. https://doi.org/10.1016/j.dadm.2017.12.001.

Mueller, C., Soysal, P., Rongve, A., Isik, A. T., Thompson, T., Maggi, S., … Veronese, N. (2019). Survival time and differences between dementia with Lewy bodies and Alzheimer's disease following diagnosis: A meta-analysis of longitudinal studies. *Ageing Research Reviews*, *50*, 72–80. https://doi.org/10.1016/j.arr.2019.01.005.

Névéol, A., Dalianis, H., Velupillai, S., Savova, G., & Zweigenbaum, P. (2018). Clinical natural language processing in languages other than English: Opportunities and challenges. *Journal of Biomedical Semantics*, *9*(1), 12. https://doi.org/10.1186/s13326-018-0179-8.

Paige, E., Barrett, J., Pennells, L., Sweeting, M., Willeit, P., Di Angelantonio, E., et al. (2017). Use of repeated blood pressure and cholesterol measurements to improve cardiovascular disease risk prediction: An individual-participant-data meta-analysis. *American Journal of Epidemiology*, *186*(8), 899–907. https://doi.org/10.1093/aje/kwx149.

Palmqvist, S., Hansson, O., Minthon, L., & Londos, E. (2009). Practical suggestions on how to differentiate dementia with Lewy bodies from Alzheimer's disease with common cognitive tests. *International Journal of Geriatric Psychiatry*, *24*(12), 1405–1412. https://doi.org/10.1002/gps.2277.

Pate, A., Emsley, R., Ashcroft, D. M., Brown, B., & van Staa, T. (2019). The uncertainty with using risk prediction models for individual decision making: An exemplar cohort study examining the prediction of cardiovascular disease in English primary care. *BMC Medicine*, *17*(1), 134. https://doi.org/10.1186/s12916-019-1368-8.

Patterson, B. W., Jacobsohn, G. C., Shah, M. N., Song, Y., Maru, A., Venkatesh, A. K., et al. (2019). Development and validation of a pragmatic natural language processing approach to identifying falls in older adults in the emergency department. *BMC Medical Informatics and Decision Making*, *19*(1), 138. https://doi.org/10.1186/s12911-019-0843-7.

Perera, G., Broadbent, M., Callard, F., Chang, C. K., Downs, J., Dutta, R., et al. (2016). Cohort profile of the South London and Maudsley NHS Foundation Trust Biomedical Research Centre (SLaM BRC) case register: Current status and recent enhancement of an electronic mental health record-derived data resource. *BMJ Open*, *6*(3). https://doi.org/10.1136/bmjopen-2015-008721, e008721.

Pons, E., Braun, L. M. M., Hunink, M. G. M., & Kors, J. A. (2016). Natural language processing in radiology: A systematic review. *Radiology*, *279*(2), 329–343. https://doi.org/10.1148/radiol.16142770.

Powell, J., Fitton, R., & Fitton, C. (2006). Sharing electronic health records: The patient view. *Informatics in Primary Care*, *14*(1), 55–57. https://doi.org/10.14236/jhi.v14i1.614.

Price, A., Farooq, R., Yuan, J. M., Menon, V. B., Cardinal, R. N., & O'Brien, J. T. (2017). Mortality in dementia with Lewy bodies compared with Alzheimer's dementia: A retrospective naturalistic cohort study. *BMJ Open*, *7*(11), e017504. https://doi.org/10.1136/bmjopen-2017-017504.

Pruitt, P., Naidech, A., Van Ornam, J., Borczuk, P., & Thompson, W. (2019). A natural language processing algorithm to extract characteristics of subdural hematoma from head CT reports. *Emergency Radiology*, *26*(3), 301–306. https://doi.org/10.1007/s10140-019-01673-4.

Rees, F., Doherty, M., Lanyon, P., Davenport, G., Riley, R. D., Zhang, W., et al. (2017). Early clinical features in systemic lupus erythematosus: Can they be used to achieve earlier aiagnosis? A risk prediction model. *Arthritis Care & Research*, *69*(6), 833–841. https://doi.org/10.1002/acr.23021.

Resnik, P., Armstrong, W., Claudino, L., Nguyen, T., Nguyen, V.-A., & Boyd-Graber, J. (2015). Beyond lda: Exploring supervised topic modeling for depression-related language in Twitter. *2nd Workshop on Computational Linguistics and Clinical Psychology: From Linguistic Signal to Clinical Reality*, *Denver, Colorado*, 99–107. https://www.aclweb.org/anthology/W15-1212.pdf.

Roth, G. A., Johnson, C., Abajobir, A., Abd-Allah, F., Abera, S. F., Abyu, G., et al. (2017). Global, regional, and national burden of cardiovascular diseases for 10 causes, 1990 to 2015. *Journal of the American College of Cardiology*, *70*(1), 1–25. https://doi.org/10.1016/j.jacc.2017.04.052.

Selby, L. V., Narain, W. R., Russo, A., Strong, V. E., & Stetson, P. (2018). Autonomous detection, grading, and reporting of postoperative complications using natural language processing. *Surgery*, *164*(6), 1300–1305. https://doi.org/10.1016/j.surg.2018.05.008.

Sheikhalishahi, S., Miotto, R., Dudley, J. T., Lavelli, A., Rinaldi, F., & Osmani, V. (2019). Natural language processing of clinical notes on chronic diseases: Systematic review. *JMIR Medical Informatics*, *7*(2), e12239. https://doi.org/10.2196/12239.

Shoenbill, K., Song, Y., Gress, L., Johnson, H., Smith, M., & Mendonca, E. A. (2020). Natural language processing of lifestyle modification documentation. *Health Informatics Journal*, *26*(1), 388–405. https://doi.org/10.1177/1460458218824742.

Sinha, S., Peach, G., Poloniecki, J. D., Thompson, M. M., & Holt, P. J. (2013). Studies using English administrative data (hospital episode statistics) to assess health-care outcomes: Systematic review and recommendations for reporting. *European Journal of Public Health*, *23*(1), 86–92. https://doi.org/10.1093/eurpub/cks046.

Soni, H., Grando, A., Murcko, A., Diaz, S., Mukundan, M., Idouraine, N., et al. (2020). State of the art and a mixed-method personalized approach to assess patient perceptions on medical record sharing and sensitivity. *Journal of Biomedical Informatics*, *101*. https://doi.org/10.1016/j.jbi.2019.103338, 103338.

## 116 Big data in psychiatry and neurology

Stapley, S. A., Rubin, G. P., Alsina, D., Shephard, E. A., Rutter, M. D., & Hamilton, W. T. (2017). Clinical features of bowel disease in patients aged <50 years in primary care: A large case-control study. *The British Journal of General Practice, 67*(658), e336–e344. https://doi.org/10.3399/bjgp17X690425.

StatsNZ. (2018). *Integrated data infrastructure.* https://www.stats.govt.nz/integrated-data/integrated-data-infrastructure/.

Stewart, R., Soremekun, M., Perera, G., Broadbent, M., Callard, F., Denis, M., et al. (2009). The South London and Maudsley NHS Foundation Trust Biomedical Research Centre (SLAM BRC) case register: Development and descriptive data. *BMC Psychiatry, 9*, 51. https://doi.org/10.1186/1471-244X-9-51.

Stinton, C., McKeith, I., Taylor, J. P., Lafortune, L., Mioshi, E., Mak, E., et al. (2015). Pharmacological management of Lewy body dementia: A systematic review and meta-analysis. *The American Journal of Psychiatry, 172*(8), 731–742. https://doi.org/10.1176/appi.ajp.2015.14121582.

Stubbs, B., Mueller, C., Gaughran, F., Lally, J., Vancampfort, D., Lamb, S. E., et al. (2018). Predictors of falls and fractures leading to hospitalization in people with schizophrenia spectrum disorder: A large representative cohort study. *Schizophrenia Research, 201*, 70–78. https://doi.org/10.1016/j.schres.2018.05.010.

Tadayon, S., Wickramasinghe, K., & Townsend, N. (2019). Examining trends in cardiovascular disease mortality across Europe: How does the introduction of a new European standard population affect the description of the relative burden of cardiovascular disease? *Population Health Metrics, 17*, 6. https://doi.org/10.1186/s12963-019-0187-7.

Tate, A. R., Dungey, S., Glew, S., Beloff, N., Williams, R., & Williams, T. (2017). Quality of recording of diabetes in the UK: How does the GP's method of coding clinical data affect incidence estimates? Cross-sectional study using the CPRD database. *BMJ Open, 7*(1). https://doi.org/10.1136/bmjopen-2016-012905, e012905.

Teixeira, P. L., Wei, W. Q., Cronin, R. M., Mo, H., VanHouten, J. P., Carroll, R. J., et al. (2017). Evaluating electronic health record data sources and algorithmic approaches to identify hypertensive individuals. *Journal of the American Medical Informatics Association, 24*(1), 162–171. https://doi.org/10.1093/jamia/ocw071.

Townsend, N., Wilson, L., Bhatnagar, P., Wickramasinghe, K., Rayner, M., & Nichols, M. (2016). Cardiovascular disease in Europe: Epidemiological update 2016. *European Heart Journal, 37* (42), 3232–3245. https://doi.org/10.1093/eurheartj/ehw334.

Toyabe, S. (2012). Detecting inpatient falls by using natural language processing of electronic medical records. *BMC Health Services Research, 12*, 448. https://doi.org/10.1186/1472-6963-12-448.

Truyers, C., Goderis, G., Dewitte, H., Van den Akker, M., & Buntinx, F. (2014). The Intego database: Background, methods and basic results of a Flemish general practice-based continuous morbidity registration project. *BMC Medical Informatics and Decision Making, 14*, 48. https://doi.org/10.1186/1472-6947-14-48.

Vann Jones, S. A., & O'Brien, J. T. (2014). The prevalence and incidence of dementia with Lewy bodies: A systematic review of population and clinical studies. *Psychological Medicine, 44*(4), 673–683. https://doi.org/10.1017/S0033291713000494.

Velupillai, S., Epstein, S., Bittar, A., Stephenson, T., Dutta, R., & Downs, J. (2019). Identifying suicidal adolescents from mental health records using natural language processing. *Studies in Health Technology and Informatics, 264*, 413–417. https://doi.org/10.3233/SHTI190254.

Verbeke, G., & Molenberghs, G. (2000). Linear mixed models for longitudinal data. *Springer series in statistics.* Springer-Verlag.

Watson, J., Nicholson, B. D., Hamilton, W., & Price, S. (2017). Identifying clinical features in primary care electronic health record studies: Methods for codelist development. *BMJ Open*, *7*(11). https://doi.org/10.1136/bmjopen-2017-019637, e019637.

Webb, P., Bain, C., & Page, A. (2020). *Essential epidemiology: An introduction for students and health professionals* (4th ed.). Cambridge University Press.

Williams, M. M, Xiong, C., Morris, J. C, & Galvin, J. E. (2006). Survival and mortality differences between dementia with Lewy bodies vs Alzheimer disease. *Neurology*, *67*(11), 1935–1941. https://doi.org/10.1212/01.wnl.0000247041.63081.98.

Wilson, P. W., D'Agostino, R. B., Levy, D., Belanger, A. M., Silbershatz, H., & Kannel, W. B. (1998). Prediction of coronary heart disease using risk factor categories. *Circulation*, *97*(18), 1837–1847. https://doi.org/10.1161/01.cir.97.18.1837.

Wolf, A., Dedman, D., Campbell, J., Booth, H., Lunn, D., Chapman, J., et al. (2019). Data resource profile: Clinical Practice Research Datalink (CPRD) aurum. *International Journal of Epidemiology*, *48*(6). https://doi.org/10.1093/ije/dyz034. 1740–1740g.

Wu, C. Y., Chang, C. K., Robson, D., Jackson, R., Chen, S. J., Hayes, R. D., et al. (2013). Evaluation of smoking status identification using electronic health records and open-text information in a large mental health case register. *PLoS One*, *8*(9). https://doi.org/10.1371/journal.pone.0074262, e74262.

Yawn, B. P., Yawn, R. A., Geier, G. R., Xia, Z., & Jacobsen, S. J. (1998). The impact of requiring patient authorization for use of data in medical records research. *The Journal of Family Practice*, *47*(5), 361–365.

Yim, W. W., Yetisgen, M., Harris, W. P., & Kwan, S. W. (2016). Natural language processing in oncology: A review. *JAMA Oncology*, *2*(6), 797–804. https://doi.org/10.1001/jamaoncol.2016.0213.

Zhang, X., Bellolio, M. F., Medrano-Gracia, P., Werys, K., Yang, S., & Mahajan, P. (2019). Use of natural language processing to improve predictive models for imaging utilization in children presenting to the emergency department. *BMC Medical Informatics and Decision Making*, *19*(1), 287. https://doi.org/10.1186/s12911-019-1006-6.

Zhou, X., Wang, Y., Sohn, S., Therneau, T. M., Liu, H., & Knopman, D. S. (2019). Automatic extraction and assessment of lifestyle exposures for Alzheimer's disease using natural language processing. *International Journal of Medical Informatics*, *130*. https://doi.org/10.1016/j.ijmedinf.2019.08.003, 103943.

## Chapter 6

# Neutrosophic rule-based classification system and its medical applications

Sameh H. Basha[*,a], Areeg Abdalla[a], and Aboul Ella Hassanien[b,c]

[a]*Faculty of Science, Cairo University, Cairo, Egypt,* [b]*Faculty of Computers and Information, Cairo University, Cairo, Egypt,* [c]*Scientific Research Group in Egypt (SRGE), Cairo, Egypt*

## 1 Introduction

An expert system, from the artificial intelligence point of view, is a computer system that emulates the human expertise in solving complex problems (Jackson, 1998). The main components of expert systems include knowledge base (KB), inference engine, and user interface. The KB is the main part of expert systems. One of the important factors in the success of the expert system to solve a particular problem depends on the accuracy and precision of the knowledge. The KB contains two parts: database (which is the data related to a particular problem at hand) and rule base (almost in the form of IF-THEN rules). The inference engine treats, acquires, and interprets the KB to achieve an accurate result. In many applications, there is a need for an interaction between the expert system and the user of such systems and this interaction is through the user interface.

Rule-based expert system is a famous expert system in which the knowledge is represented by IF-THEN rules which consist of two parts: antecedent part (IF-part) and consequent part (THEN-part) (Liu, Gegov, & Cocea, 2016; Nagori & Trivedi, 2014). The fulfillment of the rule antecedent gives rise to the execution of the consequent. Construction of rule-based systems can be divided into two types of deterministic rule-based systems based on the type of logic systems: probabilistic rule-based systems and fuzzy rule-based systems. This is based on the type of logic as fuzzy logic or probabilistic logic used as a tool to represent different forms of KB (Liu, Gegov, & Stahl, 2014; Magdalena, 2015). A rule-based expert system has been applied in different areas such as medical,

---

* Current address: Mathematics Department, Faculty of Science, Cairo University, Cairo, Egypt.

**Big Data in Psychiatry and Neurology.** https://doi.org/10.1016/B978-0-12-822884-5.00004-0
Copyright © 2021 Elsevier Inc. All rights reserved.

# 120 Big data in psychiatry and neurology

transportation, manufacturing, space and aviation, accounting, agriculture, etc. (Bhatt & Buch, 2016). Also, rule-based expert systems are used with machine learning in many problems such as classification, decision making, monitoring, process control, regression, clustering, pattern recognition, and natural language processing (NLP) ( Jadhav, Nalawade, & Bapat, 2013).

With a huge amount of data and a huge need to deal with problems in data such as imprecision, incomplete, vagueness, and inconsistency. A fuzzy rule-based system is the most famous rule-based system type and the most used and is the most important application of the fuzzy set theory introduced by Zadeh (1996). Fuzzy rule-based systems are considered an extension of classical rule-based systems in that they deal with fuzzy rules and the representation of the knowledge form is based on the fuzzy logic instead of classical logic (Alcalá, Casillas, Cordón, Herrera, & Zwir, 1999). Research on the rule-based system focuses on improving this system. The general structures of the fuzzy rule-based system contain four components (Alcalá et al., 1999; Sivanandam, Sumathi, & Deepa, 2007):

- **Knowledge base**: Contains two parts: data base and rule base.
  - *Data base*: Contains the dataset and the fuzzy membership functions used in the rule-based system that will be used in the inference engine.
  - *Rule base*: Contains fuzzy rules in the form of "IF-THEN" rules.
- **Fuzzifier**: Converts the crisp input to fuzzy one that will be used in the inference engine.
- **Inference engine**: Produces the result as the fuzzy output by using the fuzzy input and KB.
- **Defuzzifier**: Converts the fuzzy output into the crisp output.

The fuzzy rule-based system has many applications in different areas mainly in decision-making, classification tasks, and control problems (Mohammadpour, Abedi, Bagheri, & Ghaemian, 2015). For classification tasks as well as the decision-making and many other tasks, the fuzzy rule-based classification system is used by Ishibuchi, Nakashima, and Nii (2005). The main tasks in using fuzzy rule-based systems are automatically generating the fuzzy rules, learning of the fuzzy rules, and optimizing the fuzzy rules. There are several methods to generate automatically and learning the fuzzy rules as simple heuristic procedures, genetic algorithms, and neuro-fuzzy model (Mohammadpour et al., 2015). The fuzzy rule-based classification system has received great research attention in the last two decades due to its excellent performance (Ravi & Khare, 2016). There is a lot of fuzzy rule-based classification system in many areas. We will focus on their applications in the medical field.

One of the main health problems which caused a high death rate is cardiovascular diseases (CVDs). CVDs are usually caused by some blood clots that obstructs the blood flow, thereby affecting the heart. There is a high need to get an early diagnosis of these diseases.

Sanz et al. (2011) proposed a method based on a fuzzy rule-based classification system for classifying the patients with respect to the risk of suffering

CVDs. They used interval-valued fuzzy sets to model the linguistic labels of the classifier. In order to tune the amplitude of the support of the upper bound of each membership function, they used a genetic algorithm as postprocessing.

Sanz et al. (2014) provide a classifier based on a fuzzy rule-based classification system that tackles the problem of determining the risk of a patient developing CVD within the next 10 years and provide both a diagnosis and an interpretable model explaining the decision. They proposed a method that combines fuzzy rule-based classification systems with interval-valued fuzzy sets. They used a genetic tuning in order to find the best degree for every interval-valued fuzzy set as well as for finding the best values for the parameters.

Although the fuzzy rule-based classification system is a useful tool to deal with the classification problem and it has a good classification rate and a highly interpretable model, it has some shortcoming that influences the classification accuracy. It lacks dealing with incomplete and inconsistent information. Moreover, the inflexibility of the linguistic variable concept in case of dealing with complex system imposes hard restrictions on the fuzzy rule structure (Sanz, Fernández, Bustince, & Herrera, 2010; Sanz et al., 2014).

Basha, Abdalla, and Hassanien (2016) proposed a neutrosophic rule-based classification system (NRCS) which is a generalization of the fuzzy rule-based classification system, based on neutrosophic logic (NL). NL proposed by Smarandache as a generalization of fuzzy, interval-valued fuzzy, intuitionistic fuzzy, and interval-valued intuitionistic fuzzy logic (Wang, Smarandache, Sunderraman, & Zhang, 2005). In NL, each proposition is estimated to have a percentage of truth, a percentage of indeterminacy, and a percentage of falsity (Ansari, Biswas, & Aggarwal, 2013; Robinson, 2003). In contrast to fuzzy logic, NL can handle incomplete information as well as inconsistent information (Smarandache, 2003; Wang et al., 2005). The main three stages of NRCS are neutrosophication, inference engine, and deneutrosophication as shown in Fig. 1.

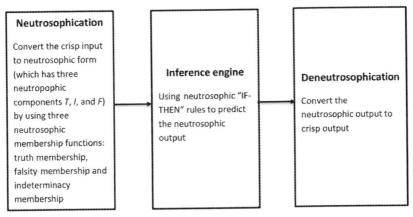

FIG. 1  The main stages of neutrosophic rule-based classification system.

122 Big data in psychiatry and neurology

In this chapter, NRCS is presented as well as some of its medical applications as a prediction system for toxicity effects assessment of biotransformed hepatic drugs, a prediction model for diabetics, and a predictive model for seminal quality.

The remaining chapter is organized as follows: Section 2 presents the theoretical background and steps of the proposed model. Three medical applications solved using the proposed NRCS model are presented in Section 3. Finally, conclusions are presented in Section 4.

## 2 Theoretical background

### 2.1 Neutrosophic logic and neutrosophic set

Neutrosophy was introduced by Smarandache in 1995, and it deals with the origin, nature, and scope of neutralities, as well as their interactions with different mental visions (Wang et al., 2005). This theory considers three concepts: (1) the idea $\langle A \rangle$, (2) its opposite $\langle Anti–A \rangle$, and (3) a spectrum of "neutralities" $\langle Neut–A \rangle$. The $\langle Neut–A \rangle$ and $\langle Anti–A \rangle$ together are referred as $\langle Non–A \rangle$ (Wang et al., 2005). The calculations of $\langle Anti–A \rangle$ and $\langle Non–A \rangle$ are used to neutralize and balance the idea $\langle A \rangle$ (Wang et al., 2005).

The fuzzy set (FS) theory was introduced by A.L. Zadeh in 1965 to handle vague, and fuzzy information and it is represented by a membership degree say $\mu_A(x)$ for each element $x$ in a set $A$, where $\mu_A(x) \in [0, 1]$ (Zadeh, 1996). FS was generalized by many theories such as interval-valued fuzzy sets (Turksen, 1986), intuitionistic fuzzy sets (Atanassov, 1989), and interval-valued intuitionistic fuzzy set (Atanassov, 1989). Each of these theories can handle only one aspect of imprecision. However, the FS theory cannot deal with incomplete and inconsistent information. For this reason, the neutrosophic set was constructed to handle incomplete information as well as inconsistent information. Instead, the neutrosophic set is a huge formal structure which generalizes the concept of all sets such as, the classic set, fuzzy set, interval-valued fuzzy set, intuitionistic fuzzy set, and interval-valued intuitionistic fuzzy set (Arora, Biswas, & Pandy, 2011).

### 2.1.1 Neutrosophic set

Fundamental concepts of neutrosophic set was introduced by Smarandache (2003) and Alblowi, Salama, and Eisa (2013). They provide a natural foundation for treating mathematically the neutrosophic phenomena for building new branches of neutrosophic mathematics.

Mathematically, an element $x(T, I, F)$ belongs to the set in the following way: it is $t$ true in the set, $i$ indeterminate in the set, and $f$ false, where $t$, $i$, and $f$ are real numbers taken from the sets $T$, $I$, and $F$, respectively, with no restriction on $T, I, F$, nor on their sum $n = t + i + f$ (Smarandache, 2003).

Let $X$ is a space of points (objects) with a generic element in $X$ denoted by $x$. A neutrosophic set $A$ in $X$ is characterized by a truth-membership function $(T_A)$, an indeterminacy-membership function $(I_A)$, and a falsity-membership function $(F_A)$. $T_A(x)$, $I_A(x)$, and $F_A(x)$ are real standard or nonstandard subsets of $]^-0, 1^+[$. That is, $T_A: X \rightarrow ]^-0, 1^+[$, $I_A: X \rightarrow ]^-0, 1^+[$, and $F_A: X \rightarrow ]^-0, 1^+[$. There is no restriction on the sum of $T_A(x)$, $I_A(x)$, and $F_A(x)$, so, $^-0 \leq supT_A(x) + supI_A(x) + supF_A(x) \leq 3^+$.

There are many ways to construct neutrosophic set operators (complement, union, intersection, containment, difference, etc.) according to the problem at hand (Arora et al., 2011; Smarandache, 2003; Wang et al., 2005).

### 2.1.2 Neutrosophic logic

NL was developed to represent mathematical models that contains uncertainty, vagueness, ambiguity, imprecision, incompleteness, inconsistency, redundancy, and contradiction (Hassanien, Basha, & Abdalla, 2018; Smarandache, 2003). NL is a logic in which each proposition is estimated to have a percentage of truth in a subset $T$, a percentage of indeterminacy in a subset $I$, and a percentage of falsity in a subset $F$, where $T, I, F$ are standard or nonstandard real subsets of $]^-0, 1^+[$ where $]^-0, 1^+[$ is a nonstandard unit interval (Ansari et al., 2013; Robinson, 2003). $T$, $I$, and $F$ are called neutrosophic components, and these components represent the truth, indeterminacy, and falsehood values, respectively, referring to neutrosophy, neutrosophic logic, neutrosophic set, neutrosophic probability, and neutrosophic statistics (Ashbacher, 2002). In real-world applications, it is easier to use standard real interval $[0, 1]$ for $T$, $I$, and $F$ instead of the nonstandard unit interval $]^-0, 1^+[$ (Ansari et al., 2013; Basha, Sahlol, El Baz, & Hassanien, 2017).

The sets $T$, $I$, and $F$ are not necessarily intervals but may be any real subunitary subsets: discrete or continuous; single element, finite, or (countable or uncountable) infinite; union or intersection of various subsets (Basha et al., 2016; Smarandache, 2003). Statically, $T, I$, and $F$ are subsets, but dynamically the components $T$, $I$, and $F$ are set-valued vector functions/operators depending on many parameters, such as time, space, etc. (Smarandache, 2003).

## 2.2 Neutrosophic rule-based classification system

The NRCS is a generalization of fuzzy rule-based classification system (FRCS) which is a rule-based expert system. NRCS uses NL for generalizing the fuzzy rule-based classification system. The antecedents and consequents of the "IF-THEN" rules in the NRCS are NL statements, instead of fuzzy logic. The NRCS has three stages:
1. **Neutrosophication**: Construction of the neutrosophic KB by converting crisp inputs using the neutrosophic three membership functions: truth membership, falsity membership, and indeterminacy membership.

2. **Inference engine**: The KB and neutrosophic "IF-THEN" rules are applied to get a neutrosophic output.
3. **Deneutrosophication**: Converts the neutrosophic output of the previous step back to a crisp value using three functions analogous to the ones used by the neutrosophication.

The used KB stores the available knowledge in the form of neutrosophic "IF-THEN" rules, and then the KB captures the neutrosophic rule semantics using neutrosophic sets. Fig. 2 shows the NRCS consisting of four phases: information extraction phase, neutrosophication phase, rules generation phase, and the classification phase. More details about each phase in the following sections.

### 2.2.1 Information extraction phase

In this phase, important information is extracted to apply NRCS by reading data files and extracting (1) the number of attributes, (2) minimum and maximum value of each attribute, (3) number of classes and their names, and (4) class labels or decisions.

### 2.2.2 Neutrosophication phase

In this phase, three membership functions, namely, truth membership, falsity membership, and indeterminacy membership are defined. These membership functions will be extracted from fuzzy-trapezoidal membership function. To represent each value for each feature in the neutrosophic form, three

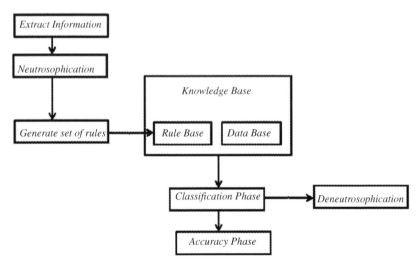

**FIG. 2** Block diagram of the proposed NRCS model.

neutrosophic components $\langle T, I, F \rangle$ are required. To get these three components, we applied three membership functions on each value in each attribute in the dataset.

### 2.2.3 Rules generation phase

The goal of this phase is to generate rules which will be used in the next phase (classification phase). Assume the data is denoted by $X = \{x_1, x_2, ..., x_n\}$, where $x_i$ is the $i$th sample and $n$ is the total number of samples. Each sample has one class label which is denoted by $c_i \in \{1, 2, ..., C\}$, where $C$ is the total number of classes. First, the dataset is divided into training data ($X_{training}$) which has class labels and testing data ($X_{testing}$) without class labels. In this phase, exact neutrosophic rules are generated from the training and testing data. In NRCS, each attribute in each neutrosophic rule has three components which describe the degree of truth, degree of indeterminacy, and degree of falsity.

### 2.2.4 Classification phase

In this phase, the testing rule matrix without class labels is constructed. For each testing rule ($x_t \in X_{testing}$), the intersection percentage is calculated between that testing rule and all training rules ($X_{training}$) (see Fig. 2), and these percentages are denoted by $P = \{p_1, p_2, ..., p_q\}$, where $q$ is the number of rules in the training set and $p_i$ is the matching percentage between $x_t$ and the training rule $x_i$. The class label of the training rule which has the maximum intersected percentage is assigned to the testing rule. If there is no intersection between training rules and the current testing rule with at least 50% ($p_i < 0.5, \forall i = 1, ..., q$), the class label is determined from the exact rules set. After that, this testing rule is added to the training rules instead of testing rules ($X_{training} = X_{training} \cup x_t$).

Finally, the testing matrix which has predicted class labels is compared with the exact matrix which has actual class labels. To evaluate our model, the confusion matrix is computed (see Fig. 3). From the confusion matrix, different measures can be calculated such as true positive (TP), true negative (TN), false positive (FP), and false negative (FN).

|  | | True class | |
|---|---|---|---|
|  | | Positive (P) | Negative (N) |
| Predicted class | True (T) | True positive (TP) | False positive (FP) |
|  | False (F) | False negative (FN) | True negative (TN) |
|  | | P=TP+FN | N=FP+TN |

**FIG. 3** An illustrative example of the 2 × 2 confusion matrix.

126  Big data in psychiatry and neurology

## 3  NRCS medical applications

### 3.1  Neutrosophic rule-based prediction system for toxicity effects assessment of biotransformed hepatic drugs

Estimating the toxicity effects is an essential step in the development process of any drug. The currently used methods for the prediction of the toxicity risks of drugs are expensive and demand high computational processes. The large-scale evaluation of drug toxicity rises the demand to implement alternative more intelligent techniques. Here, we use a relatively big dataset consisting of 553 drugs, which are biotransformed in liver. There are 31 chemical descriptors for each drug and four toxic effects; mutagenic, tumorigenic, irritant, and reproductive effect. Here, an NRCS is used in classifying unknown drugs into toxic or nontoxic.

### 3.1.1  Dataset description

In this study, the dataset is extracted from the drug bank database. Out of 6712 drugs, 1448 of these drugs are FDA-approved small molecule drugs, 131 are FDA-approved biotech (peptide/protein) drugs, 85 are nutraceuticals, and the remaining 5080 are experimental drugs. Here, 553 drugs biotransformed in the liver are used (Sander, Freyss, von Korff, & Rufener, 2015).

Table 1 illustrates the four toxic effects, the number of positive, and negative samples. Each drug is represented by 31 features, calculated using Data Warrior package (Sander et al., 2015). As shown in Table 1, the data is not balanced. Moreover, the imbalance ratio "the ratio of the number of samples of the majority class to the number of samples of the minority class" of these four classes is different. The imbalance ratio of the mutagenic,

**TABLE 1** The number of positive samples and negative samples and the imbalance ratio for each toxic effect in our dataset.

| Toxic effect | Samples in positive class | Samples in negative class | Imbalance ratio |
|---|---|---|---|
| Mutagenic effect | 90 = 16.28% | 463 = 83.73% | 5.14 |
| Tumorigenic effect | 90 = 16.28% | 463 = 83.73% | 5.14 |
| Reproductive effect | 187 = 33.82% | 366 = 66.18% | 1.96 |
| Irritant effect | 67 = 12.16% | 486 = 87.88% | 7.25 |

tumorigenic, and irritant effects has relatively high values compared with the reproductive. In these experiments, the positive class is the minor one. This may negatively affect the sensitivity measure of the NRCS model. The reproductive effect has the highest risk effect (33.82%), the effects of the mutagenic and tumorigenic risks are equal (16.28%), and the irritant effect has least effect (12.16%). It is worth mentioning here that this dataset has a lot of imprecision, incomplete, vagueness, and inconsistency.

### 3.1.2 Toxicity classification using the NRCS

The NRCS is used for predicting the four toxicity effects of the biotransformed drugs in the previous dataset. The model did not use any feature selection, nor it did any preprocessing of the data. The results of the model were compared with other conventional classifiers such as neural network (NN) (Yamany et al., 2015), $k$-nearest neighbors ($k$-NN) (Tharwat, Ghanem, & Hassanien, 2013), and linear discriminant analysis (LDA) (Tharwat, 2016). Common performance measures are used for this comparison (Table 2). We highlight here the important observations about the result for each toxicity effect.

The results of this experiment in terms of accuracy, precision, sensitivity, specificity, F1 score, and GM are summarized in Table 2.

From Table 2, the following remarks can be drawn.

- *Mutagenic effect*: Although the NRCS algorithm obtained the worst accuracy among all classifiers, it has achieved the best results for all other measures. For the sensitivity and specificity measures the NRCS outperformed the other learning algorithms with a big difference.
- *Tumorigenic effect*: While the NRCS achieved the best results in two measures, the LDA achieved the best score in the other four measures. Yet, the NRCS model was the second best among these four measures.
- *Irritant effect*: The NRCS algorithm achieved the best results in almost all measures and LDA obtained competitive results.
- *Reproductive effect*: The NRCS algorithm obtained the best results in all measures except the specificity. The LDA achieved the best specificity results.

The results showed that the NRCS model obtained high sensitivity for all toxic effects. The imbalance in the data discussed earlier reflects the low values of the sensitivity measure compared to the specificity results. In the reproductive effect, the sensitivity results were much higher than the other toxicity effects due to its minimum imbalance ratio. Overall, the NRCS algorithm outperforms the other classifiers in most of the cases. This is due to the indeterminacy term in the NL which is best treating the neutral and nonsignificant features. We conclude that the model can be used in predicting the drug toxicity in early stages of drug development.

**TABLE 2** A comparison between the results (accuracy, precision, sensitivity, specificity, F1 score, and GM) of the proposed model using NRCS and results obtained from NN, LDA, and *k*-NN classifiers.

| Metrics | Mutagenic effect | | | | Tumorigenic effect | | | | Irritant | | | | Reproductive effect | | | |
|---|---|---|---|---|---|---|---|---|---|---|---|---|---|---|---|---|
| | *NN* | *LDA* | k-NN | *NRCS* | *NN* | *LDA* | k-NN | *NRCS* | *NN* | *LDA* | k-NN | *NRCS* | *NN* | *LDA* | k-NN | *NRCS* |
| Accuracy | 83.8 | 84.2 | 85.5 | 83.4 | 78.7 | 86.6 | 83.0 | 83.5 | 85.5 | 81.9 | 90.9 | 86.6 | 63.5 | 72.9 | 70.7 | 73.3 |
| Precision | 33 | 44 | 17 | 45.7 | 29 | 50 | 33 | 45.2 | 14 | 39 | 27 | 40.9 | 43 | 54 | 53 | 65.3 |
| Sensitivity | 20 | 33 | 20 | 36.3 | 20 | 38 | 18 | 32.5 | 17 | 15 | 15 | 27.7 | 32 | 52 | 49 | 65.2 |
| Specificity | 84.3 | 89.3 | 86.6 | 92 | 81.9 | 90.7 | 84.7 | 92.7 | 89.3 | 84.9 | 93.6 | 94.9 | 70.3 | 80.3 | 78.1 | 77.5 |
| F1 score | 37.7 | 37.7 | 35.7 | 40.5 | 14.9 | 41.2 | 11.6 | 37.8 | 9.5 | 21.7 | 19.3 | 31.6 | 36.7 | 53 | 50.9 | 65.2 |
| GM | 13 | 54.3 | 13.2 | 57.8 | 28.6 | 58.7 | 24.3 | 54.9 | 25.0 | 35.7 | 37.5 | 49.4 | 47.4 | 64.6 | 61.9 | 71.08 |

## 3.2 A predictive model for diabetics using NRCS

Diabetes is considered one of the most frequent diseases worldwide, which targets the elderly people. It is caused by pancreas dysfunction in producing enough insulin to process the blood sugar, resulting in abnormal levels of glucose in the blood causing diabetes of types 1 and 2. Diabetic people face risks of developing secondary health problems with eyes, heart diseases, and nerve system. This makes diabetes a chronic and one of the deadliest diseases all over the world. In spite of being incurable, diabetes can be controlled by changing lifestyle and medication. Therefore, discovering prediabetes and treating diabetes in early stages prevent complications.

The causes of diabetes are still unknown, yet it is believed that family history or and lifestyles play a major role in having diabetes. The ordinary medical approaches of regular visits to diagnostic centers for checking fasting glucose, potassium, sodium, urea, albumin, creatinine, etc. are exhausting and expensive. This rises the need to predict the diabetes using computational models. Here, a classification of cases to be potentially diabetic or not using NRCS is proposed.

### 3.2.1 PIMA dataset description

Although diabetes is a common disease, very few datasets have been collected from patients and used in its prediction. The PIMA dataset is one of the rare diabetes datasets available publically on the UCI machine learning repository website. It has 8 features and 768 of sampling instances (500 nondiabetic and 268 diabetic instances). There are two output classes to indicate if the person is diabetic or not (Table 3).

We have chosen PIMA in our study, because it is a common benchmark in comparing the classifiers' performance. Fig. 4 shows the two interleaved classes 0 and 1. They are overlapped and their centers are very close, which makes it difficult to identify them separately.

### 3.2.2 Diabetic predictive using the NRCS

The NRCS classification system is applied to the PIMA dataset for predicting the diabetes. A comparison with its fuzzy rule-based classification counterpart system is done to show its superiority. The common performance measures in

**TABLE 3** Details of PIMA dataset.

| Dataset name | Number of sampling | Number of features | Number of classes |
|---|---|---|---|
| PIMA | 768 | 8 | 2 |

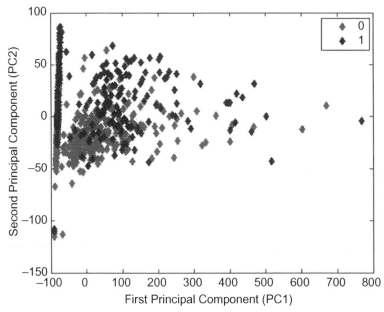

FIG. 4 PIMA dataset.

classification such as precision, sensitivity, and specificity are used. Figs. 5–7 show that the NRCS system was more accurate in classification. Moreover, the indeterminacy term in NL reduces the complexity and computations as it removes any overlaps between sets and classes.

### 3.3 A predictive model for seminal quality using NRCS

Lately, men fertility rates recorded noticeable decrease. Researchers indicated that many factors in current living conditions affect semen quality resulting in men infertility. Hence, the search for a good predictor for men fertility is potential and urgent research problem. This section introduces a predictive model of sperm quality according to the individual's habits and the surrounding environmental factors.

The NRCS is used in classifying an unknown seminal quality to either normal or abnormal. Nine input parameters are used in covering person's lifestyle. The NRCS model has been compared with three well-known classifiers such as multilayer perceptron (MLP), decision trees (DT), and support vector machines (SVM). Performance measures of accuracy, sensitivity, specificity, and GM are used for this comparison.

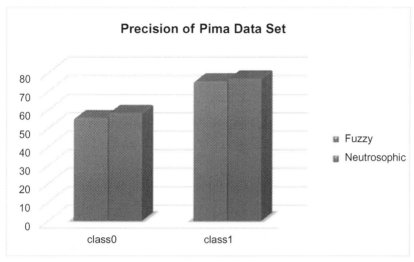

FIG. 5 Precision of PIMA dataset.

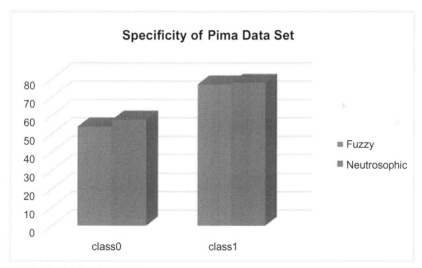

FIG. 6 Sensitivity of PIMA dataset.

### 3.3.1 The fertility dataset description

The fertility dataset, available at UCI website machine learning repository is used. In this data, 100 semen samples of individuals were collected. Each sample is characterized by nine features covering environmental, health, and lifestyle factor; season number, personage, childhood diseases, accidents or

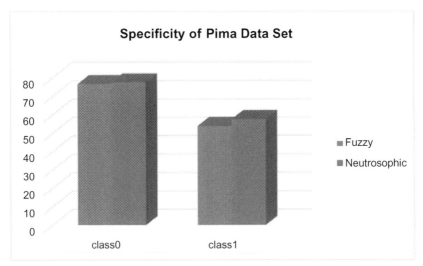

FIG. 7 Specificity of PIMA dataset.

seriously previous life event, surgery, past fever rates, the consumed rate of alcohol, sitting hours, and smoking. Out of the 100 samples, 88 were normal cases and only 12 were abnormal cases; that is, this data is imbalanced favoring the normal cases. Fig. 8 shows this high imbalance ratio.

### 3.3.2 Seminal quality classification using the NRCS model

The goal of this experiment is to use the NRCS for predicting the seminal quality. In this experiment, the results of the NRCS model was compared with conventional classifiers such as MLP, DT, and SVM. Table 4 shows the results of these comparisons according to the performance measures. It is clear that the NRCS has reached the best accuracy with remarkable differences. This is due to the use of the indeterminacy term in the NRCS algorithm.

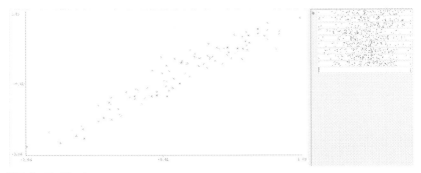

FIG. 8 Fertility dataset.

# Neutrosophic rule-based classification system Chapter | 6 133

**TABLE 4** A comparison between NRCS, MLP, DT, and SVM in terms of accuracy, sensitivity, specificity, and GM.

| Metrics | MLP | SVM | DT | NRCS |
|---|---|---|---|---|
| Accuracy | 69 | 69 | 67 | 95.6 |
| Specificity | 72.5 | 73.9 | 71.7 | 97.6 |
| Sensitivity | 25 | 12.5 | 12.5 | 66.7 |
| GM | 42.6 | 30.4 | 29.9 | 80.7 |

## 4 Conclusions and future work

This chapter presents an application of the NRCS in the medical field. The NRCS generalizes the fuzzy logic-based classification system using NL instead of fuzzy logic. The proposed NRCS gives better in determining the classes and more robust than its fuzzy counterpart. Also, the NRCS reduces the complexity and the computational of the classifier. It has been shown that it can be used in different vital medical applications. One of the major drawbacks of the NRCS is the lack of the ability to learn. It rather requires the KB to be driven from expert knowledge. The key point is to use an evolutionary learning process to automate the NRCS.

## References

Alblowi, S. A., Salama, A. A., & Eisa, M. (2013). New concepts of neutrosophic sets. *International Journal of Mathematics and Computer Applications Research*, *3*(4), 95–102.

Alcalá, R., Casillas, J., Cordón, O., Herrera, F., & Zwir, S. J. I. (1999). Techniques for learning and tuning fuzzy rule-based systems for linguistic modeling and their application. In C. Leondes (Ed.), *Knowledge* (pp. 889–941). Academic Press (Ed.).

Ansari, A. Q., Biswas, R., & Aggarwal, S. (2013). Neutrosophic classifier: An extension of fuzzy classifer. *Applied Soft Computing*, *13*(1), 563–573.

Arora, M., Biswas, R., & Pandy, U. S. (2011). Neutrosophic relational database decomposition. *International Journal of Advanced Computer Science and Applications*, *2*(8), 121–125.

Ashbacher, C. (2002). *Introduction to neutrosophic logic*. Infinite Study.

Atanassov, K. T. (1989). More on intuitionistic fuzzy sets. *Fuzzy Sets and Systems*, *33*(1), 37–45.

Basha, S. H., Abdalla, A. S., & Hassanien, A. E. (2016). NRCS: Neutrosophic rule-based classification system. In *Proceedings of SAI intelligent systems conference* (pp. 627–639).

Basha, S. H., Sahlol, A. T., El Baz, S. M., & Hassanien, A. E. (2017). Neutrosophic rule-based prediction system for assessment of pollution on Benthic Foraminifera in Burullus Lagoon in Egypt. In *2017 12th international conference on computer engineering and systems (ICCES)* (pp. 663–668).

Bhatt, M. R., & Buch, S. (2016). Application of rule based and expert systems in various manufacturing processes—A review. In S. Satapathy, & S. Das (Eds.), *Vol. 50. Proceedings*

## 134 Big data in psychiatry and neurology

*of first international conference on information and communication technology for intelligent systems: Volume 1. Smart Innovation* (pp. 459–465). Cham: Springer (Eds.).

Hassanien, A. E., Basha, S. H., & Abdalla, A. S. (2018). Generalization of fuzzy C-means based on neutrosophic logic. *Studies in Informatics and Control, 27*(1), 43–54.

Ishibuchi, H., Nakashima, T., & Nii, M. (2005). *Classification and modeling with linguistic information granules.* Berlin, Heidelberg: Springer-Verlag.

Jackson, P. (1998). *Introduction to expert systems.* Addison Wesley.

Jadhav, J. S., Nalawade, K. M., & Bapat, M. M. (2013). Rule based expert system application for crime against women law in Indian judicial system. *IOSR Journal of Humanities and Social Science, 16*(6), 19–22.

Liu, H., Gegov, A., & Cocea, M. (2016). Rule-based systems: A granular computing perspective. *Granular Computing, 1*, 259–274.

Liu, H., Gegov, A., & Stahl, F. (2014). Categorization and construction of rule based systems. In V. Mladenov, C. Jayne, & L. Iliadis (Eds.), *Vol. 459. Engineering applications of neural networks. EANN 2014. Communications in computer and information science* (pp. 183–194). Cham: Springer (Eds.).

Magdalena, L. (2015). Fuzzy rule-based systems. In J. Kacprzyk, & W. Pedrycz (Eds.), *Springer handbook of computational intelligence* (pp. 203–218). Berlin, Heidelberg: Springer Handbooks (Eds.).

Mohammadpour, R., Abedi, S., Bagheri, S., & Ghaemian, A. (2015). Fuzzy rule-based classification system for assessing coronary artery disease. *Computational and Mathematical Methods in Medicine, 2015*, 8 pp.

Nagori, V., & Trivedi, B. (2014). Types of expert system: Comparative study. *Asian Journal of Computer and Information Systems, 2*(2). Retrieved from: https://www.ajouronline.com/index.php/AJCIS/article/view/948.

Ravi, C., & Khare, N. (2016). Review of Fuzzy rule based classification systems. *Computational and Mathematical Methods in Medicine, 9*(8), 1299–1302.

Robinson, A. (2003). Non-standard analysis. In *Mathematical logic in the 20th century* (pp. 385–393). World Scientific.

Sander, T., Freyss, J., von Korff, M., & Rufener, C. (2015). DataWarrior: An open-source program for chemistry aware data visualization and analysis. *Journal of Chemical Information and Modeling, 55*(2), 460–473.

Sanz, J. A., Fernández, A., Bustince, H., & Herrera, F. (2010). Improving the performance of Fuzzy rule-based classification systems with interval-valued fuzzy sets and genetic amplitude tuning. *Information Sciences, 180*(19), 3674–3685.

Sanz, J. A., Galar, M., Jurio, A., Brugos, A., Pagola, M., & Bustince, H. (2014). Medical diagnosis of cardiovascular diseases using an interval-valued fuzzy rule-based classification system. *Applied Soft Computing, 20*, 103–111.

Sanz, J. A., Pagola, M., Bustince, H., Brugos, A., Fernández, A., & Herrera, F. (2011). A case study on medical diagnosis of cardiovascular diseases using a genetic Algorithm for tuning Fuzzy rule-based classification systems with interval-valued Fuzzy sets. In *2011 IEEE symposium on advances in type-2 Fuzzy logic systems (T2FUZZ), April* (pp. 9–15). https://doi.org/10.1109/T2FUZZ.2011.5949553.

Sivanandam, S. N., Sumathi, S., & Deepa, S. N. (2007). *Introduction to Fuzzy Logic using MATLAB.* Berlin, Heidelberg: Springer-Verlag.

Smarandache, F. (2003). *A unifying field in logics: Neutrosophic logic. Neutrosophy, neutrosophic set, neutrosophic probability: Neutrosophic logic: Neutrosophy, neutrosophic set, neutrosophic probability.* Infinite Study.

Tharwat, A. (2016). Linear vs. quadratic discriminant analysis classifier: A tutorial. *International Journal of Applied Pattern Recognition, 3*(2), 145–180.

Tharwat, A., Ghanem, A. M., & Hassanien, A. E. (2013). Three different classifiers for facial age estimation based on k-nearest neighbor. In *Proceedings of the 9th international computer engineering conference (ICENCO)* (pp. 55–60).

Turksen, I. B. (1986). Interval valued fuzzy sets based on normal forms. *Fuzzy Sets and Systems, 20*(2), 191–210.

Wang, H., Smarandache, F., Sunderraman, R., & Zhang, Y.-Q. (2005). *Vol. 5. Interval neutrosophic sets and logic: Theory and applications in computing: Theory and applications in computing*. Infinite Study.

Yamany, W., Tharwat, A., Hassanin, M. F., Gaber, T., Hassanien, A. E., & Kim, T.-H. (2015). A new multi-layer perceptrons trainer based on ant lion optimization algorithm. In *Fourth international conference on information science and industrial applications (ISI)* (pp. 40–45).

Zadeh, L. A. (1996). Fuzzy sets. In *Fuzzy sets, Fuzzy logic, and Fuzzy systems: Selected papers by Lotfi A. Zadeh* (pp. 394–432). World Scientific.

# Chapter 7

# From complex to neural networks

Nicola Amoroso[a,b] and Loredana Bellantuono[c]

[a]*Dipartimento di Farmacia—Scienze del Farmaco, Università di Bari, Bari, Italy,* [b]*Istituto Nazionale di Fisica Nucleare, Sezione di Bari, Bari, Italy,* [c]*Dipartimento Interateneo di Fisica, Università di Bari, Bari, Italy*

## 1 Big data and MRI analyses

Big data analytics is nowadays an invaluable tool to manage and extract knowledge from exponentially growing volumes of data, especially for quantitative neuroscience. By definition, for Big Data we mean high volume, velocity, and variety of information (Katal, Wazid, & Goudar, 2013); more recently, a fourth "v" has gained popularity: veracity (Rubin & Lukoianova, 2013). These attributes are far from providing an objective definition of what should/could considered "big," nevertheless it is possibly the best definition we have to explain what it is meant for. First of all, it is pretty much easy to understand why we care about volume data. Do we expect big data to include kilobytes ($10^3$), gigabytes ($10^9$), petabytes ($10^{15}$), or more like yottabytes ($10^{24}$)? When big is big enough? A pictorial representation is presented in Fig. 1.

Generally, we mean big data when a standard laptop or a personal desktop is not able to manage it; even few kilobytes can be big when the computational infrastructure we use is sufficiently old. The second "v" concerns the data usage. In several applications not only the possibility to manage high volumes of data matters, but it also becomes of paramount importance the possibility to process it fast, or even extremely fast, if you think about real-time or quasi-real-time applications. Would you be comfortable in waiting 60 s to watch your preferred YouTube video? How many seconds would it be reasonable to wait to listen to your favorite song on Spotify? Would you be happy to wait a few minutes every time you want to post a photo on Facebook or just letting your friends know you are happy about your soccer team scoring a goal? These are just few examples of everyday life applications dealing with big data and the necessity to rapidly respond to user solicitations. For science, there is not usually the same struggle for velocity, as scientific analyses can be generally

Big Data in Psychiatry and Neurology. https://doi.org/10.1016/B978-0-12-822884-5.00011-8
Copyright © 2021 Elsevier Inc. All rights reserved.

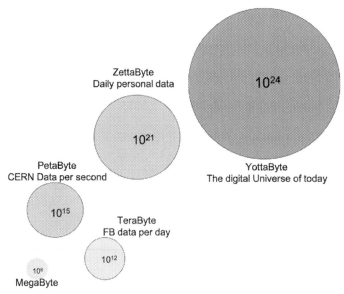

FIG. 1  A pictorial representation of Big Data.

performed in weeks or even months; however, even scientists have to face feasibility and computational burden of their analyses. Variety is the more technical attribute; in that it emphasizes the need for specific techniques to manage big data. When we think about big data, we conceive heterogeneous data including text, images, and videos, in brief we figure out data which cannot be simply stored and represented as a plain text table. This need requires the development and the design to federate and integrate information from different sources. Last but not least, there is a growing interest about data veracity because the increment of available data highlights the need to check their quality.

For what concerns neuroscience applications, neuroscience is exploiting huge worldwide neuroimaging initiatives for data collection and sharing. Just to consider a few cases, it is worth mentioning ABIDE (Di Martino et al., 2014), ADNI (Mueller et al., 2005), Brainmap (Laird, Lancaster, & Fox, 2005), ENIGMA (Schmaal et al., 2016), FBIRN (Potkin & Ford, 2009), Human Connectome (Van Essen, David, et al., 2012), OASIS (Marcus et al., 2007), PPMI (Marek et al., 2011). The idea behind these initiatives is that collecting large databases and sharing them across the scientific community is the most efficient way toward the design and development of novel, robust, and accurate algorithms to exploit their informative content, thus paving the way for advances in diagnosis, treatments, or even basic fundamental understanding of pathological pathways.

However, there is no consensus about the strategy to experimentally map the human brain and several studies in recent years have faced the challenge to design dedicated brain imaging models. This spread has also been eased by computational advances: modern supercomputing technologies and distributed computational infrastructures offer the possibility to afford the burden of more and more sophisticated models and analyses; the use of cloud computing solutions, especially in terms of Software as a Service (SaaS), provides the final user with the possibility to just investigating the data without specific technological background and without worries about installation and coding issues. Neuroscientists can focus on their data, learning, and understanding, without struggling with technicalities. Besides, the computational power offered by modern technologies outperforms the possibilities we just had few years ago; training deep neural networks with hundreds of neurons and layers is now an affordable issue even for standard resources.

The increasing possibility to access novel data sources and the enhancement provided by this data offers huge possibilities to investigate novel algorithms, to deepen our understanding of several mechanisms, both physiological and pathological ones, but these opportunities come at a cost. The primary issue is probably data heterogeneity, which requires dedicated strategies for standardization and harmonization, a problem widely studied recently (Karayumak et al., 2019; Mirzaalian et al., 2016, 2018; Yamashita et al., 2019). Different techniques are used to standardize data, making possible a comparison among them; for scientist working with MRI data the challenge is twofold, on one hand it is of fundamental importance the normalization of gray-level intensities. These inhomogeneities derive from different aspects, one of the best known is that they originated from the nonuniformity of magnetic fields, which is also called bias field. Several solutions to face this problem have been proposed in recent years, through the development of algorithms exploiting different strategies to standardize the gray-level distribution of MRI scans (Li, Gore, & Davatzikos, 2014; Liao, Lin, & Li, 2008; Tustison & Gee, 2009). Another important issue concerns spatial normalization. In this case, the intersubject differences arising from different positioning of the patient in the MRI scan or the differences in terms of image resolution are tackled in order to standardize the image representation in a common reference template. This step is fundamental as, on one hand, it ensures the spatial anatomical correspondence between different scans, so that the same anatomical district is expected to have the same spatial coordinates in scans of different patients; on the other hand, this operation is necessary for statistical considerations, as it allows the comparison of scans having initially different dimensions, which become perfectly comparable in size, once projected in the common reference space. Spatial standardization is usually referred as registration. Even in this case, there are several different algorithms and packages offering a solution (Cao et al., 2020; Dadar et al., 2018; Klein et al., 2009).

Although data standardization is probably the most important preprocessing technique for MRI analysis, other options are frequently considered: brain extraction, tissue classification, motion correction, image enhancement, artifact removal. In fact, the choice of a suitable preprocessing pipeline is demanded to the specific analysis and the desired effects under investigation; however, it should be considered that each particular step has its own specific assumptions which are not easily met by all subjects within a cohort, therefore these actions often yield bias and confounding factors. Accordingly, the development of strategies requiring minimal preprocessing or eventually using raw data has gained more attention, especially in applications exploiting deep learning approaches (Akkus et al., 2017; Cole et al., 2017; Işın, Direkoğlu, & Şah, 2016). Among the most promising strategies, complex networks deserve a particular mention. The reasons are manifold: first of all, brain connectivity is accounted to play a pivotal role in pathogenic mechanisms of several diseases, thus investigating connectivity properties paves the way for the design of novel biomarkers and deepening our comprehension; another important consideration holds specifically for big data analyses: a consolidated neuroimaging strategy consists in examining MRI voxel by voxel, namely Voxel-Based Morphometry (VBM), looking for morphological differences eventually explaining groupwise differences as, for example, between healthy controls and patients (Ashburner & Friston, 2000), see an example in Fig. 2.

VBM is computationally intensive, considering that a single MRI brain scan accounts for millions of voxels; besides, VBM suffers from a lack of statistical power which impairs the robustness of conclusions. In order to highlight a significant difference within a cohort, the voxel-wise distributions of controls and patients are compared, of course this yields millions of comparisons and statistical tests, therefore retrieving a statistical significance becomes particularly challenging. In many cases VBM remains the best and only option to consider; nevertheless, other approaches have been investigated. A valid alternative to VBM is provided by Region-of-Interest (ROI) approaches. In ROI studies brain anatomical districts are segmented so that measuring their volumes or other features characterizing their shapes allows to investigate pathological or physiological changes (Douaud et al., 2006; Etzel, Gazzola, & Keysers, 2009; Lopez-Garcia et al., 2006). More recently, between these two approaches, a third option has gained some popularity. It is an intermediate strategy where brain is not segmented in regions of interest nor its voxels are considered singularly. In particular, this approach has exploited complex network theory

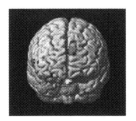 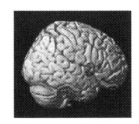 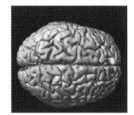

**FIG. 2** A VBM analysis represents the activity of each voxel.

(Amoroso et al., 2018a, 2018b, 2019; Kanavati et al., 2017; Kong et al., 2019). In the next paragraph a brief introduction to complex network theory is provided.

## 2 Modeling purposes: Complex networks

The complex network formalism provides an abstract representation of real-world complex systems, consisting of elementary units among which similarities or interactions exist. The system elements are modeled as nodes (also called vertices), while relations are identified by the presence of edges (or links) connecting pairs of such nodes (Newman, 2018). Typical examples of nodes are users of a social network, train stations, genes that manifest together to provide a certain anatomical feature. With reference to the earlier examples, edges can be represented by interactions between users, train routes connecting stations, and the complex interactions that influence gene expression. A sequence of links indirectly connecting two nodes A and B is called a path between A and B. Relations between nodes of a network can be asymmetric: this occurs, for example, in a public transport network in which a route allows to travel directly from A to B, but not from B to A, or in a family network with links determined by the parenthood relation. In these cases, the links of the complex network are characterized by a given direction, and the corresponding graph is said oriented. In general, the same system can be represented in different ways by the complex network formalism, since the definition of nodes and connections can change according to the specific aspect that one wants to examine. Therefore it is essential to identify the more suitable observation scale of the system, at which the interesting collective behaviors emerge, and the nature of relations that should be represented by network connections.

Fig. 3 shows the graph of a complex network formed by 18 nodes, connected by 23 edges. The importance of a node within the network can be quantified by different centrality measures of different kinds, including degree and betweenness. The degree of a node is defined as the number of links connected to it; in an oriented graph, one can distinguish the in-degree, equal to the number of connected links oriented toward the node, and the out-degree, namely the number of connected links oriented from considered node to other nodes. Complex networks that describe systems with a hierarchical nature are characterized by the presence of hubs, which are nodes characterized by a very large degree compared to the rest. Betweenness of a node is the number of shortest paths, indirectly connecting any pair of other nodes in the network, that pass through the considered node. It is worth stressing that degree and betweenness represent node centrality measures whose information content is absolutely not redundant: the degree evaluates the importance of a node in terms of number of connections, while betweenness can be considered a measure of the strategic role of the node. This difference is clear in the network of Fig. 3: node m is the most important from the point of view of connections, being characterized by degree

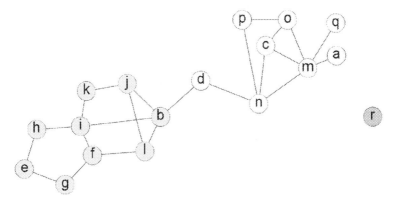

**FIG. 3** Complex network consisting of 18 nodes, connected to each other through 23 links; nodes are colored according to their community membership.

equal to 5, but its betweenness is roughly one half of that of node d, which has instead a lower degree.

A network community can be defined as a set of nodes with a tendency to interact among themselves more than with nodes in the complementary set. From a mathematical point of view, a community is defined by the condition that the number of internal links between its nodes is relevantly larger than the number of links with external elements (Newman, 2006). The network in Fig. 3 is characterized by the presence of three communities, highlighted in different colors; one of them, in particular, is formed by an isolated node, disconnected from the rest of the network. Community detection plays a paramount role in the context of network models for social sciences.

So far, we have described complex networks as objects that model binary relations, represented by edges that can either exist or not between two nodes. This picture is actually oversimplified and does not allow to describe heterogeneity in the intensity of real-world connections. For example, bounds between users in a social network can be more or less strong, can correspond to different frequencies of interaction and contact, and can even be characterized by an adversative nature. To take into account the different possible calibers of a relation, one can associate a weight to each link, namely a continuous numerical value, positive or negative, that enables to compare the intensity of different connections. Strength is the node centrality measure that generalizes the concept of degree to the case of a weighted network, quantifying both the number and the weight of connections to a given node. Considering a node with given degree and strength, very different situations can occur: connections with other nodes can have essentially all the same weight, or, instead, the connection to few nodes can be much stronger than the rest. Participation ratio is a node centrality measure that quantifies the unevenness in the distribution of the weights of its connection. For a node with $k$ connections, participation ratio is equal to

$1/k$ if edges are all characterized by the same weight, and close to 1 if the weight of a single edge is much larger than the other ones (Boccaletti et al., 2006).

Recently, a novel tool for the study of complex systems, represented by multilayer networks, has greatly developed, enabling us to model and analyze complex data with multivariate and multiscale information (Boccaletti et al., 2014; De Domenico et al., 2013; Domenico & Manlio, 2017; Kivelä et al., 2014). A multilayer network is formed by various standard complex networks, referred to as layers, with edges connecting nodes belonging to different layers. Each layer contains information on a specific aspect of the system under investigation. A typical example of multilayer network is provided by public transport, formed by.

- a layer in which nodes represent airports and edges represent flights;
- a layer in which nodes represent train stations and edges represent train routes;

and so on. In this scheme, two layers can be linked by, e.g., establishing a connection between an airport and a train station that are sufficiently close according to a distance criterion. Following the edges of a multilayer network constructed in this way, one obtains all the routes on which a public transport user can travel.

The study of human brain largely benefited from models based on multilayer networks. In this framework, the same node is usually replicated on more layers, in which the edges are determined by different connectivity properties, depending on the specific context. A multiplex is a particular topology of a multilayer network, in which nodes of each layer actually represent the same set of elements in the real system (Newman, 2018). An example of multiplex is represented by a social network, in which nodes coincide with people, and connections can be determined by kinship, friendship, work collaboration, etc. The whole relation network on an individual can be thus represented by inserting interlayer edges that connect all node copies of the same person in each layer. In the following, we will discuss how the complex network and multiplex formalism can be employed to model brain connectivity, both in structural and functional terms, providing an interesting application field of complexity to neurosciences, opened by the seminal paper by Bullmore and Sporns (2009).

Structural connectivity, ranging from neuronal to region connections, has been widely investigated by empirical neuroscientists. In particular, network analysis, derived by a well-known mathematical branch called graph theory, has offered a number of quantitative ways to evaluate and characterize brain and its anatomical patterns. According to graph theory, networks are composed of two distinct elements: nodes or vertices denoting the elements of interests characterizing the system (neurons, brain regions, anatomical districts); edges accounting for physical or mathematical connections between nodes (synapses, axonal projections, white matter fibers). Thanks to complex network it is

**144** Big data in psychiatry and neurology

possible to outline which nodes play a pivotal role within the system, therefore if for example we are comparing a cohort of patients and controls we could in principle unveil which brain regions are mostly affected by the disease. Besides, complex networks provide several measures characterizing connections too; topological distances between elements in brain can often be used to outline physiological or pathological patterns. For these purposes, centrality measures are of fundamental importance. Measuring how "central" a node or an edge is means to evaluate its importance in the network and consequently whether or not it has a strategic role in the brain. More recently, thanks to computational and technological advances, several attempts have been made to map the structural networks of the human brain, the *human connectome*. First attempts exploited information provided by structural MRI data based on patterns from cross correlations in cortical thickness or volume across individual brains. Seminal studies revealed interesting properties such as small-worldness and the existence of communities of brain regions. Then, thanks to diffusion weighted imaging (DWI), a novel approach to structural connectivity was born. This modality allows the reconstruction of white matter fiber bundles connecting different brain regions, making it possible to study structural brain connectivity by exploring physical connections instead of mathematical relationships. The process of mathematical reconstruction of white matter pathways through DWI measures is called tractography. Even in this case some intriguing properties have been outlined. DTI networks exhibit high clustering and short path length. An important drawback of tractography reconstructions is the computational burden, as they can require even 24 h of CPU time. Nevertheless, it is worth mentioning that thanks to GPUs these times can be reduced to less than 1 h. Another important limitation comes from fiber reconstruction accuracy; in this case there is not a unanimous solution and many researchers all over the world are investigating different approaches to improve the accuracy of this process.

The main alternative to structural connectivity is examining the brain from a functional perspective. While structural connectivity explores the architecture of brain physical connections and their mathematical properties, functional connectivity evaluates how different brain districts respond to a physiological stimulus or behave during resting. Functional networks aim at elucidating how brain architecture supports neurophysiological activities. There are several ways to assess brain activity, the most famous and studied approaches are based on functional MRI (fMRI), electroencephalography (EEG), and magnetoencephalography (MEG). The basic idea behind these approaches is that different brain regions which collaborate for a task should show an analogous pattern, be it an oxygenation blood level as for fMRI or an electric activity as for EEG and MEG. The signals measured by functional approaches are usually temporal series and the inherent brain connectivity relies on different strategies: structural equation modeling, dynamic causal modeling, or Granger causality. The aforementioned techniques for mapping the brain functional connectivity within the described approach differ from each other by the nature of the signal they detect and the statistical methodology they use to recognize regions displaying functional similarities. fMRI, MEG, and EEG reveal the temporal coordination of brain activity at different scales and in different states of brain activation or rest, providing complementary benefits and disadvantages (Rossini, 2019). In particular, fMRI monitors local blood flow and metabolism by evaluating temporal correlations between blood-oxygen-level-dependent (BOLD) signal fluctuations in different anatomical regions. This method allows for the achievement of excellent spatial resolutions but suffers from drawbacks related to its indirect measurement nature and provides a limited temporal resolution. On the other hand, EEG and MEG measure the electrical and magnetic signals produced by neural oscillations, respectively; these techniques allow to directly and noninvasively monitor neuronal activity at its natural temporal resolution, namely in the milliseconds range, while providing a relatively low spatial resolution, in the centimeter range.

Again, measuring functional connectivity between different regions can be used to generate a network, which can then be topologically described using graph theory. Considerable heterogeneity in the aforementioned methods leads nonetheless to a promising level of convergence between their outcomes, which appear consistent with a small-world model of brain connectivity (Bullmore & Sporns, 2009).

## 3 Learning from data

Whatever mathematical description you decide to use, the basic idea behind machine learning approaches is that it is possible to describe continuous and discrete patterns according to a number of numerical or categorical descriptors, usually called features. Think about a brain MRI scan, the possibilities are almost infinite. For example, you could use each voxel, specifically its gray-level intensity, as a mathematical feature and the whole brain will be described as a vector with millions of components. Of course, given this basic representation it is also possible to increase the data dimensionality by considering other statistical features related to voxel-wise measurements: let us consider a voxel and its neighbors (which typically are those voxels with a unitary distance from the considered voxel, but depending on the cases you could consider farther voxels), for each voxel it is possible to compute statistical moments such as mean, variance, skewness, and kurtosis. Voxel-wise features are just an example of possible mathematical features which can be associated to a brain description; however, several other features can be taken into account. Recall ROI-based approaches, in that perspective it is possible to consider volumes of anatomical regions or other geometrical features related to their shapes. Besides, it is also possible to fuse voxel-based and ROI-based mathematical features; the result is always the same, each MRI scan and therefore each patient is mathematically represented by a high dimensional vector of features. The goal of machine learning is to determine patterns characterizing different behaviors

**146** Big data in psychiatry and neurology

within the cohort: controls and patients, male and female, children and adults. Another option is to use these patterns to provide an assessment of the cohort, for example the age of subjects, the severity of a disease, the Mini-Mental State Examination (MMSE) for cognitive impairment. In the first case, we face what are called classification problems, in the second case regression problems. From a mathematical point of view, the difference is that classification problems deal with categorical variables, like control and patient, while regression problems aim at predicting continuous outputs, like age. The framework remains unchanged: on one hand a vector of mathematical features describes the different observations of the data sample, on the other hand the output or dependent variable representing a category we want to discriminate or a continuous evaluation we want to perform. Accordingly, the data can be represented as a matrix with $N$ rows and $M + 1$ columns, with $N$ being the number of observation, $M + 1$ the sum of $M$ features, and one column representing the label, categorical or continuous.

Once a tabular representation is obtained, machine learning aims at defining a model describing the data. The model learning consists of three different steps: first we formulate a hypothesis about the data, in general this hypothesis concerns the relationship between features and labels, should it be linear? Polynomial? Should it be even more complicated? In practice, formulating a hypothesis about the data consists in determining a number of parameters according to which our model will be defined. The second step consists in defining a suitable cost function. The role of the cost is evaluating the model performance as a function of the parameter values; in this way we learn whether a parameter configuration is better or worse than another one. Finally, an update function is used to incorporate the information derived from cost evaluation in model learning; searching the feature space for optimal parameter configuration is based on the assumption that according to update function it is possible to iteratively improve the model until the best configuration is achieved. Of course, this is not always true. For example, in some cases the search starts with a random initialization of parameters and this choice could significantly affect the possibility of the model to reach the best configuration. Besides, there are other aspects to consider, how many steps does the algorithm need to reach the best configuration? Is this an affordable computational cost? In any cases the answer is yes, but there is no guarantee it will definitely happen. A final consideration about what is meant for optimal configuration must be given. Consider Fig. 4.

When the model has too few parameters to capture the complexity, by definition the model is said to be biased and underfits the data; on the contrary, if too many parameters are incorporated the model is able to mimic the data adapting to data itself, in this case the model has high variance and it overfits the data. The appropriate model requires finding the optimal trade-off between bias and variance; in fact, a biased model will not be able to account for the data complexity whereas a high-variance model won't be able to reproduce findings on independent test sets. Let us consider now how different algorithms exploit this

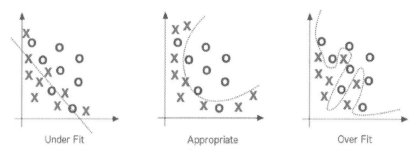

**FIG. 4** From left to right three different possible parameter configuration. From left to right three models (*dotted lines*) of increasing complexity.

learning framework. We will examine three different algorithms, which can be considered the state of the art: Random Forests, Support Vector Machines, and Artificial Neural Networks.

Random Forests (Breiman, 2001) is in most cases the first option you would consider. Easy to tune, as it depends on just two parameters, and accurate in several domains (Gentzkow, Kelly, & Taddy, 2019; Georganos et al., 2019; Steyerberg, 2019; Yang et al., 2019). This algorithm is an optimal choice also because of its robustness, thanks to an internal validation yielded by the so-called out-of-bag examples, random forest predictions usually provide a good approximation of generalization error. Let us examine in further detail the random forest algorithm. As for bagging algorithms (Breiman, 1996), random forests first of all maximally exploit the informative content of available data by bootstrapping all observations and therefore obtaining a number of replicas of the initial data which can be used for learning. This number is the first parameter of the model and corresponds to the number of trees that will grow the forest. The great intuition of Breiman, the inventor of Random Forests, was about the high correlation of bootstrapped trees; in order to avoid this correlation and improve the overall classifier strength he suggested to inject some randomness within this framework: each tree in fact is grown with different features. The basic idea of a classification tree is that, according to available features, it is possible to split the data in classes using each feature; in random forests data is split several times using a random subsample of features, until optimal separation is held. The number of features used at each split is the second and last parameter of Random Forests, it is usually set to one-third of all features $f$ for regression and for classification.

Random Forests are usually adopted when a large number of features is available, in order to exploit its robustness and strength. It is an extremely robust algorithm, making it suitable for problems where generalization is an important issue; when this is not the case other algorithms can obtain better accuracy. This is the case, for example, of Support Vector Machines (Bennett & Demiriz, 1999; Cauwenberghs & Poggio, 2001; Suykens & Vandewalle, 1999). Consider a two-class classification problem, in principle this classification problem is solved when an optimal hyperplane separating the observations according to their class exists and is determined. Support Vector Machines postulate that it can be easier and more efficient to determine the optimal hyperplane in a feature space of dimension higher than the original one (see Fig. 5).

As an example, on the left we see a clearly circular pattern, the two classes could be easily separated by a circumference with a suitable radius. Now, let us introduce an auxiliary feature; in the novel three-dimensional space $(x, y, z)$ the two patterns are linearly separable! Notice however that this comes for free, as the third dimension is a function of the previous ones and the distance in the novel space can be easily rewritten in terms of original features (kernel trick): Of course, in real applications finding the optimal separation is not so easy; however, given this high flexibility SVM is usually a good choice for both model accuracy and versatility in finding discriminating patterns.

More recently another approach has gained wide popularity in scientific communities: deep learning. However, the truth is that deep learning techniques derive from another machine learning algorithm which has been studied and used for decades, far before Random Forests and SVM, Artificial Neural Networks (Bishop, 1995). For our purposes it does not make sense to emphasize the differences between neural networks and deep neural networks; it is true that some differences, specifically for what concerns convolutional and recurrent neural networks could be outlined, the interested reader can find an exhaustive discussion elsewhere (Goodfellow, Bengio, & Courville, 2016). Neural networks were designed to mimic human reasoning, specifically human brain physiology. In a neural network the basic processing unit is the neuron, just as in our brain! The input features describing an observation are passed to the input layer which consists of a number of neurons equal to the number of input features; then several options can be considered. In any case the idea is that thanks to suitable vector of weights and a bias term the input features are summed up and their effect is propagated to hidden inner layers of the network until the final layer, also called output layer, is reached (see Fig. 6).

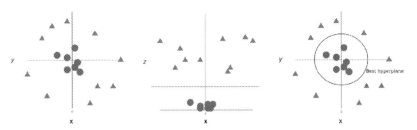

**FIG. 5** A figurative example of kernel trick. A problem not linearly separable in the original feature space becomes linearly separable in another space with higher dimensionality.

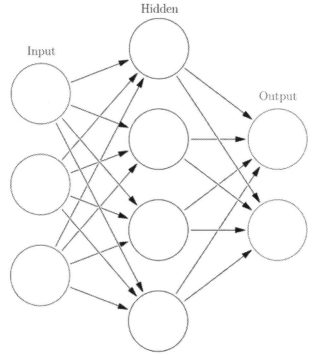

**FIG. 6** An illustrative architecture of neural networks: the three distinctive elements are input, hidden, and output layers.

Nowadays, for deep learning it is meant a network architecture which contemplates at least two hidden layers.

Neural networks are an extremely powerful tool to learn intricate patterns, the main drawback being the outnumbering values of parameter you need to tune to find the optimal configuration. The number of hidden layers, the number of neurons, the activation functions, the number of epochs, and the learning rate are just few examples of all parameters determining a neural network model. Besides, when dealing with deep neural network the number of parameters exponentially grows with the network size. Thanks to modern computer data centers the computational burden required by large neural networks, but we can call now deep, is affordable and this is one of the main reasons neural networks have experienced a so huge comeback. In fact, there are more specific reasons deep learning models have gained popularity, but they are too technical for the purposes of this book. However, let us mention at least one aspect related to Big Data; the fact that huge databases are now publicly available requires a dramatic effort by human experts to label them (otherwise supervised learning would not be possible). A great advantage of deep learning techniques is that these models can also be used in an unsupervised way, thus they are the best

**150**    Big data in psychiatry and neurology

option for huge amounts of unlabeled data. Despite all these considerations, the reader must always take into account that it is not possible to establish a priori which is the best learning algorithm to use, according to the "no free-lunch theorem." A good recommendation is to start with a robust and easily tunable algorithm as Random Forests to get a first idea of what is concealed within your data, then evaluate and compare other models until you find the one which is the best for your application.

## 4 A multiplex model to diagnose neurodegenerative diseases and anomalous aging

In this section we present a case study, in which a multiplex model is used to schematize the structural aspects of brain connectivity (Amoroso et al., 2018a, 2018b, 2019). In our approach, MRI brain scans of different subjects are recorded in a common template, known as MNI152. Next, each scan is segmented into a three-dimensional grid of rectangular boxes (patches), consisting of a fixed number of three-dimensional pixels, called voxels. Patches represent nodes in an undirected, subject-specific complex network, where a link that connects a given patch pair is weighted based on Pearson's correlation between their gray-level distributions. Thus link weights quantify the morphological similarity among the brain districts. Since spatial distributions of white matter, gray matter, and cerebrospinal fluid in the brain are affected by age and pathological conditions, the entire cohort of MRI scans is modeled as a multiplex, in which each layer contains a subject-specific brain network. For each layer of the multiplex, a set of nodal metrics are calculated: strength and inverse participation, and their conditional means at fixed node degree. In order to monitor the differences between the brain networks of different subjects, we build the aggregate adjacency matrix of the cohort and calculate the multiplex versions of the aforementioned nodal metrics, weighing each contribution of the node with its aggregate degree. The single-layer and multiplex metrics, extracted as described before for each subject, are then used as features to train appropriate machine learning or deep learning algorithms to perform the following tasks:

1. classification, to distinguish normal controls (NC) from people affected by neurodegenerative disorders; in particular, specific models were developed to recognize the following pathological conditions: Alzheimer's disease (AD), mild cognitive impairment (MCI), or Parkinson's disease (PD);
2. regression, to predict aging.

The proposed framework employs cross-validation techniques, which enable to test the developed algorithms across the entire dataset and ensure reliability in performance assessment. A schematic overview of such pipeline is displayed in Fig. 7. This approach is not irreparably affected by possible errors in the recording and segmentation steps, as it only requires a rough spatial overlap between MRI scans of different individuals, and not prior segmentation operations. The

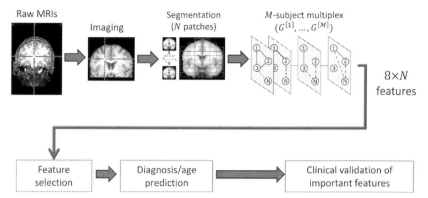

**FIG. 7** Workflow employed in our multiplex model, to achieve NC/AD, NC/MCI, and NC/PD classifications and to predict brain age, from structural MRI scans.

procedure is repeated multiple times, looking for the configuration providing the most accurate performance for the classification and regression tasks; this optimization operation is carried out by changing patch sizes and tuning a number of parameters, specific to each classification or regression algorithm.

In our study, we identified optimal spatial dimension scales for observation and detection of different neurodegenerative pathological conditions and anomalous brain aging. In particular, AD and MCI classification performances are maximized by parceling out brain scans in patches of volume 3000 mm$^3$, while PD effects are most visible in 125 mm$^3$ volume patches. As concerns regression for brain age prediction, the configuration that provides the best results is the one that uses 3000 mm$^3$ volume patches. In all of our classification models, nodal metrics undergo a feature selection process, based on a specific Random Forest wrapper algorithm. In AD and MCI subject recognition tasks, classification is performed by an additional Random Forest block. On the other hand, in the case of NC/PD classification, the performance of different algorithms was evaluated and compared: Support Vector Machine, Random Forest, Naive Bayes, and Neural Network classifiers, each optimized by tuning its internal characteristic parameters; the results appear robust regardless of the specific algorithm choice, although the Support Vector Machine provides slightly better performance.

It should be noted that the network approach discussed here provides, in all cases examined, more accurate results than those obtainable with standard methods, which consist of voxel-based morphometry techniques and Regions of Interest (ROI) determination (Ashburner & Friston, 2000; Fischl, 2012). As for the regression task for brain age prediction, in our study we considered Deep Learning and several Machine Learning algorithms (Ridge and Lasso regression, Random Forest, and Support Vector Machine). The optimal configuration of each regressor was identified through the systematic exploration of its parameter space; the best performances were achieved through the Deep Learning algorithm.

**152** Big data in psychiatry and neurology

The use of the multiplex model looks very promising, as the accuracy obtained in both classification and regression tasks compares favorably with the outcomes of other state-of-the-art methods (Franke et al., 2012). In addition, the multiplex framework allows to determine the importance of the proposed features in classification and regression tasks. Because each feature concerns a specific patch, and therefore a brain district, the model is able to detect morphological changes occurring in a specific region at the outbreak of neurodegenerative diseases or during aging processes, and these indications are a valuable tool for identifying appropriate biomarkers of these conditions.

## References

Akkus, Z., et al. (2017). Deep learning for brain MRI segmentation: State of the art and future directions. *Journal of Digital Imaging, 30*(4), 449–459.

Amoroso, N., et al. (2018a). Complex networks reveal early MRI markers of Parkinson's disease. *Medical Image Analysis, 48*, 12–24.

Amoroso, N., et al. (2018b). Multiplex networks for early diagnosis of Alzheimer's disease. *Frontiers in Aging Neuroscience, 10*, 365.

Amoroso, N., et al. (2019). Deep learning and multiplex networks for accurate modeling of brain age. *Frontiers in Aging Neuroscience, 11*, 115.

Ashburner, J., & Friston, K. J. (2000). Voxel-based morphometry—The methods. *NeuroImage, 11*(6), 805–821.

Bennett, K. P., & Demiriz, A. (1999). Semi-supervised support vector machines. *Advances in Neural Information Processing Systems*, 368–374.

Bishop, C. M. (1995). *Neural networks for pattern recognition*. Oxford University Press.

Boccaletti, S., et al. (2006). Complex networks: Structure and dynamics. *Physics Reports, 424*, 175–308.

Boccaletti, S., et al. (2014). The structure and dynamics of multilayer networks. *Physics Reports, 544*(1), 1–122.

Breiman, L. (1996). Bagging predictors. *Machine Learning, 24*(2), 123–140.

Breiman, L. (2001). Random forests. *Machine Learning, 45*(1), 5–32.

Bullmore, E. T., & Sporns, O. (2009). Complex brain networks: Graph theoretical analysis of structural and functional systems. *Nature Reviews Neuroscience, 10*, 186–198.

Cao, X., et al. (2020). Image registration using machine and deep learning. In *Handbook of medical image computing and computer assisted intervention* (pp. 319–342). Academic Press.

Cauwenberghs, G., & Poggio, T. (2001). Incremental and decremental support vector machine learning. *Advances in Neural Information Processing Systems, 13*, 409–415.

Cole, J. H., et al. (2017). Predicting brain age with deep learning from raw imaging data results in a reliable and heritable biomarker. *NeuroImage, 163*, 115–124.

Dadar, M., et al. (2018). A comparison of publicly available linear MRI stereotaxic registration techniques. *NeuroImage, 174*, 191–200.

De Domenico, M., et al. (2013). Mathematical formulation of multilayer networks. *Physical Review X, 3*(4), 041022.

Di Martino, A., et al. (2014). The autism brain imaging data exchange: Towards a large-scale evaluation of the intrinsic brain architecture in autism. *Molecular Psychiatry, 19*(6), 659–667.

Domenico, D., & Manlio. (2017). Multilayer modeling and analysis of human brain networks. *GigaScience, 6*, 1–8.

Douaud, G., et al. (2006). Distribution of grey matter atrophy in Huntington's disease patients: A combined ROI-based and voxel-based morphometric study. *NeuroImage, 32*(4), 1562–1575.

Etzel, J. A., Gazzola, V., & Keysers, C. (2009). An introduction to anatomical ROI-based fMRI classification analysis. *Brain Research, 1282*, 114–125.

Fischl, B. (2012). FreeSurfer. *NeuroImage, 62*(2), 774–781.

Franke, K., et al. (2012). Brain maturation: Predicting individual BrainAGE in children and adolescents using structural MRI. *NeuroImage, 63*(3), 1305–1312.

Gentzkow, M., Kelly, B., & Taddy, M. (2019). Text as data. *Journal of Economic Literature, 57*(3), 535–574.

Georganos, S., et al. (2019). Geographical random forests: a spatial extension of the random forest algorithm to address spatial heterogeneity in remote sensing and population modelling. *Geocarto International*, 1–16.

Goodfellow, I., Bengio, Y., & Courville, A. (2016). *Deep learning*. MIT press.

Işın, A., Direkoğlu, C., & Şah, M. (2016). Review of MRI-based brain tumor image segmentation using deep learning methods. *Procedia Computer Science, 102*, 317–324.

Kanavati, F., et al. (2017). Supervoxel classification forests for estimating pairwise image correspondences. *Pattern Recognition, 63*, 561–569.

Karayumak, S. C., et al. (2019). Retrospective harmonization of multi-site diffusion MRI data acquired with different acquisition parameters. *NeuroImage, 184*, 180–200.

Katal, A., Wazid, M., & Goudar, R. H. (2013). Big data: Issues, challenges, tools and good practices. In *2013 Sixth international conference on contemporary computing (IC3)*IEEE.

Kivelä, M., et al. (2014). Multilayer networks. *Journal of Complex Networks, 2*(3), 203–271.

Klein, A., et al. (2009). Evaluation of 14 nonlinear deformation algorithms applied to human brain MRI registration. *NeuroImage, 46*(3), 786–802.

Kong, Y., et al. (2019). Iterative spatial fuzzy clustering for 3D brain magnetic resonance image supervoxel segmentation. *Journal of Neuroscience Methods, 311*, 17–27.

Laird, A. R., Lancaster, J. J., & Fox, P. T. (2005). Brainmap. *Neuroinformatics, 3*(1), 65–77.

Li, C., Gore, J. C., & Davatzikos, C. (2014). Multiplicative intrinsic component optimization (MICO) for MRI bias field estimation and tissue segmentation. *Magnetic Resonance Imaging, 32*(7), 913–923.

Liao, L., Lin, T., & Li, B. (2008). MRI brain image segmentation and bias field correction based on fast spatially constrained kernel clustering approach. *Pattern Recognition Letters, 29*(10), 1580–1588.

Lopez-Garcia, P., et al. (2006). Automated ROI-based brain parcellation analysis of frontal and temporal brain volumes in schizophrenia. *Psychiatry Research: Neuroimaging, 147*(2–3), 153–161.

Marcus, D. S., et al. (2007). Open access series of imaging studies (OASIS): Cross-sectional MRI data in young, middle aged, nondemented, and demented older adults. *Journal of Cognitive Neuroscience, 19*(9), 1498–1507.

Marek, K., et al. (2011). The Parkinson progression marker initiative (PPMI). *Progress in Neurobiology, 95*(4), 629–635.

Mirzaalian, H., et al. (2016). Inter-site and inter-scanner diffusion MRI data harmonization. *NeuroImage, 135*, 311–323.

Mirzaalian, H., et al. (2018). Multi-site harmonization of diffusion MRI data in a registration framework. *Brain Imaging and Behavior, 12*(1), 284–295.

Mueller, S. G., et al. (2005). The Alzheimer's disease neuroimaging initiative. *Neuroimaging Clinics, 15*(4), 869–877.

Newman, M. E. J. (2006). Modularity and community structure in networks. *Proceedings of the National Academy of Sciences of the United States of America, 103*(23), 8577–8582.

**154** Big data in psychiatry and neurology

Newman, M. E. J. (2018). *Networks*. Oxford University Press.

Potkin, S. G., & Ford, J. M. (2009). Widespread cortical dysfunction in schizophrenia: The FBIRN imaging consortium. *Schizophrenia Bulletin, 35*(1), 15–18.

Rossini, P. M. (2019). Methods for analysis of brain connectivity: An IFCN-sponsored review. *Clinical Neurophysiology, 130*(10), 1833–1858.

Rubin, V., & Lukoianova, T. (2013). Veracity roadmap: Is big data objective, truthful and credible? *Advances in Classification Research Online, 24*(1), 4.

Schmaal, L., et al. (2016). Subcortical brain alterations in major depressive disorder: Findings from the ENIGMA major depressive disorder working group. *Molecular Psychiatry, 21*(6), 806–812.

Steyerberg, E. W. (2019). *Clinical prediction models*. Springer International Publishing.

Suykens, J. A. K., & Vandewalle, J. (1999). Least squares support vector machine classifiers. *Neural Processing Letters, 9*(3), 293–300.

Tustison, N., & Gee, J. (2009). N4ITK: Nick's N3 ITK implementation for MRI bias field correction. *Insight Journal, 9*.

Van Essen, D. C., et al. (2012). The human connectome project: A data acquisition perspective. *NeuroImage, 62*(4), 2222–2231.

Yamashita, A., et al. (2019). Harmonization of resting-state functional MRI data across multiple imaging sites via the separation of site differences into sampling bias and measurement bias. *PLoS Biology, 17*(4), e3000042.

Yang, K.-. C., et al. (2019). Arming the public with artificial intelligence to counter social bots. *Human Behavior and Emerging Technologies, 1*(1), 48–61.

Chapter 8

# The use of Big Data in Psychiatry—The role of administrative databases

Manuel Gonçalves-Pinho[a,b,c] and Alberto Freitas[a,b]

[a]Department of Community Medicine, Information and Health Decision Sciences, Faculty of Medicine, University of Porto, Porto, Portugal, [b]Center for Health Technology and Services Research (CINTESIS), Porto, Portugal, [c]Department of Psychiatry and Mental Health, Centro Hospitalar do Tâmega e Sousa, Penafiel, Portugal

This chapter is organized into four sections. Sections 1 and 2 describe the concepts of Big Data and Administrative Database use in Mental Health. Section 3 details the advantages and disadvantages of using Administrative Databases in Mental Health research and the appropriate methodological steps a researcher must follow when using Administrative Databases. Section 4 is the conclusion of the chapter with the main keypoints.

## 1  Introduction

Since Hippocratic times to modern days, medicine and medical doctors relied on the creation and use of medical records in order to document and summarize diseases and critical information regarding patients' health. With increasing access to medical services, imaging and analytical examinations, outpatient and inpatient episodes, there is a day-to-day exponential increase in medical information (Cottle et al., 2013; Raghupathi, 2010). This data previously kept on physical format (paper based) is currently being shifted toward a digital form with electronic health records (EHRs). The great amount of information gathered in digital databases gave birth to the concept of *Big Data* in healthcare, where the quantity of data is so vast that it requires specific software and/or hardware to be analyzed and managed (Raghupathi & Raghupathi, 2014).

*Big Data* is usually linked to the *five V's rule*. Volume—regarding the vast amount of data and cases recorded, often gathering health information from entire hospitals, regions, or even countries; Velocity—allowing real-time analysis and real-time access through a friendly user software/interface;

Big Data in Psychiatry and Neurology. https://doi.org/10.1016/B978-0-12-822884-5.00009-X
Copyright © 2021 Elsevier Inc. All rights reserved.

**156** Big data in psychiatry and neurology

Variety—different variables available, from clinical information (medical diagnosis, medical and/or surgical procedures, hospitalization outcome, etc.) to administrative information (sex, patient's residence, marital status, employment information, etc.); Variability—data whose meaning is constantly changing, most information is often kept in structured fields with limited and constant filling options. Nevertheless, there is an increasing interest in natural language processing methodologies to extract information from open text fields; Veracity—accuracy in the information kept in databases (Stewart & Davis, 2016). *Big Data* use is accompanied with a series of challenges in data capturing, analysis, storage, searching, sharing, visualization, transferring, and concerns regarding privacy violations (Anuradha, 2015).

Administrative databases (AD) are conceived for primary purposes other than research (Schwartz, Gagnon, Muri, Zhao, & Kellogg, 1999). Usually they collect information regarding sociodemographic characteristics and clinical information of the diagnosis, procedures, and prescriptions associated to an inpatient or outpatient visit of a patient to the hospital (Stewart & Davis, 2016). This information is then used for hospital charging and administrative purposes (Byrne, Regan, & Howard, 2005).

Mental disorders are treated in an ambulatory (consultations, day hospital, etc.) and inpatient level of care. Due to the chronicity of care needed in a significant number of patients, the amount of data available is vast. Nonetheless, there are few databases specifically conceived to collect and analyze mental disorder-related data and its use depends on the quality of the variables registry (Stewart & Davis, 2016). Inpatient treatment is one of the cornerstones of mental disorders therapy; often long admissions are required which increases its importance for epidemiological studies and also rises the amount of information available in hospitalization administrative databases. As so, secondary data originated from administrative databases arise as an important asset in mental health.

## 2 Big Data, administrative databases, and mental health

Psychiatry is the field of medicine that acts on preventing, diagnosing, and treating mental disorders. The importance of analyzing and registering patient's characteristics in psychiatry was always acknowledged by psychiatrists. During a normal consultation the mental health professional registers information from the moment the patient enters the office until the moment he leaves. With multiple types of information (e.g., imaging and pathologic exams, medical registers, among other) and the increasing access to medical care in most countries, the amount of data available is rising exponentially.

From the old notebook used in the XIX century to nowadays electronic medical records (EMRs) used in most hospital and psychiatric departments, medical information and the way health professionals organize it has changed considerably throughout time (Cushman, 1997; Munk-Jørgensen, Okkels, Golberg,

Ruggeri, & Thornicroft, 2014). EMRs are associated with benefits in savings, maximization of information with rapidity of access, multidisciplinary coordination of medical care, increased quality of care, evidenced-based medical care, education, and research (Kaufman & Hyler, 2005). Even though the list of advantages is significant there are some cons that EMRs may have in mental healthcare. The fact that a psychiatric evaluation is a holistic exam with multiple components makes difficult to establish objective inputs when a clinician wants to translate uncertainty, more easily transmitted through free-text entries (Eason & Waterson, 2014; Morrison, Fernando, Kalra, Cresswell, & Sheikh, 2014; Whooley, 2010). The fact that EMRs may give a more objective structure of input does not make the recorded information more valid or accurate per se. When psychiatrists use EMRs they will be influenced by the digital display where they are entering the input, shifting the focus from the patient to the computer. Eye contact is of great importance in medicine but of even greater importance when evaluating a patient in a psychiatric consultation where one must access the patient body movements, monitoring facial expressions and general appearance leading to a clear change in the psychiatrist–patient relationship (Kaufman & Hyler, 2005).

EMRs are usually aggregated in larger electronic databases that allow an easy and ready access to the global information they contain. Administrative databases are repositories of information pertaining to various levels of healthcare services. From medical diagnosis to sociodemographic characteristics, the amount and type of information kept in administrative databases depends greatly of the purpose for which the databases were conceived. In general, administrative databases are designed to organize hospital discharge notifications, billing for procedures, or prescriptions (Blumenthal, 2009; Stewart & Davis, 2016). AD may originate from all levels of medical care, from primary care, emergency departments, and ambulatory/inpatient hospital care. The diagnoses, procedures, and outcomes of patients' visits are documented in databases conceived for reimbursement or managing purposes (Golinvaux et al., 2014). There are various examples of AD used in mental health research, from the *Danish Psychiatric Central Register* which has allowed the study of rare events and disorders to the South Verona Case Psychiatric Registers in Italy, but mostly conceived in Scandinavian countries (Erlangsen, Mortensen, Vach, & Jeune, 2005; Munk-Jørgensen & Mortensen, 1997; Ruggeri et al., 2005).

## 3 Pros and cons of administrative databases research in mental health

The exponential growth of digital information in AD provides one of the key advantages of AD-related research. The easy and simple access to clinical and sociodemographic information and the systematic collection processes used in AD-related research allow to better represent a specific population by increasing the precision of the measured variables and by collecting information regardless of the influence of other variables (sex, race, marital status, employment, insurance, etc.). AD representation may vary from administrative information of a single hospital to information collected at a national or even continental level (Gonçalves-Pinho, Bragança, & Freitas, 2020). In psychiatry, the use of AD is of great importance due to this broad and complete representation of patients and contact episodes. Both in ambulatory and at an inpatient level of care, information regarding patient's contact with the health system allows to estimate mental disorder's prevalence, incidences, prescription patterns, and to indirectly evaluate the quality of care given. Some examples may be given with recent paper describing suicide patterns in medically ill patients or the mortality assessment of anorexia nervosa patients in acute care hospitals (Edakubo & Fushimi, 2020; Ishikawa et al., 2020; Kessler et al., 2020; Shen, Lo-Ciganic, Segal, & Goodin, 2020; Stewart & Davis, 2016).

Another quality of AD is that the information portrayed represents real-world data from an existing clinical setting in opposition to what occurs in clinical trials where data is originated in a research setting. Real-world data is originated from heterogeneous sources and may lead to real-world evidence when it generates medical/scientific evidence. This effect is even of great importance in psychiatry, where the effect of all the methodologic "theater" behind a clinical trial may influence the outcomes and the sense of improving in patients with mental disorders. Medication side effects or social and functional evaluations of patients with schizophrenia are examples of specific research areas where the real-life setting works as an important variable that is often ignored in clinical trials (Cascade, Kalali, & Kennedy, 2009; Cascade, Kalali, Mehra, & Meyer, 2010; Menendez-Miranda et al., 2015).

Nonetheless, their incomparable representativity and large data sample, AD-related research faces numerous challenges regarding the quality of information it contains. The possible errors present on the data entering process may affect the quality of the clinical information in AD (Cruz-Correia et al., 2009).

Although AD clinical diagnoses are usually systematically organized accordingly to an international classification system (e.g., ICD—*International Classification of Diseases* (Organization, 1978) or DSM—*Diagnostic and Statistical Manual of Mental Disorders* (Association, 2013)), the quality of AD information depends on the following: the clinical accuracy of the physician, the correct register of the input variables (depending on the professional responsible for the registering and codification of the input variables), purpose of the AD (usually AD which are conceived for billing and insurance payments have a more detailed and rigorous classification).

Different input variables of an AD may vary in terms of data quality. Sociodemographic variables (age, sex, identification number, etc.) are usually cross-checked between different AD and governmental databases, guaranteeing a more accurate and valid input when compared to clinical variables (diagnoses, procedures, comorbidities, allergies, etc.) that might be extracted from EHRs or directly recorded by a physician.

The use of Big Data in Psychiatry Chapter | 8    **159**

An AD may be representative when it contains the total amount of hospitalization episodes of a particular disease or the total amount of emergency visits in a determined time period, but if the clinical variables are wrongly recorded it may lead to selection bias and a compromise of internal and/or external validity of a study that uses an AD.

In order to use AD in mental health research, there are important methodological processes that a researcher must perform to avoid bias or the misuse of administrative information. We will now describe in detail the relevant steps one must follow to use AD in mental health research (Alonso et al., 2020; Freitas, Silva-Costa, Marques, & Costa-Pereira, 2010; van Walraven & Austin, 2012).

(a) *Description of the datasets:* AD-related research gathers information from a dataset of administrative episodes (e.g., hospitalizations, emergency visits, ambulatory contacts, prescriptions) and not from a direct population sample. As Schneeweiss said (Schneeweiss, 2007), to understand how a database was generated one must detail the official documentation and regulations behind it but also the mundane realities of how healthcare encounters translate into standardized codes. We are not analyzing individuals but administrative contacts with patients. Even if it is one of the easiest steps to perform in order to understand an AD, the description of the databases is rarely and poorly performed (Byrne et al., 2005). As so, studies using AD should detail the specific characteristics of the database regarding a variety of topics:

- the type of administrative information;
- the inclusion and exclusion criteria of the episodes used;
- how information is inserted and generated in the database;
- the professionals responsible for managing and adding information in the database.

(b) *Accuracy of diagnostic and procedural codes:*

Most AD-related research uses diagnosis and procedures codes in order to select and identify the cases/episodes they want to include. However, they rarely measure or reference the association of the diagnostic/procedure code with the entity it supposedly represented (van Walraven, Bennett, & Forster, 2011).

In psychiatry the most used diagnostic classifications are ICD—*International Classification of Diseases* and DSM—*Diagnostic and Statistical Manual of Mental Disorders* in their various editions. The first is worldly used to report health statistics and the latter is used in clinical and research settings (Kupfer, First, & Regier, 2008). The use of two different classifications, that may be dissonant between each other, increases the discrepancy between research findings and administrative derivative health statistics and estimates of burden of diseases (Kupfer et al., 2008). Coding accuracy depends on the accuracy of the clinical diagnosis that depends, in most cases, on the clinical accuracy of a psychiatrist or mental health professional (diagnostic error). There are a variety of methods and studies that analyze this step on clinical research that depends greatly on clinical training and classification standardization (Aboraya, 2007; Basco et al., 2000; Eack, Singer, & Greeno, 2008; Snowden et al., 2011; Souza, Santos, Lopes, & Freitas, 2019; Takwoingi, Riley, & Deeks, 2015; Whooley, 2010).

After the clinical diagnosis there is another step influencing the diagnostic/procedure accuracy in AD, i.e., the association of the code with the documented diagnosis or procedure which may be linked to administrative errors (van Walraven & Austin, 2012). From the individual pathologic identity of a single patient to a diagnostic code from a standardized classification goes a stairway of generalization. The diagnostic or procedure code that represents a clinical identity must be able to be associated to it. The use of a specific code should be accompanied, when possible, with a validation study of that same code. The importance of validation studies is well established and mostly shows positive conclusions about validity in AD. Nevertheless, methodological literature is sparse and validity measures vary substantially in their conceptual design (Byrne et al., 2005). Most validation studies about diagnoses analyze few psychiatric disorders (mostly schizophrenia). Also, the evaluation of an AD may lead to very different quality levels according to the diagnosis/procedure; in the Danish Psychiatric Case Register, an accuracy of 94% for affective disorders was found and only 50% for substance misuse diagnoses (Hansen et al., 2000; Kessing, 1998). Psychotic disorders seem to be linked to higher diagnostic accuracy as they are usually well described and categorized (Davis, Sudlow, & Hotopf, 2016). Validity cannot be inferred for the entire register if only demonstrated in one specific domain/diagnosis (Byrne et al., 2005).

The possible validation studies are as follows:

- *1-Ecological studies*—where one compares AD clinical data with a more "reliable" prevalence/incidence data source. This kind of study is susceptible to the ecological bias as the information from the more reliable source might not translate data to an individual level.
- *2-Abstraction studies*—methodological analysis where the medical record abstraction process is repeated and then the code status of the two processes is compared to generate agreement statistics (van Walraven & Austin, 2012).
- *3-Gold-standard studies*—comparative studies of the code use in an AD with a gold-standard determination of the disease. This kind of validation study entails the highest level of confidence in validating AD codes.

In mental health, these validation studies might not always be possible to perform due to a lack of population-based disease registries. The output of the coding accuracy studies may vary but generally the statistical measure most used is

The use of Big Data in Psychiatry Chapter | 8  **161**

the PPV—positive predictive value, indicating the probability of a code representing the disease/procedure they intend to describe. However, PPV pose some important statistical concerns once its validity varies with the prevalence of the disease/procedure they intend to represent. Diseases with low prevalence, often present in psychiatry due to the vast diagnostic ramification present in DSM and ICD that may divide the diagnostic groups and cause a dilution of the number of patients/episodes of each diagnosis, may be underrepresented with the use of PPV (Aboraya, 2007). Sensitivity/specificity and a receiver operating characteristic (ROC) curve analysis (specifically, area under the curve [AUC]) may be used in these cases once they do not vary with disease prevalence (Brenner & Gefeller, 1997; Mandic, Go, Aggarwal, Myers, & Froelicher, 2008).

**(c)** *Clinical significance vs statistical significance:*

AD normally gather large data samples and with large data samples comes statistical significance. The concept of statistical significance does not always come to hand with the one of clinical significance. For instance, even if we find *p-values* of less than 0.001 in a specific comparison, that finding might not be clinically relevant if it entails a very small relative or absolute difference.

Precision increases with the number of episodes included in a hypothesis test, as so it is comprehensible that when analyzing mental disorders in AD one might find more "attractive" *P values* when studying broad generic diagnostic groups such as depressive and anxiety disorders compared to specific ICD or DSM diagnostic codes. The use of confidence intervals (CI) instead of isolated *P values* gives more information regarding the relative/absolute differences between the tested groups.

**(d)** *Time-dependent nature of variables:*

AD normally are originated from an endpoint of a specific event (hospitalization, consultation, emergency department visit, etc.) and some variables collected in AD may change during observation. For instance, the use of a diagnostic code of depression in a hospitalization episode may reflect a depression episode that started previous or during the hospitalization. As so, the use of variables that do not specify their onset may lead to biased conclusions if interpreted as if they were known at the start of the episode—a bias present in both AD- and non-AD-related studies.

The use of time-dependent covariates may help to unveil the changes that variables suffer over time and understand the real impact of time in the studied variables (Fisher & Lin, 1999).

**(e)** *Clustered data and study outcomes:*

AD is often used to analyze and compare the impact of administrative variables of healthcare providers (physicians, mental health departments, hospitals, etc.) in patients' outcomes. For instance, differences regarding the impact of the differentiation level of a hospital on patient's outcome may be one of the examples where clustering analysis may be performed using variables available in an AD. This may be performed by a clustering

**162** Big data in psychiatry and neurology

methodological process that groups patients/episodes by a determined variable (hospital, medical doctor, public/private sector). Clustering allows to determine the effect of a specific variable on the outcome that one is studying. This process needs to be performed with analytical methods that take in account the clustering variables (Rocha et al., 2019; van Walraven & Austin, 2012).

## 4 Conclusions

Often, medical records in psychiatrist consultations vary tremendously between professionals and even between different time periods in the same professional contributing to the high heterogeneity recorded in registries derived AD (Davis et al., 2016).

Not only the quantity of information is vast but also the characteristics of the information are heterogeneous between mental health professionals, hospitals, and cultural backgrounds. Psychiatry as a medical specialty needs to record narrative reports that usually do not fit on closed forms of EMRs.

AD arise as a great tool to extract clinical and administrative information about mental disorders and patients with mental disorders. The strengths of AD-related research are the large number of episodes/patients that they may contain, increasing the representativity of an AD-related study. This representativity is accomplished without the usual monetary expenses linked to clinical trials. The fact that AD represents real-world data and not data derived from a research and clinical trial setting helps to extrapolate the conclusions to a clinical setting.

Regardless of their numerous advantages, AD have some characteristics that may lead to biased results and conclusions. One must not forget that AD's primary purpose is to register administrative encounters and not clinical consultations or encounters.

It is of utmost importance to describe in detail the AD characteristics and the population from which its data is derived in order to be able to correctly generalize the AD-related studies' conclusions.

AD contain information about clinical and sociodemographic characteristics that are susceptible to registration errors both at the beginning when the clinician or mental health professional is registering the clinical diagnosis and at the end due to administrative errors.

Each diagnosis is an entirely different universe inside an AD and its behavior may shift greatly from other diagnostic categories/groups. Administrative data is variable in its accuracy for diagnosis, and it may not be possible to generalize from one data source to another. It is important to be able to understand the data, be aware of AD limitations and clearly discuss them, considering the aims of a particular study; in fact, the same data can be of high quality for one purpose but of low quality for one other. Although

The use of Big Data in Psychiatry Chapter | 8  **163**

administrative errors are a weakness of AD, the clinical errors are described to be the biggest source of error (Davis et al., 2016).

The use of AD research in mental health is of great importance once it allows to derive prevalence and incidence information on mental disorders, allowing to develop detailed health policies and reallocate resources accordingly. Its use should be performed with adequate caution and accompanied by the methodological steps described in this chapter.

## References

Aboraya, A. (2007). The reliability of psychiatric diagnoses: POINT—Our psychiatric diagnoses are still unreliable. *Psychiatry (Edgmont)*, *4*(1), 22.

Alonso, V., Santos, J. V., Pinto, M., Ferreira, J., Lema, I., Lopes, F., et al. (2020). Health records as the basis of clinical coding: Is the quality adequate? A qualitative study of medical coders' perceptions. *Health Information Management*, *49*(1), 28–37. https://doi.org/10.1177/1833358319826351.

American Psychiatric Association. (2013). *Diagnostic and statistical manual of mental disorders (DSM-5®)*. American Psychiatric Pubishing.

Anuradha, J. (2015). A brief introduction on Big Data 5Vs characteristics and Hadoop technology. *Procedia Computer Science*, *48*, 319–324.

Basco, M. R., Bostic, J. Q., Davies, D., Rush, A. J., Witte, B., Hendrickse, W., et al. (2000). Methods to improve diagnostic accuracy in a community mental health setting. *American Journal of Psychiatry*, *157*(10), 1599–1605.

Blumenthal, D. (2009). Stimulating the adoption of health information technology. *West Virginia Medical Journal*, *105*(3), 28–30.

Brenner, H., & Gefeller, O. (1997). Variation of sensitivity, specificity, likelihood ratios and predictive values with disease prevalence. *Statistics in Medicine*, *16*(9), 981–991.

Byrne, N., Regan, C., & Howard, L. (2005). Administrative registers in psychiatric research: A systematic review of validity studies. *Acta Psychiatrica Scandinavica*, *112*(6), 409–414.

Cascade, E., Kalali, A. H., & Kennedy, S. H. (2009). Real-world data on SSRI antidepressant side effects. *Psychiatry (Edgmont)*, *6*(2), 16.

Cascade, E., Kalali, A. H., Mehra, S., & Meyer, J. M. (2010). Real-world data on atypical antipsychotic medication side effects. *Psychiatry (Edgmont)*, *7*(7), 9.

Cottle, M., Hoover, W., Kanwal, S., Kohn, M., Strome, T., & Treister, N. (2013). *Transforming Health Care through Big Data Strategies for leveraging big data in the health care industry*. Institute for Health Technology Transformation. http://ihealthtran.com/big-data-in-healthcare.

Cruz-Correia, R. J., et al. (2009). Data quality and integration issues in electronic health records. *Information discovery on electronic health records*. Chapman and Hall; CRC Press. In press.

Cushman, R. (1997). Serious technology assessment for health care information technology. *Journal of the American Medical Informatics Association*, *4*(4), 259–265.

Davis, K. A., Sudlow, C. L., & Hotopf, M. (2016). Can mental health diagnoses in administrative data be used for research? A systematic review of the accuracy of routinely collected diagnoses. *BMC Psychiatry*, *16*(1), 263.

Eack, S. M., Singer, J. B., & Greeno, C. G. (2008). Screening for anxiety and depression in community mental health: The beck anxiety and depression inventories. *Community Mental Health Journal*, *44*(6), 465–474.

**164** Big data in psychiatry and neurology

Eason, K., & Waterson, P. (2014). Fitness for purpose when there are many different purposes: Who are electronic patient records for? *Health Informatics Journal, 20*(3), 189–198.

Edakubo, S., & Fushimi, K. (2020). Mortality and risk assessment for anorexia nervosa in acute-care hospitals: A nationwide administrative database analysis. *BMC Psychiatry, 20*(1), 19.

Erlangsen, A., Mortensen, P. B., Vach, W., & Jeune, B. (2005). Psychiatric hospitalisation and suicide among the very old in Denmark: Population-based register study. *The British Journal of Psychiatry, 187*(1), 43–48.

Fisher, L. D., & Lin, D. Y. (1999). Time-dependent covariates in the Cox proportional-hazards regression model. *Annual Review of Public Health, 20*(1), 145–157.

Freitas, J., Silva-Costa, T., Marques, B., & Costa-Pereira, A. (2010). Implications of data quality problems within hospital administrative databases. In *Paper presented at the XII mediterranean conference on medical and biological engineering and computing 2010*.

Golinvaux, N. S., Bohl, D. D., Basques, B. A., Fu, M. C., Gardner, E. C., & Grauer, J. N. (2014). Limitations of administrative databases in spine research: A study in obesity. *The Spine Journal, 14*(12), 2923–2928.

Gonçalves-Pinho, M., Bragança, M., & Freitas, A. (2020). Psychotic disorders hospitalizations associated with cannabis abuse or dependence: A nationwide big data analysis. *International Journal of Methods in Psychiatric Research, 29*(1), e1813.

Hansen, S. S., Munk-Jørgensen, P., Guldbaek, B., Solgård, T., Lauszus, K., Albrechtsen, N., et al. (2000). Psychoactive substance use diagnoses among psychiatric in-patients. *Acta Psychiatrica Scandinavica, 102*(6), 432–438.

Ishikawa, T., Obara, T., Kikuchi, S., Kobayashi, N., Miyakoda, K., Nishigori, H., et al. (2020). Antidepressant prescriptions for prenatal and postpartum women in Japan: A health administrative database study. *Journal of Affective Disorders, 264*, 295–303. https://doi.org/10.1016/j.jad.2020.01.016.

Kaufman, K. R., & Hyler, S. E. (2005). Problems with the electronic medical record in clinical psychiatry: A hidden cost. *Journal of Psychiatric Practice, 11*(3), 200–204.

Kessing, L. (1998). Validity of diagnoses and other clinical register data in patients with affective disorder. *European Psychiatry, 13*(8), 392–398.

Kessler, R. C., Bauer, M. S., Bishop, T. M., Demler, O. V., Dobscha, S. K., Gildea, S. M., et al. (2020). Using administrative data to predict suicide after psychiatric hospitalization in the Veterans Health Administration System. *Frontiers in Psychiatry, 11*, 390.

Kupfer, D. J., First, M. B., & Regier, D. A. (2008). *A research agenda for DSM V*. American Psychiatric Publishing.

Mandic, S., Go, C., Aggarwal, I., Myers, J., & Froelicher, V. F. (2008). Relationship of predictive modeling to receiver operating characteristics. *Journal of Cardiopulmonary Rehabilitation and Prevention, 28*(6), 415–419.

Menendez-Miranda, I., Garcia-Portilla, M. P., Garcia-Alvarez, L., Arrojo, M., Sanchez, P., Sarramea, F., et al. (2015). Predictive factors of functional capacity and real-world functioning in patients with schizophrenia. *European Psychiatry, 30*(5), 622–627.

Morrison, Z., Fernando, B., Kalra, D., Cresswell, K., & Sheikh, A. (2014). National evaluation of the benefits and risks of greater structuring and coding of the electronic health record: Exploratory qualitative investigation. *Journal of the American Medical Informatics Association, 21*(3), 492–500.

Munk-Jørgensen, P., & Mortensen, P. B. (1997). The Danish psychiatric central register. *Danish Medical Bulletin, 44*(1), 82–84.

Munk-Jørgensen, P., Okkels, N., Golberg, D., Ruggeri, M., & Thornicroft, G. (2014). Fifty years' development and future perspectives of psychiatric register research. *Acta Psychiatrica Scandinavica, 130*(2), 87–98.

The use of Big Data in Psychiatry Chapter | 8  **165**

Raghupathi, W. (2010). Data mining in health care. In *Healthcare informatics: Improving efficiency and productivity* (pp. 211–223).

Raghupathi, W., & Raghupathi, V. (2014). Big data analytics in healthcare: Promise and potential. *Health Information Science and Systems, 2*(1), 3.

Rocha, A., et al. (2019). Internal deterministic record linkage using indirect identifiers for matching of same-patient hospital transfers and early readmissions after acute coronary syndrome in a nationwide hospital discharge database: a retrospective observational validation study. *BMJ Open, 9*(12). https://doi.org/10.1136/bmjopen-2019-033486.

Ruggeri, M., Nosè, M., Bonetto, C., Cristofalo, D., Lasalvia, A., Salvi, G., et al. (2005). Changes and predictors of change in objective and subjective quality of life: Multiwave follow-up study in community psychiatric practice. *The British Journal of Psychiatry, 187*(2), 121–130.

Schneeweiss, S. (2007). Understanding secondary databases: A commentary on "Sources of bias for health state characteristics in secondary databases". *Journal of Clinical Epidemiology, 60*(7), 648–650.

Schwartz, R. M., Gagnon, D. E., Muri, J. H., Zhao, Q. R., & Kellogg, R. (1999). Administrative data for quality improvement. *Pediatrics, 103*(Suppl. E1), 291–301.

Shen, Y., Lo-Ciganic, W.-H., Segal, R., & Goodin, A. J. (2020). Prevalence of substance use disorder and psychiatric comorbidity burden among pregnant women with opioid use disorder in a large administrative database, 2009–2014. *Journal of Psychosomatic Obstetrics and Gynecology, 18*, 1–7. https://doi.org/10.1080/0167482X.2020.1727882.

Snowden, J. S., Thompson, J. C., Stopford, C. L., Richardson, A. M., Gerhard, A., Neary, D., et al. (2011). The clinical diagnosis of early-onset dementias: Diagnostic accuracy and clinicopathological relationships. *Brain, 134*(9), 2478–2492.

Souza, J., Santos, J. V., Lopes, F., & Freitas, A. (2019). Quality of coding within clinical datasets: A case-study using burn-related hospitalizations. *Burns, 45*(7), 1571–1584.

Stewart, R., & Davis, K. (2016). 'Big data' in mental health research: Current status and emerging possibilities. *Social Psychiatry and Psychiatric Epidemiology, 51*(8), 1055–1072.

Takwoingi, Y., Riley, R. D., & Deeks, J. J. (2015). Meta-analysis of diagnostic accuracy studies in mental health. *Evidence-Based Mental Health, 18*(4), 103–109.

van Walraven, C., & Austin, P. (2012). Administrative database research has unique characteristics that can risk biased results. *Journal of Clinical Epidemiology, 65*(2), 126–131.

van Walraven, C., Bennett, C., & Forster, A. J. (2011). Administrative database research infrequently used validated diagnostic or procedural codes. *Journal of Clinical Epidemiology, 64* (10), 1054–1059.

Whooley, O. (2010). Diagnostic ambivalence: Psychiatric workarounds and the diagnostic and statistical manual of mental disorders. *Sociology of Health & Illness, 32*(3), 452–469.

World Health Organization. (1978). *International classification of diseases:[9th] ninth revision, basic tabulation list with alphabetic index.* World Health Organization.

Chapter 9

# Predicting the emergence of novel psychoactive substances with big data

## Robert Todd Perdue[a] and James Hawdon[b]

[a]Department of Sociology & Anthropology, Elon University, Elon, NC, United States, [b]Center for Peace Studies and Violence Prevention, Virginia Polytechnic Institute and State University, Blacksburg, VA, United States

## 1 Introduction

Substance abuse among adolescents in the United States is common, with about a quarter of youth using at least one illicit drug in the past year (Monitoring the Future, 2019). Use has significant consequences, as young drug users are more likely to engage in high-risk sexual behaviors, perform poorly in school, and have behavior and emotional problems. Moreover, mortality and morbidity rates are substantially elevated for this group. While well-deserved attention is given to the abuse of prescription drugs and heroin, an underexamined, but very serious emerging threat is posed by novel psychoactive substances (NPS). These drugs are being developed and introduced at alarming rates, posing significant challenges to law enforcement and healthcare workers who often have little understanding of the origin, makeup, and toxicity of these substances (Khey, Stogner, & Miller, 2013). NPS include newly synthesized drugs that mimic existing drugs (e.g., bath salts or synthetic cannabinoids—see Vandrey, Dun, Fry, & Girling, 2012), newly discovered or rediscovered psychoactive botanicals in fortified or strengthened form (e.g., *Salvia divinorum*—see González, Riba, Bouso, Gómez-Jarabo, & Barbanoj, 2006), and new and innovative means of administering drugs (e.g., "dabbing" marijuana or butane hash oil—see Stogner & Miller, 2015—or purple drank—see Agnich, Stogner, Miller, & Marcum, 2013).

As with other recreational drugs, these substances can pose serious risks to their users, ranging from anxiety, depression, and cognitive impairment to acute psychotic reactions and fatal overdoses (see Cohen et al., 2017; Hermanns-Clausen, Kneisel, Szabo, & Auwärter, 2013; Lozier et al., 2015). For example,

Big Data in Psychiatry and Neurology. https://doi.org/10.1016/B978-0-12-822884-5.00014-3
Copyright © 2021 Elsevier Inc. All rights reserved.

**167**

**168** Big data in psychiatry and neurology

the number of emergency room visits involving synthetic cannabinoids increased from 11,406 visits in 2010 to 28,531 visits in 2011 (Bush & Woodwell, 2014, p. 1), and based on the most recent data, synthetic cathinones or bath salts were associated with nearly 23,000 emergency room visits in 2011 (Drug Abuse Warning Network, 2013). The number of NPS being discovered and produced is believed to be at an all-time high, with estimates of more than one hundred NPS introduced every year (European Monitoring Centre for Drugs and Drug Addiction, 2015; Khey et al., 2013). The variety of these drugs, the speed with which they are developed and introduced, and the lack of scientific data concerning their potential dangers pose significant challenges for law enforcement officials trying to control access to these drugs, healthcare providers trying to understand the dangers of these drugs, and rehabilitation workers trying to develop effective treatment protocols for these drugs. As such, researchers are wrestling with this multifaceted and growing social problem. Fundamental to the effort to understand these drugs are two critical questions: Is there any way we can track trends in novel drug use? And, is there any way we can predict trends in novel drug abuse?

Here, our primary aim is to assess if Google Trends data can serve as a useful proxy measure to track and evaluate NPS trends. If this free source of real-time data proves relatively accurate, those concerned with NPS will be able to more quickly act on this information. We begin by comparing Google Trends data with Monitoring the Future (MTF) data. MTF has surveyed American secondary school students since 1975 and it is designed to collect a nationally representative sample that can provide an accurate description of youth drug use. The data are also extremely useful for tracking changes in use over time (NIDA, 2019). In making this comparison between Google Trends data and the "gold standard" of drug use data, we aim to address the issue of poor-quality data that hampers the efforts of healthcare professionals, law enforcement agents, and policymakers attempting to deal with NPS. Valid drug data has always been difficult to collect, as the stigma of being a user and the possible legal consequences associated with use pose significant challenges. Other problems that plague all data collection efforts that rely on traditional social science methods are also barriers for collecting reliable drug use data. For example, survey methodologies are time consuming and expensive, and sampling issues often result in omitting data relevant to the heaviest drug users. While hospital records can provide insights into these users, the fact that these data are frequently reported in a manner that combines effects can prevent the necessary parsing of the data. Finally, other official data sources such as police records are problematic in that these often reflect political calculations concerning which drugs are to be targeted for enforcement and which areas are going to be most heavily surveilled (see Perdue, Hawdon, & Thames, 2018 for a similar argument).

In addition to these well-documented issues which hinder data collection efforts on drug use trends, technical and strategic barriers also make collecting robust drug data extremely difficult. First, researchers and research organizations often use different metrics when collecting data, thereby making comparisons impossible. Second, multiple drugs are frequently combined into broad categories. For example, MTF reports figures for "bath salts," which refer to a long list of various synthetic cathinones. Spatial and temporal issues also hinder our efforts. Many research bodies collect data infrequently and irregularly, leaving critical gaps in our data that harms our ability to predict trends, as is the case with the DAWN (Drug Abuse Warning Network, 2013) data. Even those who collect data more systematically, such as MTF, do so annually, at best. As with all collected data, there is the inevitable lag between data collection and when the data are available for analysis. With respect to problems associated with physical space, at least two weaknesses exist: a lack of granularity and the data divide between urban and rural areas. Drug abuse data are frequently collected at scales that deny useful information to those working in localized areas. This problem becomes increasingly acute as one moves from urban to rural areas. For instance, the Substance Abuse and Mental Health Services Administration (SAMSHA) collects data at the state, US Census Bureau Region, and Metro Areas. These areas fail to provide useful information because these scales are exceedingly large and diffuse. Again, this problem becomes especially acute in rural areas, which is extremely problematic given the distinctiveness of rural addiction.

Because existing drug data are plagued by these weakness, we assess the utility of using big data to ameliorate some of these limitations. Specifically, we ask: Can Internet search queries fill gaps in United States drug abuse data? To begin, we review scholarly attempts to use Internet searches to address a wide array of research questions.

## 2 Internet search queries as data

Commercial search engines are vital to knowledge dissemination, as individual users rely on these almost exclusively as fundamental sources of information (Trevisan, 2014; Van Couvering, 2008). Because search engines track individuals' Internet explorations, scholars have used these searches as behavioral measures of an individual's interest in a topic (Granka, 2009). According to this line of research, the salience of an issue to a specified population can be inferred based on the volume of Internet search queries related to the issue (Zheluk, Quinn, Hercz, & Gillespie, 2013). Researchers note several advantages of Internet search trend analysis over more traditional methods such as surveys. First, data are available at no cost to researchers (e.g., Google Trends). Second, these data are real-time insights into the public's concern about a given issue, reflecting dynamic behavior at the aggregate level (see Trevisan, 2014). Third, because individuals do these inquiries as part of their everyday life, searches are unobtrusive data that avoid issues such as telescoping, memory decay, and respondents providing socially desirable answers. Finally, as unobtrusive data these searches avoid potential Human Subject issues related to possible issues

**170** Big data in psychiatry and neurology

of linkage between responses and the individual providing the responses, while also making it impossible to link an individual and their response.

An increasingly popular tool for scientific inquiry, analysis of Internet search queries is being utilized across a wide range of disciplines, including economics, political science, and public health epidemiology. Econometricians, for instance, have successfully forecasted trends in unemployment rates in the United States, Germany, Israel, and Italy (Askitas & Zimmerman, 2009; Choi & Varian, 2009; D'Amuri, 2009; D'Amuri & Marcucci, 2009; Suhoy, 2009) and have predicted consumer behaviors such as automobile and real estate sales (Choi & Varian, 2012). Political science researchers have used Internet search trends to measure public attention to global warming, terrorism, and healthcare in the United States (Ripberger, 2001). Additionally, Internet search query analysis uncovered the negative effect of racial animus on President Barak Obama's 2008 presidential campaign, which had remained hidden to analysts using traditional survey methods (Stevens-Davidowitz, 2012).

The most sustained and widespread use of Internet search data may be found in public health and epidemiological research, with studies effectively tracking and forecasting the spread of infectious diseases, such as influenza (Eysenbach, 2006; Ginsberg et al., 2009; Hulth, Rydevik, & Linde, 2009; Polgreen, Cehn, Pennock, Nelson, & Weinstein, 2008), dengue fever (Chan, Sahai, Conrad, & Brownstein, 2011), listeriosis (Wilson & Brownstein, 2009), and Lyme disease (Seiffer, Schwarzwalder, Geis, & Aucott, 2010). Search query analysis has also been successfully utilized in explorations of the implications of governmental policy related to healthcare, such as the substitution of electronic nicotine delivery systems in the wake of increases in the cigarette tax (Ayers, Ribisl, & Brownstein, 2011) and international patterns in abortion rates relative to laws regulating abortion (Reis & Brownstein, 2010).

More recently, researchers have investigated the efficacy of Internet search query data in understanding patterns of the manufacture and use of NPS. This methodology proved particularly effective in the study of illicitly made desomorphine—the injectable drug known as "krokodil" in Russia (Zheluk, Quinn, & Meylakhs, 2014). Not only was search query data found to be a valuable proxy measure of the production and use of krokodil, but the volume of search queries related to the drug substantially increased immediately following the amplification of media reports and government demonstrations framing krokodil as a significant social problem, while significantly decreasing immediately following the implementation of governmental restrictions on the drug. Similarly, a study by Kapitany-Foveny and Demetovics (2017) indicates that the volume of Internet entries—including search queries, articles published online, and online advertisements—related to the NPS mephedrone is inversely related to its legal status. That is, in the period of time after mephedrone was banned, Internet activity related to the drug increased. The authors note the increased Internet activity surrounding mephedrone was significant, while measures of activity related to drugs such as cocaine, heroin, and ecstasy remained consistent during the same period (Kapitany-Foveny & Demetovics, 2017).

Predicting the emergence of novel psychoactive substances  Chapter | 9  **171**

Taken together, these findings highlight both the utility of search query analysis in the study of drug use patterns and the influence of governmental regulations on these same patterns.

The vast majority of Internet search trend studies discussed here have utilized Google's free, publicly available Google Trends data. The market leader, Google represents 79.29% of all desktop-based search queries to-date in 2017 and 96.17% of all mobile-based search queries to-date in 2017 (Netmarketshare.com, 2017). Google Trends (https://www.google.com/trends/) offers global tracking of Google search queries that can be organized by country and various subnational units. Here, our primary aim is to assess the viability of Google Trends data as a proxy measure to evaluate NPS trends in the United States. As such, we begin by comparing Google Trends data with that collected by Monitoring the Future (MTF), a well-respected study that has been ongoing since 1975.

# 3 Methods

To assess the extent to which Google Trends can provide valid insights into NPS trends, we first identified drugs included in MTF data that could be considered novel. MTF is an ongoing study of American secondary school students and young adults that has surveyed 12th grade students annually since 1975. MTF is designed to collect a nationally representative sample that can provide an accurate description of youth drug use in a given year, while documenting changes over time (NIDA, 2019). As such, while some "traditional" drugs such as alcohol, marijuana, and cocaine are always included in the study, other drugs come and go. We identified five drugs that fit our criteria and had been in the study long enough to make correlations valid (see Perdue et al., 2018). For each drug, we collected monthly Google Trends data for each year MTF data were available. The years these MTF data were available varied, with the starting dates ranging from 2009 for "Adderall" and "*Salvia divinorum*," to 2012 for "bath salts." Google Trends search data were then collected from those start dates until 2016 (see Table 1).

**TABLE 1** Search one—NPS included in Google Trends searches.

| Drug search term | Years |
| --- | --- |
| Adderall | January 1, 2009–December 31, 2016 |
| *Salvia divinorum* | January 1, 2009–December 31, 2016 |
| Snus | January 1, 2011–December 31, 2016 |
| Synthetic Marijuana | January 1, 2011–December 31, 2016 |
| Bath Salts | January 1, 2012–December 31, 2016 |

**172** Big data in psychiatry and neurology

**TABLE 2** Search two—NPS included in Google Trends searches.

| Drug search term | Years |
|---|---|
| Adderall | January 1, 2009–December 31, 2019 |
| *Salvia divinorum* | January 1, 2009–December 31, 2019 |
| Snus | January 1, 2011–December 31, 2019 |
| Synthetic Marijuana | January 1, 2011–December 31, 2019 |
| Bath Salts | January 1, 2012–December 31, 2019 |

This study builds on Perdue et al., 2018 by adding three more years of data, allowing for correlations with the most recent MTF data (see Table 2). We used the same search terms and start dates and limited the study to the United States.

We then use Pearson's correlations to determine the relationships between the MTF data and the Google Trends data. Google explains the relative manner of their data collection thusly:

*The numbers that appear show total searches for a term relative to the total number of searches done on Google over time. A line trending downward means that a search term's relative popularity is decreasing. But that doesn't necessarily mean the total number of searches for that term is decreasing (Google, 2017).*

Because monthly data are not available in MTF (an example of the limitations of the data we discussed before), we were forced to calculate the average relative popularity of the search term for each year by aggregating the monthly Google Trends data over the year and dividing by 12.

## 4   Results

Looking at Table 3, there is evidence that Google Trends and MTF data are highly correlated, and we argue these correlations support the argument that Google Trends is useful for detecting drug use and tracking trends in use. Four of the five correlations between the Google Trends and MTF data are positive and statistically significant at the 0.05 level. The lone exception is for snus, which is significant using a more liberal level ($P = 0.099$). Obtaining statistical significance is surprising given the small samples on which the correlations are based. Importantly, these correlations achieve statistical significant because they are very strong, ranging from 0.707 to 0.950. Thus, while the sample sizes for these correlations are well below what is typically desired, the results are suggestive and encouraging.

# Predicting the emergence of novel psychoactive substances Chapter | 9 **173**

**TABLE 3** Correlations between MTF rates of use and Google Trends searches for novel drugs ending 2016.

| Drug (N = years) | Pearson's correlation (significance) |
| --- | --- |
| Adderall | 0.707 |
| (8) | (0.050) |
| *Salvia divinorum* | 0.936 |
| (8) | (0.001) |
| Snus | 0.731 |
| (6) | (0.099) |
| Synthetic Marijuana | 0.950 |
| (6) | (0.004) |
| Bath salts | 0.929 |
| (5) | (0.022) |

**TABLE 4** Correlations between MTF rates of use and Google Trends searches for novel drugs ending 2019.

| Drug (N = years) | Pearson's correlation (significance) |
| --- | --- |
| Adderall | −0.330 |
| (11) | (0.050) |
| *Salvia divinorum* | 0.918 |
| (11) | (0.001) |
| Snus | 0.725 |
| (9) | (0.099) |
| Synthetic Marijuana | 0.956 |
| (9) | (0.004) |
| Bath salts[a] | 0.454 |
| (7) | (0.022) |

[a]*Monitoring the Future did not collect data on bath salts in 2019, therefore this drug has only 2 years of new data rather than three like the others under study.*

We then update the study with the most recent data available from MTF (see Table 4).

The results of this latter analysis are significant for all drugs at the 0.099 level, while four of the five are significant at 0.05. However, two of the drugs, Adderall and bath salts, are no longer exhibiting the predictive power previously seen. Bath salts have dropped from a robust 0.929 to just 0.454, while Adderall is now showing an inverse relationship between Google Trends and MTF data. These findings speak to the broader efficacy and challenges of using big data as a proxy for more traditional methodologies, discussed later.

## 5 Discussion and conclusion

The results found here tell us much, especially those related to Adderall and bath salts. First, we contend that because Adderall is a legal drug often prescribed to adolescents, searches on Google Trends are likely to behave in a different manner than would those for illegal drugs and especially NPS. While Adderall is certainly abused for recreational purposes, Internet searches are also very likely to include significant numbers driven by reasons unrelated to recreational abuse. For example, parents whose children are prescribed Adderall are likely to search for information regarding the drug. In short, it is unlikely that Adderall meets the definition of a novel drug. As such, it is very possible that Google Trends would not be a useful tool for tracking its use and abuse. Similarly, "bath salts" are really no longer "novel." These drugs were becoming widely used in Europe by 2008, and by 2010 they had emerged on the streets in the United States, showing up in poison control centers. Bath salts were added to MTF in 2012, and their use among high school seniors peaked at 1.0% in 2013. MTF stopped reporting data for bath salt use in 2019, thus, this once novel drug is now nearly a decade old.

The choice to include these two drugs in this study is based solely on the fact that they are one of only a handful of drugs included in MTF that is not a "traditional" drug such as alcohol, marijuana, LSD, heroin, or cocaine. Nevertheless, their inclusion, and these findings, are valuable, as they suggest that Google Trends is likely a good proxy for little known novel drugs, a moderately useful proxy for well-known illegal drugs, and a poor proxy for well-known legal drugs like Adderall.

This model implies that the more well known a drug is, the less percentage of searches are conducted by drug users (or potential drug users) and the greater percentage of searches are conducted by moral entrepreneurs or concerned nonusers (Becker, 1963). This is likely the case for bath salts, as searches increased dramatically following a series of high-profile attacks supposedly caused by the drug, including one involving cannibalism (Herald, 2012). As such, exaggerated concern from nonusers in response to sensational events may reduce the viability of Google Trends. In short, potential users are unlikely to search the Internet for, say, marijuana, as they are aware of its effects and are likely to have access to the drug. Conversely, if potential users become aware of a new drug just hitting the streets, they may turn to their computer to find out more.

Predicting the emergence of novel psychoactive substances Chapter | 9  **175**

A supportive example is the emergence of isotonitazene. According to one of the very few academic studies available, this drug, known as "Iso," is "a potent NPS opioid and the first member of the benzimidazole class of compounds being available in online markets" (Blanckaeret et al., 2019). Iso is similar to other NPS such as synthetic marijuana and bath salts in that it mimics the effects of well-established illicit drugs, but aims to escape legislation by employing chemical structures similar, but slightly different from, those drugs (Blanckaeret et al., 2019). What makes Iso different from synthetic marijuana and bath salts and especially troubling, however, is that it is a synthetic opioid. Very few substances mimicking opioid activity have emerged in dark web markets, and the elevated risk of overdose and fatality such drugs pose demands immediate attention (Coopman, Blanckaert, Van Parys, Van Calenbergh, & Cordonnier, 2016; Degreef, Blanckaert, Berry, van Nuijs, & Maudens, 2019). Exploring Google Trends for searches of "isotonitazene" reveals the novelty of this synthetic opioid, as we find *zero* results until March 2019, when a search score of 14 is recorded, followed by a steep increase up to 100 in January 2020 (see Fig. 1). This maps closely to what we know from academic work, with the first published report documenting Iso as an NPS on the black market being published in November 2019 by Blanckaert et al.

As isotonitazene searches highlight, Google Trends can point to NPS before they are widely known and data has been collected and made available by more traditional sources. Indeed, accurate pictures of the size and contours of illicit drug use are a prerequisite for informed public debate and policymaking. Unfortunately, drug abuse researchers are unable to paint these pictures because of inadequate data, allowing misconceptions to develop and resources to be misallocated (Miech et al., 2017, p. 6). Google Trends appears to be a useful data

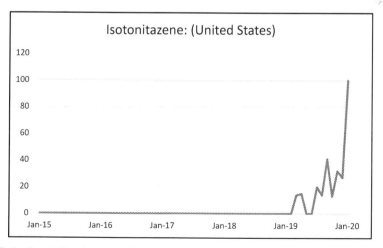

**FIG. 1** Google Trend searches of isotonitazene.

source regarding NPS because Google is the first place many go for information on topics of which they are unfamiliar, possibly allowing prediction of future trends. Google Trends also provides finer grained temporal and spatial data, and may be able to shed light on how NPS differ along rural and urban divides.

Given the current findings and the significant public health threats posed by NPS, continued research in the utility of Google Trends to forecast NPS and other drug abuse prevalence is warranted. Consistent with the growing body of literature utilizing Internet search query data to forecast structural and behavioral outcomes, we find Google Trends analysis to be a valuable tool in the exploration of public interest in and use of NPS. Trends in the majority of NPS search queries are highly correlated with data collected by MTF surveys, suggesting that Google Trends are capable of providing a proxy measure of NPS use prevalence. We do find, however, that this methodology is likely to be more useful with drugs that are truly novel and may be less valuable in relation to traditional illicit drug use and abuse. Nevertheless, we do not find this to be overly problematic. Traditional data can track trends in well-known drugs, and public officials do not necessarily need information about these trends in the same way as they do about novel drugs.

Offering data at no cost in near real time, Google Trends, in conjunction with existing drug use data, may provide researchers with an enhanced understanding of current public interest in NPS and may be particularly useful in predicting NPS use prevalence over time and across regions. For example, tracking novel drugs such as Iso, where rates of use appear to be increasing rapidly, can allow law enforcement and health professionals to intervene in a more timely manner.

While our research suggests the potential utility of Google Trends data in tracking the use of NPS, ironically, the limitations of existing data that motivated our study also prevent us from making definitive data comparisons and statements. Not only is MTF limited in the types of drugs that are reported, the data are also available only annual, while Google Trend data are far more temporally refined. Moreover, MTF is an excellent source of data on the use patterns of youth and young adults, but is limited to just this population and cannot be used to draw conclusions about older Americans. This is extremely problematic, particularly for research designed to inform policy. As such, future research should continue exploring the relationship between existing data sources and Google Trends, but additional drug data sources are needed. Researchers should consider correlating Google Trends and overdose death data reported by the CDC, while data on emergency room visits, drug seizures, and drug arrests could also be compared to Google Trend data.

Yet, even if these studies are conducted, the problems of existing data sources will continue to hinder their utility in establishing the accuracy of Google Trend data. To do this adequately, researchers will need to gather survey data related to the use of NPS and other illicit drugs on a large sample, and they will have to collect these data systematically several times over several months or (hopefully) years. Such a data collection effort would be extremely time consuming and expensive, but it is such an effort that is needed to offer more complete comparisons with Google Trends data. While we believe Google Trend data is useful, we must also be aware of possible distortions in Google Trends data. Search queries may be influenced by moral panics or even by misspellings in search bars. Nevertheless, while recognizing Google Trends has limitations, these may be a promising complement to existing data that can inform our responses to drug trends, allow us to act more quickly, and ultimately, to save lives.

## References

Agnich, L. E., Stogner, J. M., Miller, B. L., & Marcum, C. D. (2013). Purple drank prevalence and characteristics of misusers of codeine cough syrup mixtures. *Addictive Behaviors, 38*(9), 2445–2449.

Askitas, N., & Zimmerman, K. F. (2009). Google econometrics and unemployment forecasting. *Econometrics Quarterly, 55*(2), 107–120.

Ayers, J. W., Ribisl, K., & Brownstein, J. S. (2011). Using search query surveillance to monitor tax avoidance and smoking cessation following the United States' 2009 "SCHIP" cigarette tax increase. *PLoS One, 6*(3), e16777.

Becker, H. (1963). *Outsiders: Studies in the sociology of deviance*. New York: Free Press.

Blanckaeret, P., Cannaert, A., Uytfanghe, V., Hulpia, F., Deconinck, E., Van Calenbeergh, S., et al. (2019). Report on a novel emerging class of highly potent benzimidazole NPS opioids: Chemical and in vitro functional characterization of isotonitazene. *Drug Test and Analysis, 12*(4), 422–430.

Bush, D. M., & Woodwell, D. A. (2014). *Update: Drug-related emergency department visits involving synthetic cannabinoids*. Substance Abuse and Mental Health Services Administration. https://www.samhsa.gov/data/sites/default/files/SR-1378/SR-1378.pdf.

Chan, E. H., Sahai, V., Conrad, C., & Brownstein, J. S. (2011). Using web search query data to monitor dengue epidemics: A new model for neglected tropical disease surveillance. *PLoS Neglected Tropical Diseases, 5*(5), e1206.

Choi, H., & Varian, H. (2009). *Predicting claims for unemployment benefits* (pp. 1–5). Google Inc.

Choi, H., & Varian, H. (2012). Predicting the present with Google Trends. *The Economic Record, 88* (1), 2–9.

Cohen, K., Kapitány-Fövény, M., Mama, Y., Arieli, M., Rosca, P., Demetrovics, Z., et al. (2017). The effects of synthetic cannabinoids on executive function. *Psychopharmacology, 234*, 1121–1134.

Coopman, V., Blanckaert, P., Van Parys, G., Van Calenbergh, S., & Cordonnier, J. (2016). A case of acute intoxication due to combined use of fentanyl and 3,4-dichloro-N-[2-(dimethylamino) cyclohexyl]-N-methylbenzamide (U-47700). *Forensic Science International, 266*, 68–72.

D'Amuri, F. (2009). *Predicting unemployment in short samples with internet job search query data*. Germany: University Library of Munich.

D'Amuri, F., & Marcucci, J. (2009). *The predictive power of Google data: New evidence on US unemployment* (p. 16). VoxEU. org.

Degreef, M., Blanckaert, P., Berry, E. M., van Nuijs, A. L. N., & Maudens, K. E. (2019). Determination of ocfentanil and W-18 in a suspicious heroin-like powder in Belgium. *Forensic Toxicology, 37*(2), 474–479.

**178** Big data in psychiatry and neurology

Drug Abuse Warning Network. (2013). *Bath salts were involved in over 20,000 drug-related emergency department visits in 2011.* The DAWN Report *https://www.samhsa.gov/data/sites/default/files/spot117-bath-salts.*

European Monitoring Centre for Drugs and Drug Addiction. (2015). Available: https://europa.eu/european-union/about-eu/agencies/emcdda_en.

Eysenbach, G. (2006). Infodemiology: Tracking Flu-related searches on the Web for syndromic surveillance. In *AMIA Annual Symposium Proceedings* (pp. 244–248). American Medical Informatics Association.

Ginsberg, J., Mohebbi, M. H., Patel, R. S., Brammer, L., Smolinski, M. S., & Brilliant, L. (2009). Detecting influenza epidemics using search engine query data. *Nature, 457*(7232), 1012–1014.

González, D., Riba, J., Bouso, J. C., Gómez-Jarabo, G., & Barbanoj, M. J. (2006). Pattern of use and subjective effects of Salvia divinorum among recreational users. *Drug and Alcohol Dependence, 85*(2), 157–162.

Google. (2017). Google Trends. *http://google.com/trends/.*

Granka, L. (2009). Inferring the public agenda from implicit query data. In *Proceedings of the workshop on understanding the user—logging and interpreting user interactions in information search and retrieval. SIGIR-2009, Boston, MA, July 19–23 2009.*

Herald, M. (2012). *Friends of cannibal Rudy Eugene say he was not 'a face-eating zombie monster' as police reveal first picture of homeless victim in bizarre Miami attack.* https://www.heraldsun.com.au/news/law-order/friends-of-cannibal-rudy-eugene-say-he-was-not-a-face-eating-zombie-monster-as-police-reveal-first-picture-of-homeless-victim-in-bizarre-miami-attack/news-story/d7523e56f12676e6f7e4e2b22cf2c3a2.

Hermanns-Clausen, M., Kneisel, S., Szabo, B., & Auwärter, V. (2013). Acute toxicity due to the confirmed consumption of synthetic cannabinoids: Clinical and laboratory findings. *Addiction, 108*(3), 534–544.

Hulth, A., Rydevik, G., & Linde, A. (2009). Web queries as a source for syndromic surveillance. *PLoS One, 4*(2), e4378.

Kapitany-Foveny, M., & Demetovics, Z. (2017). Utility of Web search query date in testing theoretical assumptions about mephedrone. *Human Psychopharmacology, 32*, e2620.

Khey, D. N., Stogner, J., & Miller, B. (2013). *Emerging trends in drug use and distribution. Vol. 12.* London: Springer.

Lozier, M. J., Boyd, M., Stanley, C., Ogilvie, L., King, E., Martin, C., et al. (2015). Acetyl fentanyl, a novel fentanyl analog, causes 14 overdose deaths in Rhode Island, March–May 2013. *Journal of Medical Toxicology, 11*(2), 208.

Miech, R. A., Johnston, L. D., O'Malley, P. M., Bachman, J. G., Schulenberg, J. E., & Patrick, M. E. (2017). Monitoring the future national survey results on drug use, 1975–2016. In *Vol. 1. Secondary school students.* Ann Arbor: Institute for Social Research, The University of Michigan. http://monitoringthefuture.org/pubs.html#monographs.

Netmarketshare. (2017). *Search engine market share.* http://netmarketshare.com.

NIDA. (2019). *Monitoring the future survey: High school and youth trends.* https://www.drugabuse.gov/publications/drugfacts/monitoring-future-survey-high-school-youth-trends.

Perdue, R. T., Hawdon, J., & Thames, K. (2018). Can big data predict the rise of novel drugs? *Journal of Drug Issues, 48*(4), 508–518.

Polgreen, P. M., Cehn, Y., Pennock, D. M., Nelson, F. D., & Weinstein, R. A. (2008). Using internet searches for influenza surveillance. *Clinical Infectious Diseases, 47*(11), 1443–1448.

Reis, B. Y., & Brownstein, J. S. (2010). Measuring the impact of health policies using internet search patterns: The case of abortion. *BMC Public Health, 10*, 514.

Predicting the emergence of novel psychoactive substances **Chapter | 9 179**

Ripberger, J. T. (2001). Capturing curiosity: Using internet search trends to measure public attentiveness. *Policy Studies Journal, 39*(2), 239–259.

Seiffer, A., Schwarzwalder, A., Geis, K., & Aucott, J. (2010). The utility of Google Trends for epidemiological research: Lyme disease as an example. *Geospatial Health, 4*(2), 135–137.

Stevens-Davidowitz, S. (2012). The effects of racial animus on a Black presidential candidate: Using Google search data to find what surveys miss. *SSRN Journal*, 1–52.

Stogner, J., & Miller, B. (2015). The dabbing dilemma: A call for research on butane hash oil and other alternate forms of cannabis use. *Substance Abuse, 36*, 393–395.

Suhoy, T. (2009). *Query indices and a 2008 downturn: Israeli data.* Bank of Israel.

Trevisan, F. (2014). Search engines: From social science objects to academic inquiry tools. *First Monday, 19*(11).

Van Couvering, E. (2008). *The history of the Internet search engine: Navigational media and the traffic of commodity* (pp. 177–206). Web Search. Multidisciplinary Perspectives (Information Science and Knowledge Management).

Vandrey, R., Dun, K., Fry, J., & Girling, E. (2012). A survey study to characterize use of Spice products (synthetic cannabinoids). *Drug and Alcohol Dependence, 120*(1–3), 238–241.

Wilson, K., & Brownstein, J. S. (2009). Early detection of disease outbreaks using the Internet. *CMAJ, 180*(8), 829–831.

Zheluk, A., Quinn, C., Hercz, D., & Gillespie, J. A. (2013). Internet search patterns of human immunodeficiency virus and the digital divide in the Russian Federation: Infoveillance study. *Journal of Medical Internet Research, 15*(11), e256.

Zheluk, A., Quinn, C., & Meylakhs, P. (2014). Internet search and krokodil in the Russian Federation: An infoveillance study. *Journal of Medical Internet Research, 16*(9), e112.

Chapter 10

# Hippocampus segmentation in MR images: Multiatlas methods and deep learning methods

**Hancan Zhu[a], Shuai Wang[b], Liangqiong Qu[c], and Dinggang Shen[d,e,f]**

[a]*School of Mathematics Physics and Information, Shaoxing University, Shaoxing, China,* [b]*Imaging Biomarkers and Computer-Aided Diagnosis Laboratory, Radiology and Imaging Sciences, National Institutes of Health Clinical Center, Bethesda, MD, United States,* [c]*Department of Biomedical Data Science, Stanford University, Stanford, CA, United States,* [d]*School of Biomedical Engineering, ShanghaiTech University, Shanghai, China,* [e]*Shanghai United Imaging Intelligence Co., Ltd., Shanghai, China,* [f]*Department of Artificial Intelligence, Korea University, Seoul, Republic of Korea*

## 1 Introduction

The hippocampus is a structure in the brain of humans and other vertebrates, which presents in a deeply embedded pair into the temporal lobe of each cerebral cortex in humans and other mammals (Andersen, Morris, Amaral, Bliss, & O'Keefe, 2006), as shown in Fig. 1. As one of the key components of the limbic system, the hippocampus plays an important role in the consolidation of information from short-term to long-term memory, emotions, and learning (Anand & Dhikav, 2012). The hippocampus is one of the most thorough areas of the mammalian central nervous system ( Johnston & Amaral, 2004), since it is affected in many neuropsychiatric diseases. For example, a change will occur in the morphology of the hippocampus in diseases such as Alzheimer's disease (AD), depression, schizophrenia, hypertension, and Cushing's disease (Anand & Dhikav, 2012; Bast, 2011; Dhikav & Anand, 2007; Dhikav & Anand, 2011; Lanke, Moolamalla, Roy, & Vinod, 2018). This change in the hippocampus morphology, especially in AD, has been used as one of the main hallmarks (Zhao et al., 2020). Therefore the study of the hippocampus is of great significance in the early diagnosis, progression monitoring, and disease-modifying treatment evaluation of these neuropsychiatric diseases (Cummings, Ritter, & Zhong, 2018).

Magnetic Resonance Imaging (MRI) is currently the most used modality to monitor and access the morphological change of hippocampus in a noninvasive manner. In order to analyze the changes in the size and shape of the

*Big Data in Psychiatry and Neurology. https://doi.org/10.1016/B978-0-12-822884-5.00019-2*
Copyright © 2021 Elsevier Inc. All rights reserved.

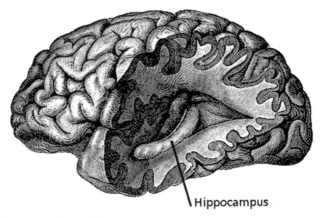

FIG. 1  Hippocampus (Gray, 1924).

hippocampus, the manual segmentation of the MR images is highly demanding, as it may need up to 2 h for an expert radiologist to identify the hippocampus slice by slice (Khlif et al., 2019). Furthermore, such manual segmentation is susceptible to both high intraoperator and interobserver variability. These bottlenecks restrict the application of this technique when dealing with larger exams. To this end, automatic methods that can accurately segment the hippocampus within a reasonable time are highly desired to alleviate the radiologists' manual efforts and remove the subjectivity in the segmentation.

Although the automatic and accurate segmentation of the hippocampus in MRI is greatly promising in the clinical applications, it still represents an open problem due to the following challenges. (1) The small structure size: compared with the whole brain structure, the hippocampus occupies only a very small part (Chen, Shi, Wang, Zhang, et al., 2017), which leads to a class imbalance problem when building the automatic model. (2) The irregular shape: despite the overall hippocampus shape resembling that of a sea horse, there are still large differences in the shape between individuals, which makes it hard to capture the shape prior information. (3) The blurred boundaries between the hippocampus and its surrounding structures, as shown in Fig. 2.

In order to overcome these challenges and alleviate the burden on radiologists, many studies were devoted to the segmentation of the hippocampus in MRI (Dill, Franco, & Pinho, 2015). The early works were initially based on semiautomatic methods, in which some spatial constraints were manually provided to guide the segmentation algorithm in the process initialization (Bartlett et al., 1994; Freeborough, Fox, & Kitney, 1997; Shen, Moffat, Resnick, & Davatzikos, 2002). For example, for most deformable model-based methods for the hippocampus segmentation, an initial shape prior is usually needed to guide the deformation process (Ghanei, Soltanian-Zadeh, & Windham, 1998; Pizer et al., 2003). These methods are *not only* sensitive to the initialization

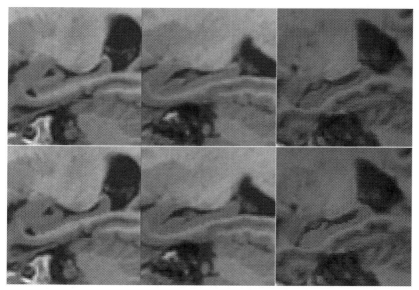

FIG. 2 MR brain images *(top row)* with hippocampus labels *(bottom row)*.

*but also* have poor application flexibility. The most commonly used automatic methods can be divided into two categories: (1) multiatlas-based methods and (2) learning-based methods.

The multiatlas-based methods have been successfully used to solve many medical image segmentation problems. These methods assume that the target image for segmentation has similar anatomical structures to those of the atlases (templates with manually labeled masks) and they can be transformed into each other by performing a spatial transformation. To segment the target image (Landman, Lyu, Huo, & Asman, 2020), the atlases are first registered onto the target image and the label maps of the atlases are propagated into the target image space based on the spatial transformations to segment the target image. Generally, there are three key steps for multiatlas-based segmentation methods: atlas selection, image registration, and label fusion (Zhu & He, 2020). For the atlas selection, a subset of all the atlases that are most similar to the target image is selected based on a predefined similarity metric, which aims to improve the robustness and also reduce the time cost of registration. For the image registration, the spatial transformation between the target image and each atlas is estimated, and the atlases and their corresponding label maps are aligned to the target image space. Since the anatomical structures usually have complex differences, the image registration is often time consuming if pursuing high accuracy. The last step, label fusion, is a core component to achieve the final segmentation results of the target image by fusing the propagated atlas labels. Since the existing multiatlas-based methods are typically characterized by their

**184** Big data in psychiatry and neurology

label fusion strategies, we will mainly focus on introducing the different label fusion methods by presenting the recent works in this field.

Unlike the multiatlas-based methods, the learning-based methods usually regard the segmentation problem as a classification or regression problem (Wang et al., 2019). The conventional learning-based methods make use of handcrafted features and train the classifier/regressor in a separate manner, which may result in a suboptimal performance. Moreover, due to the limited representation ability of the handcrafted features, the conventional learning-based methods are not suitable to address the challenging hippocampus segmentation task. Recently, deep learning has achieved leading performance in many medical image segmentation tasks and become widely used for hippocampus segmentation (Shen, Wu, & Suk, 2017). The hierarchical feature learning from the data and end-to-end training manner achieve a significantly better performance compared with the conventional learning methods. Among the various variants of deep learning architectures, fully convolutional networks (FCNs) are more efficient and can make a dense prediction of the whole image. However, deep learning-based methods also have their bottleneck in the high reliance on large training datasets, which are usually unavailable for the MRI volumes of the hippocampus. Therefore some pretrained and partially trained approaches were designed, which achieved an outstanding performance.

Within this context, this chapter mainly introduces the evolution of the multiatlas-based methods and deep learning-based methods for the hippocampus segmentation in MRI in Section 2 and Section 3, respectively. For multiatlas-based methods, several label fusion strategies combined with relevant works are presented in detail, and for deep learning-based methods, two different approaches using deep learning are introduced. Section 4 includes a summary of these methods and presents a conclusion proposing future research directions in this area.

## 2 Patch-based multiatlas labeling for Hippocampus segmentation

The multiatlas segmentation method first selects the most similar atlas images to the target image and then registers them to the target image, propagates the corresponding segmentation labels to the target space, and finally fuses the warped labels to get the estimated segmentation. Fig. 3 shows a flowchart for hippocampus segmentation with multiatlas-based methods.

Let $I$ be the target image and $\tilde{A}_i = \left( \tilde{I}_i, \tilde{L}_i \right), i = 1, 2, \ldots, \tilde{N}$ represent $\tilde{N}$ atlases, where $\tilde{I}_i$ is the $i$th image and $\tilde{L}_i$ is its segmentation label map with 1 indicating hippocampus and 0 indicating background. First, we align all the images to a template space using the linear image registration with affine transformation and identify a bounding box for both the left and right hippocampi to cover the hippocampus of the unseen target image. In particular, we scan all the atlases to find the minimum and maximum $x, y, z$ positions of the hippocampus

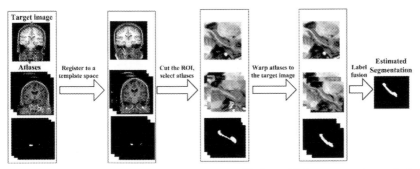

FIG. 3 The flowchart for segmenting hippocampus with the multiatlas segmentation methods (Zhu, Cheng, Yang, Fan, & Initiative, 2017).

and add several voxels in each direction to cover the hippocampus of the unseen testing images (Hao et al., 2014; Zhu et al., 2017). For each target image, we select the $N$ most similar atlases based on the normalized mutual information (NMI) between the target image and the atlas images within the bounding box. After the atlas selection, we register each atlas image to the target image using a nonlinear image registration algorithm, which results in $N$ warped atlases $A_i = (I_i, L_i)$, $i = 1, 2, \ldots, N$. Finally, we compute the label of each target voxel using the warped atlases, referred to as the label fusion, which is the core of the multiatlas segmentation. The following subsections introduce several patch-based label fusion methods.

## 2.1 Weighted voting label fusion

For the weighted voting label fusion methods, the label of the target voxel $x$ can be computed as follows:

$$L(x) = \frac{\sum_{i=1}^{N} \sum_{y \in \Omega} w_i(x, y) L_i(y)}{\sum_{i=1}^{N} \sum_{y \in \Omega} w_i(x, y)}, \quad (1)$$

where $w_i(x, y)$ is the weight between the target voxel $x$ and the selected voxel $y$ in the $i$th atlas image, and $\Omega$ denotes the searching region centered at the target voxel $x$. Based on the setting of $\Omega$, the weighted voting label fusion methods can be local and nonlocal.

Local weighted voting label fusion methods select only the corresponding voxel in each atlas, i.e., the searching region $\Omega$ in Eq. (1) only contains one voxel with the same location as the target voxel $x$. These methods use different similarity functions to compute the weight, including the inverse function (LW-INV) (Artaechevarria, Munoz-Barrutia, & Ortiz-de-Solorzano, 2009) and

**186** Big data in psychiatry and neurology

Gaussian function (LW-GU) (Sabuncu, Yeo, Van Leemput, Fischl, & Golland, 2010). The simplest local weighted voting label fusion method is the majority voting label fusion method (MV) which uses a constant value for all the weighting coefficients (Heckemann, Hajnal, Aljabar, Rueckert, & Hammers, 2006; Rohlfing, Brandt, Menzel, & Maurer Jr, 2004).

The nonlocal weighted voting label fusion methods extract the training samples in the searching region $\Omega$ with size $(2r_s + 1) \times (2r_s + 1) \times (2r_s + 1)$ in each atlas image and construct a training library of the image patches $D = [p_1, p_2, ..., p_n]$, where $n = N \cdot (2r_s + 1)^3$ denotes the number of selected patches (Coupé et al., 2011; Rousseau, Habas, & Studholme, 2011). The segmentation label of the center voxel of each image patch is used as the label of this patch denoted by $l_i$, $i = 1, 2, ..., n$. Thus a training dataset can be constructed as $\Delta = \{(p_i, l_i) | i = 1, 2, ..., n\}$, where $p_i$ is the $i$th image patch in the patch library $D$ and $l_i$ is the label of its center voxel. The voting weights are computed according to the similarity between the target patch and the patches in the patch library using a similarity function. For example, the Gaussian function is used to compute the weights (NLW-GU) (Coupé et al., 2011; Rousseau et al., 2011) as follows:

$$w_i(x, y) = e^{\frac{-\|p_t - p_i\|_2^2}{h}},$$

where $p_t$ is a vectorized target patch centered at $x$, $p_i$ is the $i$th patch in the patch library, and $h$ is a parameter.

Two types of methods were proposed under the framework of nonlocal weighted voting label fusion methods: (1) the sparse representation label fusion methods and (2) the joint label fusion methods (Liao, Gao, Lian, & Shen, 2013; Wang et al., 2013). The sparse representation label fusion methods obtain the voting weights by solving the following equation (Liao et al., 2013):

$$argmin_w \frac{1}{2} \|Dw - p_t\|_2^2 + \lambda \|w\|_1, \tag{2}$$

where $D$ denotes the patch library, $p_t$ is the target patch, and $\lambda$ is a parameter.

On the other hand, the joint label fusion method (JLF) gets the voting weights from the following optimization problem (Wang et al., 2013):

$$argmin_w \vec{w}^T M \vec{w}, \ s.t. \sum w_i = 1, \tag{3}$$

where $M$ denotes the dependency matrix, given by $M_{i,j} = [|p_t - p_i|^T |p_t - p_j|]^\beta$. The joint label fusion method jointly considers the atlases by explicitly modeling the probability of two atlases making a segmentation error (Wang et al., 2013).

In fact, most weighted voting label fusion methods can be written as Eq. (3). For the sparse representation-based label fusion method, letting the representation coefficients $w$ satisfy $\sum w_i = 1$, we have

$$p_t = \sum w_i p_t = [p_t, p_t, ..., p_t] w \triangleq P_t w,$$

where $P_t \triangleq [p_t, p_t, \ldots, p_t]$. Then, the representation term in the sparse representation method Eq. (2) can be rewritten as follows:

$$\|p_t - Dw\|_2^2 = \|P_t w - Dw\|_2^2 = \|(P_t - D)w\|_2^2 = w^T (P_t - D)^T (P_t - D)w \triangleq w^T \widetilde{M} w,$$

where $\widetilde{M} \triangleq (P_t - D)^T (P_t - D)$. Compared with Eq. (3), the representation term has the same form as the joint label fusion method with a similar dependency matrix $\widetilde{M}$.

Using different types of the dependency matrix, other weighted voting label fusion methods can also be rewritten as Eq. (3), i.e., identity matrix for the majority voting method or diagonal matrix for the local or nonlocal weighted voting label fusion methods that use a similarity function to compute the voting weights. Thus the following unified formula can be used in the weighted voting label fusion methods (Zhu & He, 2020):

$$argmin_w \, w^T \Phi w + R(w), \, s.\,t. \sum w_i = 1,$$

where $\Phi$ denotes the dependency matrix, and $R(w)$ is a regularization term. The dependency matrix $\Phi$ should be a symmetrical positive definite matrix. Under this unified formula, we argue that a weighted voting label fusion method jointly considers the pair-wise dependency between the atlases when its dependency matrix is not a diagonal one.

## 2.2 Local learning-based label fusion

In local learning-based label fusion (LLL) methods (Bai, Shi, Ledig, & Rueckert, 2015; Hao et al., 2014; Zhu, Tang, Cheng, Wu, & Fan, 2019), a classifier is built using the training dataset $\Delta$ ($l_i = -1$ for indicating background in this subsection), which is then used to classify the target voxel. For example, the L1-regularized support vector machine (SVM) classifier is trained by solving an optimization problem (Hao et al., 2014) as follows:

$$\min_{\vec{w}} \left\| \vec{w} \right\|_1 + C \sum_i \left( \max \left( 0, \, 1 - l_i \vec{w}^T \vec{f}_i \right) \right)^2, \tag{4}$$

where $\|\bullet\|_1$ denotes the L1 norm, and $C$ is a parameter.

Unlike the weighted voting label fusion methods, in LLL methods, a set of image features is extracted to capture the texture information, instead of directly using image patches. In the work of Hao et al. (2014), several types of image features are extracted for each voxel, including the intensities in its neighborhood (image patch), the first-order difference filters (FODs), the second-order difference filters (SODs), 3D Hyperplane filters, 3D Sobel filters, Laplacian filters, and Range difference filters (Toriwaki & Yoshida, 2009). All these features are concatenated to form a feature vector $\vec{f}_i$.

**188** Big data in psychiatry and neurology

After obtaining $\vec{w}$ by optimizing Eq. (4), the label of the target image voxel $x$ can be computed as follows:

$$L(x) = \text{sgn}\left(\vec{w}^T \vec{f}\right).$$

## 2.3 Supervised metric learning for label fusion

Learning a distance/similarity metric from the given training samples is an important machine learning topic, with many proposed methods (Xing, Jordan, Russell, & Ng, 2002). Among them, learning a Mahalanobis distance metric for the k-nearest neighbor classification has received considerable research interest and been successfully applied to many computer vision problems (Guillaumin, Verbeek, & Schmid, 2009; Wang, Zuo, Zhang, Meng, & Zhang, 2015). In the work of Zhu et al. (2017), a nonlocal weighted voting with metric learning method (NLW-ML) was proposed, which adopted a supervised metric learning method to learn a Mahalanobis distance metric from the training dataset and was used for the label fusion.

Given any two samples of image patches and their labels $(p_i, l_i)$ and $(p_j, l_j)$ from the training dataset $\Delta$, we obtain a doublet $(p_i, p_j)$ with a label $h$, where $h = -1$ if $l_i = l_j$, and $h = 1$ otherwise. For each training sample $p_i$, we find its $m_1$ nearest similar neighbors and $m_2$ nearest dissimilar neighbors, such that $m_1 = \{p_{i,1}^s, ..., p_{i,m_1}^s\}$ and $m_1 = \{p_{i,1}^d, ..., p_{i,m_2}^d\}$, and construct the $(m_1 + m_2)$ doublets:

$$\left\{ \left(p_i, p_{i,1}^s\right), ..., \left(p_i, p_{i,m_1}^s\right), \left(p_i, p_{i,1}^d\right), ..., \left(p_i, p_{i,m_2}^d\right) \right\}.$$

The collection of all the possible doublets builds a doublet set, denoted by $\{z_1, ..., z_{N_d}\}$, where $z_j = (p_{j,1}, p_{j,2}), j = 1, 2, ..., N_d$ and the label of $z_j$ is denoted by $h_j$. Given the doublet set $\{z_1, ..., z_{N_d}\}$, we use a kernel method to learn a classifier as follows:

$$g(z) = \text{sgn}\left(\sum_j h_j \alpha_j K(z_j, z) + b\right),$$

where $z_j$ is the $j$th doublet, $h_j$ is its label, $z = (p_{k_1}, p_{k_2})$ is a testing doublet, and $K(\bullet, \bullet)$ is a degree-2 polynomial kernel, which is defined as:

$$K(z_i, z_j) = \left[(p_{i,1} - p_{i,2})^T (p_{j,1} - p_{j,2})\right]^2.$$

Then, we have

$$\sum_j h_j \alpha_j K(z_j, z) + b = (p_{k_1} - p_{k_2})^T M(p_{k_1} - p_{k_2}) + b,$$

Hippocampus segmentation in MR images Chapter | 10 **189**

where $M = \sum_j h_j \alpha_j (p_{j,1} - p_{j,2})(p_{j,1} - p_{j,2})^T$ is the matrix to be learned in the Mahalanobis distance metric. Once we obtain $M$, the kernel decision function $g(z)$ can be used to determine whether the patches $p_{k_1}$ and $p_{k_2}$ are similar or dissimilar to each other.

In order to learn $M$ in the Mahalanobis metric, an SVM model is adopted:

$$\min_{M,b,\xi} \frac{1}{2}\|M\|_F^2 + C\sum_j \xi_j, \tag{5}$$

*s. t.* $h_j((p_{j,1} - p_{j,2})^T M(p_{j,1} - p_{j,2}) + b) \geq 1 - \xi_j, \; \xi_j \geq 0, \; \forall \, j,$

where $\|\bullet\|_F$ is the Frobenius norm. The Lagrange dual problem of the earlier doublet-SVM model is:

$$\max_\alpha -\frac{1}{2}\sum_{i,j} \alpha_i \alpha_j h_i h_j K(z_i, z_j) + \sum_i \alpha_i, \tag{6}$$

*s. t.* $0 \leq \alpha_l \leq C, \; \forall \, l,$

$$\sum_l \alpha_l h_l = 0,$$

which can be solved by SVM solvers, such as LibSVM (Chang & Lin, 2011).

To ensure that $M$ is a positive semidefinite matrix, we compute a singular value decomposition of $M = U\Lambda V$ and preserve only the positive singular values in $\Lambda$ to form another diagonal matrix $\Lambda_+$. Then, we use $M_+ = U\Lambda_+ V$.

Using the learned Mahalanobis distance metric $M$, we obtain a new metric space by introducing a norm $\|\bullet\|_M$: $\|x\|_M = \sqrt{x^T M x}$. The distance between two samples is defined as $d(x, y) = \|x - y\|_M$.

Given the target patch $p_x$ and the training patches $p_i, i = 1, 2, ..., n$, we compute their distances as follows:

$$d_i = d(p_x, p_i) = \sqrt{(p_x - p_i)^T M(p_x - p_i)}, \; i = 1, 2, ..., n.$$

According to these distances, we select the $k$ nearest training samples $\{(p_{s_j}, l_{s_j}) | j = 1, 2, ..., k\}$ to form the nearest neighborhood set $\mathcal{N}_k(p_x)$ and set their similarity weights to one while setting the others to zero:

$$w(p_x, p_i) = \begin{cases} 1, & p_i \in \mathcal{N}_k(p_x) \\ 0, & p_i \notin \mathcal{N}_k(p_x) \end{cases}.$$

Then, the weighted voting formula,

$$\hat{L}(x) = \frac{\sum\limits_{i=1}^{n} w(p_x, p_i) l_i}{\sum\limits_{i=1}^{n} w(p_x, p_i)},$$

**190** Big data in psychiatry and neurology

is used to compute the label of the target voxel. Finally, the estimated label of $\hat{L}(x)$ is thresholded to obtain a hard labeling as follows:

$$L(x) = \begin{cases} 1, & \hat{L}(x) > 0.5 \\ 0, & \hat{L}(x) < 0.5 \end{cases}.$$

## 2.4 An evaluation of different patch-based multiatlas labeling methods

In this subsection, we present an evaluation of different patch-based multiatlas labeling methods, including the methods of MV (Heckemann et al., 2006; Rohlfing et al., 2004), LW-INV (Artaechevarria et al., 2009), LW-GU (Sabuncu et al., 2010), NLW-GU (Coupé et al., 2011; Rousseau et al., 2011), LLL (Hao et al., 2014), JLF (Wang et al., 2013), and NLW-ML (Zhu et al., 2017). These algorithms were validated for hippocampus segmentation using the first release of the EADC-ADNI dataset, consisting of MRI scans and their corresponding hippocampus labels of 100 subjects (available at www. hippocampal-protocol.net). These images were obtained from the Alzheimer's Disease Neuroimaging Initiative (ADNI) database (adni.loni.usc.edu/), and the subjects were grouped into three diagnostic groups, including the normal controls (NC), mild cognitive impairment patients (MCI), and patients with Alzheimer's disease (AD).

The ADNI database (Principal Investigator: Michael W. Weiner, MD, VA Medical Center and University of California-San Francisco) is the result of efforts of many coinvestigators from a broad range of academic institutions and private corporations. The subjects have been recruited from over 50 sites across the U.S. and Canada. The initial goal of ADNI was to recruit 800 adults, aged 55 to 90, to participate in the research, approximately 200 cognitively normal older individuals for a 3-year follow-up, 400 people with MCI for a 3-year follow-up, and 200 people with early AD for a 2-year follow-up. For up-to-date information, see www.adni-info.org.

Each MRI scan was manually labeled according to a harmonized protocol (Boccardi et al., 2015). All the images were processed using a standard preprocessing protocol, including the alignment along the line passing through the anterior and posterior commissures of the brain (AC-PC line) and bias field correction, and they were warped into the MNI152 template space using the linear image registration with affine transformation. We randomly selected 40 subjects as the training set, used to optimize the parameters of the algorithms, and 60 other subjects as the testing set, used to evaluate the segmentation performance. The clinical scores and demographic information of these subjects are summarized in Table 1.

The parameters of all the methods were optimized based on the training dataset according to 40 leave-one-out cross-validation experiments with different combinations of the parameters. For LW-GU (Sabuncu et al., 2010), we

## TABLE 1 Demographic data and clinical scores of the subjects.

| Group | NC | MCI | AD |
|---|---|---|---|
| Subject size | 29 | 34 | 37 |
| Age (years): mean ± std | 75.79 ± 6.73 | 74.23 ± 7.67 | 73.93 ± 8.18 |
| Males/Females | 16/13 | 20/14 | 20/17 |
| MMSE: mean ± std | 28.92 ± 1.02 | 26.59 ± 2.72 | 21.81 ± 4.09 |

need to determine the patch radius $r_p$ and $\sigma_x$ in the Gaussian similarity metric. Using cross-validation, the optimal value of $r_p$ was determined to be 2 selected from $\{1,2,3\}$, and $\sigma_x$ was adaptively set as $\sigma_x = min_{x_i} \{ \|P(x) - P(x_i)\|_2 + \varepsilon \}$, $i = 1, ..., N$, where $\varepsilon$ is a small constant to ensure the numerical stability with a value of $1e\text{-}20$. LW-INV (Artaechevarria et al., 2009) has 2 parameters: the patch radius $r_p$ and $\gamma$ in the inverse function model. The optimal values were determined to be $r_p = 2$ and $\gamma = -3$, obtained from the ranges of $\{1, 2, 3\}$ and $\{-0.5, -1, -2, -3\}$, respectively. NLW-GU (Coupé et al., 2011; Rousseau et al., 2011) has 3 parameters: the searching radius $r_s$, patch radius $r_p$, and $\sigma_x$ in the Gaussian similarity metric model. The searching radius $r_s$ was set to 1, since a nonlinear image registration algorithm was used to warp the atlas images to the target image. With cross-validation, the optimal value of $r_p$ was determined to be 1 selected from $\{1, 2, 3\}$, and $\sigma_x$ was adaptively set as $\sigma_x = min_{x_{s,j}} \{ \|P(x) - P(x_{s,j})\|_2 + \varepsilon \}, s = 1, ....N, j \in V$, where $\varepsilon$ is a small constant to ensure the numerical stability with a value of $1e\text{-}20$.

The only difference between the methods of NLW-GU (Coupé et al., 2011; Rousseau et al., 2011) and LW-GU (Sabuncu et al., 2010) is the used image patches. Particularly, NLW-GU (Coupé et al., 2011; Rousseau et al., 2011) uses nonlocal image patches, i.e., the searching radius $r_s > 0$ is used to obtain the image patches. In contrast, LW-GU (Sabuncu et al., 2010) uses local image patches, i.e., the used searching radius $r_s = 0$. Since both of the NLW-GU (Coupé et al., 2011; Rousseau et al., 2011) and NLW-ML (Zhu et al., 2017) methods use nonlocal image patches, the only difference between them is the distance metric to measure the similarity between the image patches. The conducted experiment revealed that the multipoint estimate strategy was better than the single-point strategy in all these label fusion methods. Thus the reported results are only those obtained with the multipoint strategy.

Similar to NLW-GU (Coupé et al., 2011; Rousseau et al., 2011), the searching radius $r_s$ for LLL (Hao et al., 2014) and JFL (Wang et al., 2013) was set to be 1. The other parameters of these two methods were optimized based on the same training set with the same parameter selection as the proposed method. For LLL (Hao et al., 2014), the patch radius $r_p$ and the number of training samples $K$ need

**192** Big data in psychiatry and neurology

to be determined. With cross-validation, the optimal parameters were determined to be $r_p = 3$ and $K = 300$, selected from $\{1, 2, 3\}$ and $\{300, 400, 500\}$, respectively. In addition, sparse linear SVM classifiers with default parameter ($C = 1$) were built to fuse the labels, and the single-point label fusion strategy was used (Hao et al., 2014). For JLF, the patch radius $r_p$ and the parameter $\beta$ in the pairwise joint label difference term need to be determined. With cross-validation, the optimal parameters were found to be $r_p = 1$ and $\beta = 1$, selected from $\{1, 2, 3\}$ and $\{0.5, 1, 1.5, 2\}$, respectively.

The method of NLW-ML (Zhu et al., 2017) has the following parameters: the patch radius $r_p$, searching radius $r_s$, regularization parameter $C$ in SVM, the numbers of the nearest similar and dissimilar neighbors $m_1, m_2$ to construct the doublets, and the number of the nearest neighbors $k$ to select the most similar samples for the label fusion. According to (Wang et al., 2015), the values were set as follows: $C = 1$, $m_1 = m_2 = 1$. The searching radius was fixed at $r_s = 1$. The other two parameters, $r_p$ and $k$, were empirically determined from $\{1, 2, 3\}$ and $\{3, 9, 27\}$, respectively. The results indicated that the optimal segmentation performance could be obtained with $r_p = 1$ and $k = 9$.

Nine segmentation evaluation measures were adopted to evaluate the image segmentation results (Jafari-Khouzani, Elisevich, Patel, & Soltanian-Zadeh, 2011). Letting $A$ be the manual segmentation, $B$ be the automated segmentation, and $V(X)$ denote the volume of the segmentation result $X$, these evaluation measures can then be defined as follows:

**(1)** $\text{Dice} = 2\frac{V(A \cap B)}{V(A) + V(B)}$,

**(2)** $\text{Jaccard} = \frac{V(A \cap B)}{V(A \cup B)}$,

**(3)** $\text{Precision} = \frac{V(A \cap B)}{V(B)}$,

**(4)** $\text{Recall} = \frac{V(A \cap B)}{V(A)}$,

**(5)** $\text{MD} = \text{mean}_{e \in \partial A}(min_{f \in \partial B} d(e, f))$,

**(6)** $\text{HD} = \max(\text{H}(A, B), \text{H}(B, A))$, where $\text{H}(A, B) = max_{e \in \partial A}(min_{f \in \partial B} d(e, f))$,

**(7)** $\text{ASSD} = \frac{\left(\text{mean}_{e \in \partial A}\left(min_{f \in \partial B} d(e, f)\right) + \text{mean}_{e \in \partial B}\left(min_{f \in \partial A} d(e, f)\right)\right)}{2}$,

**(8)** $\text{RMSD} = \frac{\sqrt{D_A^2 + D_B^2}}{\text{card}\{\partial A\} + \text{card}\{\partial B\}}$, where $D_A^2 = \sum_{e \in \partial A}\left(min_{f \in \partial B} d^2(e, f)\right)$,

$D_B^2 = \sum_{e \in \partial B}(min_{f \in \partial A} d^2(e, f))$,

**(9)** HD95: similar to HD, except that 5% of the data points with the largest distance are removed before calculation.

In the above-mentioned definitions, $\partial A$ denotes the boundary voxels of $A$, $d(\bullet, \bullet)$ is the Euclidian distance between two points, and $\text{card}\{\bullet\}$ is the cardinality of a set.

Table 2 summarizes 9 index values (mean $\pm$ std) of the segmentation results obtained on the testing images using the segmentation methods under

**TABLE 2** Nine index values (mean ± std) of the segmentation results using different label fusion methods (The best mean index values are shown in bold).

| | | MV | LW-INV | LW-GU | NLW-GU | LLL | JLF | NLW-ML |
|---|---|---|---|---|---|---|---|---|
| *Dice* | *L* | *0.856 ± 0.031* | *0.868 ± 0.026* | *0.868 ± 0.025* | *0.877 ± 0.028* | *0.878 ± 0.025* | *0.880 ± 0.024* | *0.881 ± 0.026* |
| | **R** | **0.860 ± 0.033** | **0.872 ± 0.025** | **0.872 ± 0.025** | **0.881 ± 0.026** | **0.882 ± 0.024** | **0.884 ± 0.023** | 0.885 ± 0.024 |
| Jaccard | L | 0.750 ± 0.047 | 0.767 ± 0.039 | 0.768 ± 0.038 | 0.782 ± 0.043 | 0.784 ± 0.039 | 0.786 ± 0.037 | **0.788 ± 0.040** |
| | R | 0.755 ± 0.048 | 0.775 ± 0.038 | 0.774 ± 0.037 | 0.788 ± 0.040 | 0.790 ± 0.038 | 0.794 ± 0.036 | **0.795 ± 0.037** |
| Precision | L | 0.861 ± 0.048 | 0.873 ± 0.035 | 0.873 ± 0.033 | 0.878 ± 0.036 | 0.879 ± 0.035 | 0.879 ± 0.032 | **0.880 ± 0.035** |
| | R | 0.864 ± 0.052 | 0.876 ± 0.038 | 0.875 ± 0.037 | **0.885 ± 0.039** | 0.883 ± 0.038 | 0.882 ± 0.036 | 0.884 ± 0.037 |
| Recall | L | 0.854 ± 0.049 | 0.865 ± 0.040 | 0.865 ± 0.039 | 0.878 ± 0.045 | 0.880 ± 0.040 | 0.882 ± 0.036 | **0.884 ± 0.039** |
| | R | 0.859 ± 0.044 | 0.871 ± 0.032 | 0.871 ± 0.031 | 0.879 ± 0.038 | 0.883 ± 0.034 | **0.889 ± 0.029** | **0.889 ± 0.032** |
| HD | L | 3.157 ± 0.853 | 3.086 ± 0.862 | **3.038 ± 0.855** | 3.205 ± 0.909 | 3.057 ± 0.906 | 3.076 ± 0.784 | 3.069 ± 0.831 |
| | R | 3.255 ± 0.894 | 3.019 ± 0.877 | 3.038 ± 0.867 | 3.215 ± 0.902 | **3.005 ± 0.703** | 3.227 ± 1.100 | 3.238 ± 1.095 |
| HD95 | L | 1.345 ± 0.478 | 1.178 ± 0.465 | 1.145 ± 0.419 | 1.222 ± 0.476 | 1.145 ± 0.375 | **1.093 ± 0.352** | 1.114 ± 0.381 |
| | R | 1.332 ± 0.441 | 1.163 ± 0.340 | 1.141 ± 0.287 | 1.246 ± 0.386 | 1.163 ± 0.256 | **1.101 ± 0.237** | 1.177 ± 0.289 |
| MD | L | 0.284 ± 0.054 | 0.261 ± 0.037 | 0.263 ± 0.037 | 0.239 ± 0.046 | 0.241 ± 0.042 | 0.252 ± 0.048 | **0.238 ± 0.045** |
| | R | 0.278 ± 0.063 | 0.253 ± 0.047 | 0.257 ± 0.048 | 0.227 ± 0.050 | 0.230 ± 0.047 | 0.237 ± 0.051 | **0.226 ± 0.048** |
| ASSD | L | 0.334 ± 0.077 | 0.294 ± 0.060 | 0.290 ± 0.056 | 0.287 ± 0.069 | 0.278 ± 0.055 | **0.265 ± 0.052** | 0.270 ± 0.059 |
| | R | 0.328 ± 0.071 | 0.285 ± 0.048 | 0.284 ± 0.047 | 0.279 ± 0.056 | 0.273 ± 0.044 | **0.260 ± 0.043** | 0.265 ± 0.048 |
| RMSD | L | 0.632 ± 0.123 | 0.582 ± 0.106 | 0.577 ± 0.099 | 0.581 ± 0.119 | 0.563 ± 0.094 | **0.551 ± 0.090** | 0.556 ± 0.103 |
| | R | 0.628 ± 0.110 | 0.572 ± 0.078 | 0.570 ± 0.074 | 0.576 ± 0.095 | 0.557 ± 0.064 | **0.550 ± 0.074** | 0.556 ± 0.080 |

comparison. For each index, the best value is shown in bold. These results indicate that the NLW-ML method proposed by Zhu et al. (2017) achieves the best overall performance. Specifically, the Wilcoxon signed-rank test indicates that NLW-ML (Zhu et al., 2017) performs significantly better than MV (Heckemann et al., 2006; Rohlfing et al., 2004), LW-INV (Artaechevarria et al., 2009), LW-GU (Sabuncu et al., 2010), NLW-GU (Coupé et al., 2011; Rousseau et al., 2011), LLL (Hao et al., 2014) ($p < 0.001$), and JLF (Wang et al., 2013) ($p < 0.05$) in terms of the Dice and Jaccard index values of the segmentation results. The results also demonstrate that the method of NLW-GU (Coupé et al., 2011; Rousseau et al., 2011) performs better than LW-GU (Sabuncu et al., 2010), indicating that the nonlocal weighted voting methods achieve better performance than the local weighted voting methods, which only adopt the corresponding image patches for label fusion (Coupé et al., 2011; Rousseau et al., 2011).

Fig. 4 shows box plots of the Dice and Jaccard indices values of the segmentation results obtained by different methods, indicating that NLW-ML (Zhu et al., 2017) performs consistently better than the other label fusion methods. The superior performance of NLW-ML is also confirmed by the visualization results obtained using different methods, as shown in Fig. 5.

## 3 Deep learning-based methods for Hippocampus segmentation

In recent years, deep neural network (DNN) models have achieved remarkable progress in the medical image segmentation tasks including the segmentation of the hippocampus (Cao et al., 2018; Carmo, Silva, Yasuda, Rittner, & Lotufo, 2020; Chen, Shi, Wang, Sun, et al., 2017; Dinsdale, Jenkinson, & Namburete, 2019; Kim, Wu, & Shen, 2013; Ronneberger, Fischer, & Brox, 2015;

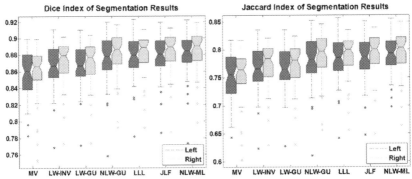

**FIG. 4** Comparison of different methods for the segmentation of the left hippocampus and the right hippocampus with respect to the Dice and Jaccard indices. In each box, the central mark is the median, and the edges are the 25th and 75th percentiles (Zhu et al., 2017).

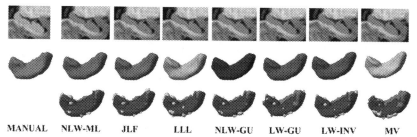

**FIG. 5** The hippocampal segmentation results obtained by different methods tested on one subject that was randomly chosen from the dataset. The first row shows the segmentation results produced by different methods, the second row shows their corresponding surface rendering results, and the difference between the results of manual and automatic segmentation methods is shown in the third row (Zhu et al., 2017).

Roy, Conjeti, Navab, Wachinger, & Initiative, 2019; Wang et al., 2020; Yu & Koltun, 2015; Zhu, Adeli, Shi, & Shen, 2020; ZHU et al., 2019). The research on hippocampus segmentation using deep learning techniques mainly focuses on exploring advanced overall methodologies (Dinsdale et al., 2019; Kim et al., 2013; Wu et al., 2018; Zhu et al., 2020), advanced network architectures (Carmo et al., 2020; Chen, Shi, Wang, Zhang, et al., 2017; Roy et al., 2019; Zhu, Shi, et al., 2019), data augmentation strategies (Roy et al., 2019; Thyreau, Sato, Fukuda, & Taki, 2018), or multitask models (Cao et al., 2018).

The deep learning-based methods for hippocampus segmentation can be roughly classified into two categories according to the applied methodologies: multiatlas-based deep learning methods and fully automatic end-to-end deep learning methods. The multiatlas-based deep learning methods apply the deep learning techniques to the traditional multiatlas labeling fusion tasks (Fang et al., 2019; Kim et al., 2013; Yang, Sun, Li, Wang, & Xu, 2018), with the advantage of being independent from the manual feature extraction schemes. For example, Kim et al. (2013) integrated a two-layer stacked convolutional Independent Subspace Analysis (ISA) network into a multiatlas-based segmentation framework for hippocampus segmentation in 7.0 T magnetic resonance (MR) images. Instead of using handcrafted features, the ISA network helps to automatically extract the hierarchical feature representation from the images. Yang and colleagues (Yang et al., 2018) formulated the multiatlas segmentation in a deep learning framework, by integrating the feature extraction and the non-local patch-based label fusion in a single deep network architecture, while Fang et al. (2019) introduced a multiatlas-guided FCN by incorporating the atlas information within the network learning process.

Unlike the multiatlas-based deep learning methods that still require the users to provide the atlas, the fully automatic end-to-end deep learning methods take a subject as the input and directly apply a DNN to learn the mapping function between the subject and its segmentation maps. For example, Chen, Shi, Wang, Sun, et al. (2017) stacked a 2D U-net with a long short-term memory (LSTM) to promote the interslice consistency in hippocampus segmentation. Roy et al. (2019) introduced QuickNAT, a multiview 2D U-shape network for hippocampus segmentation. Note that QuickNAT is composed of three 2D U-shape subnetworks, respectively, operating on coronal, axial, and sagittal views, and the results from these three views are then aggregated.

In the following subsections, a multiatlas-based deep learning confidence estimation method for hippocampus segmentation is presented. This method detects the potential errors in the warped atlas labels (Zhu et al., 2020), which are then corrected, and two label fusion schemes are used to fuse the corrected labels to obtain the final segmentation. Next, a fully automatic and end-to-end deep learning method for hippocampus segmentation is presented, in which a dilated dense network is embedded into the residual U-net. With each type of methods, we also present an evaluation to demonstrating their effectiveness.

## 3.1 Multiatlas-based deep learning method for hippocampus segmentation

Image registration is one of the core steps in the multiatlas-based methods (described in Section 2), since the registration accuracy has a direct effect on the final labeling performance. However, due to the intersubject anatomical variations, registration errors are inevitable. This subsection describes the deep learning-based confidence estimation method that was recently proposed by Zhu et al. (2020), to alleviate the potential effects of the registration errors. In this work, an FCN with residual connections was proposed to learn the relationship between the image patch pair (i.e., patches from the target subject and the atlas) and the related label confidence patch. With the obtained label confidence patch, the potential errors in the warped atlas labels can be identified and corrected. Then, two label fusion methods were used to fuse the corrected atlas labels. Fig. 6 shows the general framework of the proposed method.

### 3.1.1 FCN-based confidence estimation

Given an image patch $p_t$ in the target image and the corresponding image patch $p_a$ in a warped atlas image, an FCN (Long, Shelhamer, & Darrell, 2015; Ronneberger et al., 2015) is proposed to model the relationship $f(p_t, p_a)$ between the image patch pair $(p_t, p_a)$ and the label confidence $C$ (see Fig. 7),

$$C = f(p_t, p_a), \tag{7}$$

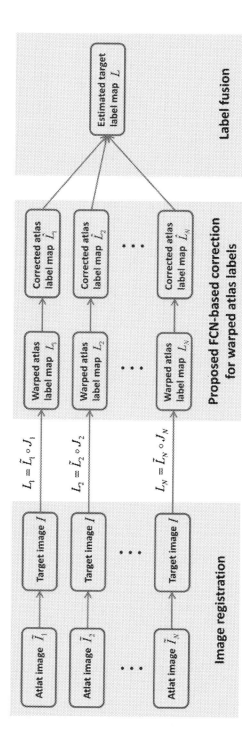

**FIG. 6** The general framework of the multiatlas-based deep learning method. $\tilde{L}_i$ is the $i$th atlas label map, and $J_i$ is the deformation field obtained by registering the $i$th atlas image to the target image (Zhu et al., 2020).

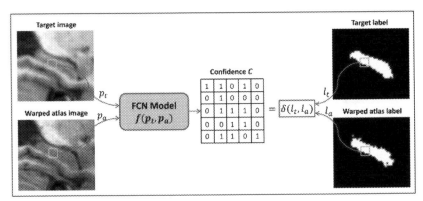

FIG. 7  An illustration of the FCN-based confidence learning (Zhu et al., 2020).

where $C$ is a patch with the same size as $p_t$ and $p_a$, which indicates whether $p_t$ and $p_a$ have the same segmentation label as follows:

$$C(x) = \delta(l_t(x), l_a(x)) = \begin{cases} 1, & \text{if } l_t(x) = l_a(x), \\ 0, & \text{if } l_t(x) \neq l_a(x), \end{cases} \quad (8)$$

where $l_t(x)$ denotes the label of the target image voxel $p_t(x)$, and $l_a(x)$ denotes the label of the warped atlas image voxel $p_a(x)$; hence, $C$ can serve as the confidence map.

Fig. 8 shows the structure of the FCN model, termed as ResUNet. Similar to U-net, ResUNet consists of encoder and decoder paths. The encoder path contains a $3 \times 3 \times 3$ convolution layer, two $2 \times 2 \times 2$ max-pooling operations with a stride of 2, and three residual blocks. Correspondingly, the decoder path contains three residual blocks, two $4 \times 4 \times 4$ deconvolution layers with a stride of 2, and one $1 \times 1 \times 1$ convolution layer. Each $3 \times 3 \times 3$ convolution is followed by a batch normalization and a rectified linear unit (ReLU). In order to retain the spatial and localization details in the decoder pathway, padded convolution layers are used in the network. The feature maps in the encoder path are connected to the corresponding features in the decoder one through element-wise summation. These skip connections concatenate the features of the same level in the encoder/decoder path, which helps to recover the lost image information and allows an uninterrupted gradient flow from the deeper layer to the shallow one. The number of possible outputs $k$ in the last $1 \times 1 \times 1$ convolution layer defines the number of classes, which is set to 2 in Zhu et al. (2020) (with '1' representing the same labels and '0' representing different ones).

The residual block consists of two $3 \times 3 \times 3$ convolution layers, each followed by a batch normalization layer and a ReLU. The residual connections here are used to connect the input features to the output feature maps of the last convolution through an element-wise summation operation. Formally, the residual block can be expressed as follows (He, Zhang, Ren, & Sun, 2016a):

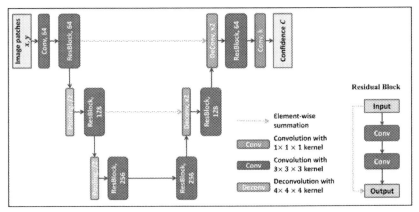

FIG. 8 An illustration of the proposed ResUNet structure. The number of kernels is denoted in each convolution operation rectangle (Zhu et al., 2020).

$$\eta = \varphi(\xi) + \xi, \tag{9}$$

where $\xi$ denotes the input feature maps, $\eta$ denotes the output feature maps, and $\varphi(\bullet)$ is the residual function. As previously investigated, the residual connections alleviate the vanishing gradient problem, promote the information propagation, and accelerate the convergence (He, Zhang, Ren, & Sun, 2016b).

A softmax loss (Gu et al., 2017) is used to train the ResUNet:

$$L_{Softmax} = -\sum_{i=1}^{m}\sum_{j=0}^{1} 1\{C(x_i) = j\} \log \frac{e^{z_{j,i}}}{\sum_{h=0}^{1} e^{z_{h,i}}}, \tag{10}$$

where $z_{j,\,i}$ is the $j$th output of the last network layer for the $i$th voxel, $C(x_i) \in \{0, 1\}$ is the ground-truth confidence at the location of voxel $x_i$, and $m$ denotes the number of voxels in the input patch.

### 3.1.2 Label fusion with FCN-based confidence estimation

Giving the target patch $p_t$, the corresponding atlas image patch $p_i$ and atlas label patch $l_i$ can be extracted from the $i$th warped atlas $i = 1, 2, ..., N$. We compute the confidence $C_i = f(p_t, p_i)$ for each patch pair $(p_t, p_i)$, $i = 1, 2, ..., N$, with the confidence estimation model (ResUNet). Then, we correct the label values in each label patch $l_i$ according to the obtained confidence $C_i$ as follows:

$$\hat{l}_i(x) = \begin{cases} l_i(x), & \text{if } C_i(x) = 1; \\ 1 - l_i(x), & \text{if } C_i(x) = 0. \end{cases} \tag{11}$$

After the label correction, we use two label fusion methods to compute the label values of the target patch, including the majority voting (Heckemann et al., 2006; Rohlfing et al., 2004) and joint label fusion (Wang et al., 2013).

**200** Big data in psychiatry and neurology

With the majority voting label fusion, the target label patch $l_t$ is determined by.

$$l_t(x) = \text{argmax}_l \sum_{i=1}^{N} \left( \hat{l}_i(x) == l \right), l \in \{0, 1\}. \tag{12}$$

With the joint label fusion, the target label patch $l_t$ can be computed by.

$$l_t(x) = \text{argmax}_l \sum_{i=1}^{N} w_i(\xi_i(x)) \left( \hat{l}_i(\xi_i(x)) == l \right), l \in \{0, 1\}, \tag{13}$$

where $\xi_i(x)$ represents the local search correspondence map between the $i$th atlas and the target image, and $w_i(\xi_i(x))$ represents the weight for the $i$th atlas. We denote $\vec{w}_x = [w_1(\xi_1(x)); w_2(\xi_2(x)); \ldots; w_N(\xi_N(x))]$. Then, $\vec{w}_x$ is determined by.

$$\text{argmin}_{\vec{w}_x} \vec{w}_x^t (M_x + \alpha I) \vec{w}_x, \tag{14}$$

$$s.t. \sum_{i=1}^{N} w_i(\xi_i(x)) = 1,$$

where $t$ stands for transpose, $I$ is an identity matrix, $\alpha$ is a parameter ($\alpha = 0.1$), and $M_x$ is a pairwise dependency matrix (Wang et al., 2013).

Since a patch-wise label fusion is used, a different label is computed for each target voxel patch instead of only taking the center voxel as the representative. The majority voting strategy can hence be used to determine the labels of the overlapping voxels of neighboring patches.

### 3.1.3 An evaluation of the multiatlas-based deep learning methods

In this subsection, we present an evaluation of the methods of FCN-MV and FCN-JLF with two widely used label fusion methods, MV (Heckemann et al., 2006; Rohlfing et al., 2004) and JLF (Wang et al., 2013), and also with a deep learning segmentation method with a 3D deeply supervised network (DSN) (Dou et al., 2017).

Forty subjects were randomly selected as atlases, and a twofold cross-validation strategy was used to evaluate the segmentation performance on the remaining 60 subjects. Specifically, 30 subjects were used to train the confidence estimation model, and the remaining 30 subjects were used to evaluate the model performance. In the training phase, 3 of 30 subjects were randomly selected for validation.

In order to reduce the computational cost, the algorithm was run on the cropped hippocampus box, which was determined by finding the minimum and maximum positions of the left and right hippocampi on the training atlases. Specifically, the obtained box was, respectively, enlarged by 7 voxels in each direction to form the cropping boxes for the left and right hippocampi, and thus they were big enough to cover the hippocampi of the unseen testing subjects. All the cropped images were normalized to have similar intensity levels using a histogram matching method. A nonlinear, cross-correlation-driven image registration algorithm (Avants, Epstein, Grossman, & Gee, 2008) was used to register the cropped atlas images to each cropped target image.

To further reduce the computational cost, the majority voting label fusion method in Heckemann et al. (2006) was used to obtain an initial segmentation of the target image. Then, the proposed method was only applied to the voxels without 100% votes for either the hippocampus or the background in the majority voting method. The image patch pairs were randomly extracted from each training image and its warped atlas images centered at the locations that did not achieve 100% votes in the majority voting-based initial segmentation. The patch size was set to $8 \times 8 \times 8$, optimally selected from $4 \times 4 \times 4$, $8 \times 8 \times 8$, $12 \times 12 \times 12$, and $16 \times 16 \times 16$, in the methods of FCN-MV and FCN-JLF.

The networks were implemented using Caffe ( Jia et al., 2014) and optimized using Adam. The batch size was set to 20, the weight decay to 0.0005, the momentum to 0.9, and the learning rate to 0.0001 (decreased by a factor of $\gamma = 0.1$ every 10,000 iterations). A weight decay of 0.0005 and a momentum of 0.9 were used in all the networks. The training process was stopped after 60,000 iterations. It is worth mentioning that the training set was separately constructed and then the separate ResUNet models were, respectively, trained for the left and right hippocampi.

The same settings of the FCN-based methods were used for the two label fusion methods (i.e., the same set of 40 atlases, same nonlinear registration and same patch-wise label fusion fashion). Atlas selection was conducted based on the NMI to select the top 20 most similar atlases from the atlas set (Hao et al., 2014; Zhu et al., 2017). The optimal hyperparameters of JLF were $r_p = 1$ and $\beta = 1$, selected from $\{1, 2, 3\}$ and $\{0.5, 1, 1.5, 2\}$, respectively, using a grid-search strategy based on the atlas dataset with 40 leave-one-out cross-validation experiments. For DSN, the same 40 subjects used as atlases in the FCN-based methods were selected as the training set and 60 other subjects as the testing set. During training, 4 subjects were randomly selected for validation. Due to the restriction of the GPU memory, the image patches were used as input to the network, instead of using the whole images. The patch size was set to $16 \times 16 \times 16$, optimally selected from $8 \times 8 \times 8$, $16 \times 16 \times 16$, and $24 \times 24 \times 24$.

Similar to Section 2.4, the ADNI database (http://adni.loni.usc.edu/) was used for the evaluation with nine segmentation evaluation metrics, including the Dice coefficient, Jaccard index, precision, recall, MD, HD, HD95, ASSD, and RMSD ( Jafari-Khouzani et al., 2011). The same image preprocessing (such as the bias field correction and linear image registration) on the ADNI database was also applied.

Table 3 lists the nine index values of the segmentation results using different segmentation methods. It shows that the methods of FCN-MV and FCN-JLF obtain the best results. Compared with the MV method, FCN-MV improves

**TABLE 3** Nine index values (mean±std) of the hippocampus segmentation results using different methods (Zhu et al., 2020).

| | MV | JLF | DSN | FCN-MV | FCN-JLF |
|---|---|---|---|---|---|
| Dice | $0.856 \pm 0.031^{a,b}$ | $0.880 \pm 0.024^{a,b}$ | $0.869 \pm 0.022^{a,b}$ | $0.883 \pm 0.022$ | **0.884 ± 0.020** |
| (L/R) | $0.860 \pm 0.033^{a,b}$ | $0.884 \pm 0.023^{a,b}$ | $0.871 \pm 0.024^{a,b}$ | $0.888 \pm 0.021^{b}$ | **0.891 ± 0.019** |
| Jaccard | $0.750 \pm 0.047^{a,b}$ | $0.786 \pm 0.037^{a,b}$ | $0.769 \pm 0.035^{a,b}$ | $0.792 \pm 0.034$ | **0.793 ± 0.032** |
| (L/R) | $0.755 \pm 0.048^{a,b}$ | $0.794 \pm 0.036^{a,b}$ | $0.772 \pm 0.036^{a,b}$ | $0.800 \pm 0.033^{b}$ | **0.803 ± 0.030** |
| Precision | $0.861 \pm 0.048^{a,b}$ | $0.879 \pm 0.032^{a}$ | $0.866 \pm 0.033^{a,b}$ | **0.896 ± 0.029** | $0.879 \pm 0.029^{a}$ |
| (L/R) | $0.864 \pm 0.052^{a,b}$ | $0.882 \pm 0.036^{a,b}$ | $0.870 \pm 0.034^{a,b}$ | **0.902 ± 0.033** | $0.889 \pm 0.0031^{a}$ |
| Recall | $0.854 \pm 0.049^{a,b}$ | $0.882 \pm 0.036^{a}$ | $0.874 \pm 0.032^{b}$ | $0.872 \pm 0.033^{b}$ | **0.890 ± 0.030** |
| (L/R) | $0.859 \pm 0.044^{a,b}$ | $0.889 \pm 0.029^{a}$ | $0.873 \pm 0.034^{b}$ | $0.876 \pm 0.028^{b}$ | **0.894 ± 0.027** |
| HD | $3.157 \pm 0.853^{a,b}$ | $3.076 \pm 0.784^{a}$ | $7.324 \pm 10.672^{a,b}$ | **2.843 ± 0.770** | $2.951 \pm 0.827$ |
| (L/R) | $3.255 \pm 0.894^{a,b}$ | $3.227 \pm 1.100$ | $5.250 \pm 8.067$ | **3.013 ± 0.878** | $3.057 \pm 0.993$ |
| HD95 | $1.345 \pm 0.478^{a,b}$ | $1.093 \pm 0.352$ | $1.934 \pm 4.532^{a,b}$ | $1.054 \pm 0.271$ | **1.028 ± 0.167** |
| (L/R) | $1.332 \pm 0.441^{a,b}$ | $1.101 \pm 0.237$ | $1.322 \pm 0.294^{a,b}$ | $1.099 \pm 0.244$ | **1.070 ± 0.196** |
| MD | $0.284 \pm 0.054^{a,b}$ | $0.252 \pm 0.048^{a,b}$ | $0.317 \pm 0.296^{a,b}$ | **0.218 ± 0.032** | $0.238 \pm 0.033^{a}$ |
| (L/R) | $0.278 \pm 0.063^{a,b}$ | $0.237 \pm 0.051^{a,b}$ | $0.258 \pm 0.058^{a,b}$ | **0.205 ± 0.044** | $0.221 \pm 0.043^{a}$ |
| ASSD | $0.334 \pm 0.077^{a,b}$ | $0.265 \pm 0.052^{b}$ | $0.353 \pm 0.145^{a,b}$ | $0.255 \pm 0.042$ | **0.252 ± 0.035** |
| (L/R) | $0.328 \pm 0.071^{a,b}$ | $0.260 \pm 0.043^{a,b}$ | $0.322 \pm 0.049^{a,b}$ | $0.249 \pm 0.036^{b}$ | **0.242 ± 0.034** |
| RMSD | $0.632 \pm 0.123^{a,b}$ | $0.551 \pm 0.090^{a,b}$ | $0.874 \pm 0.910^{a,b}$ | $0.533 \pm 0.073$ | **0.527 ± 0.058** |
| (L/R) | $0.628 \pm 0.110^{a,b}$ | $0.550 \pm 0.074^{a,b}$ | $0.679 \pm 0.293^{a,b}$ | $0.534 \pm 0.067^{b}$ | **0.523 ± 0.063** |

The best results in each row are marked in bold.
[a]Indicates that FCN-MV achieves a significant improvement over the corresponding method in the Wilcoxon signed rank tests with P < 0.05.
[b]Indicates that FCN-JLF achieves a significant improvement over the corresponding method in the Wilcoxon signed rank tests with P < 0.05.

the Dice scores by 2.7% and 2.8% for the left and right hippocampus segmentation results, respectively. This improvement is achieved by the FCN-based confidence estimation, which potentially compensates for the registration error. FCN-JLF can further improve the Dice scores by 0.1% and 0.3% for the left and right hippocampus segmentation results, respectively, compared with FCN-MV. This improvement is achieved through the use of the more advanced label fusion method, JLF, to fuse the corrected label maps. It can also be observed that JLF improves MV by 2.4% both for the left and right hippocampi, while FCN-JLF improves FCN-MV only by 0.1% and 0.3% for the left and right hippocampi, respectively. This demonstrates that the FCN-based label correction method can effectively correct the registration errors. With label correction, even the simplest majority voting label fusion can achieve better segmentation results than the state-of-the-art JLF method. The FCN-based methods also obtain better segmentation results compared with the deep learning segmentation method (DSN).

Fig. 9 shows the box plots of the segmentation results based on the nine evaluation measures. It can be seen that the methods of FCN-MV and FCN-JLF perform consistently better than other methods. Regarding the HD measure,

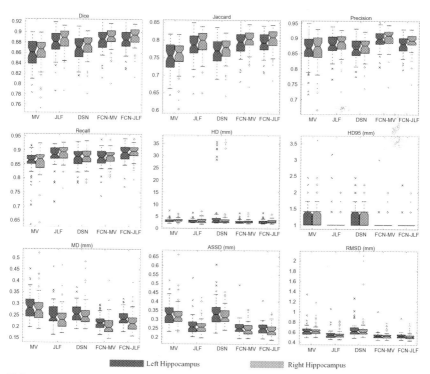

**FIG. 9** Box plots of the segmentation results based on nine evaluation measures. In each box, the central mark is the median, and the edges are the 25th and 75th percentiles (Zhu et al., 2020).

several severe outliers can be observed in the segmentation results obtained by DSN. For the HD95 measure, the figure shows how the boxes turn into lines for both left and right hippocampus segmentation results obtained by FCN-MV and FCN-JLF, which means that the HD95 values at the 25th and 75th percentiles reach the same value, indicating the high robustness of the methods. Similar results are obtained by the JLF method.

Fig. 10 shows examples of the confidence maps and corrected (warped) atlas label maps. In the confidence maps, the dark voxels denote the confidence values of 0, which means that the registration errors may happen at these voxels. The corrected atlas label maps are obtained by changing the label values at the voxels with confidence values of 0. It can be observed that the corrected (warped) atlas label maps are more similar to the target label compared with the original warped atlas labels. Meanwhile, some artifacts can also be found in the second corrected atlas label, which makes the label unsmooth. Interestingly, most of these artifacts can be perfectly eliminated by the label fusion, resulting in a smooth segmentation, which can be observed in the FCN-MV segmentation in the figure. Fig. 11 shows the sagittal view and 3D rendering of the left hippocampus segmentations for a randomly selected subject. These observations show that the methods of FCN-MV and FCN-JLF produce the most accurate segmentation results.

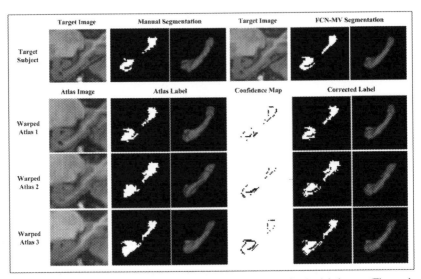

**FIG. 10** Examples of the confidence maps and corrected (warped) atlas label maps (Zhu et al., 2020).

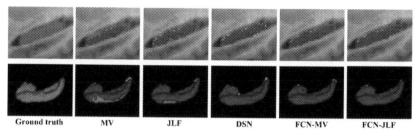

**FIG. 11** Sagittal view *(top row)* and 3D rendering *(bottom row)* of the left hippocampus segmentations for a randomly selected subject (Zhu et al., 2020).

## 3.2 End-to-end dilated residual dense U-net for hippocampus segmentation

This subsection describes a fully automatic deep learning-based method for hippocampal segmentation. This method embeds a dilated dense network in the residual U-net, namely ResDUnet. The resulting network can generate multiscale features while keeping a high spatial resolution, which is useful in fusing the low-level features in the contracting path with the high-level features in the expanding path.

### 3.2.1 The dilated residual dense U-net

The encoder-decoder like 3D ResDUNet takes a subject as input, and then predicts the posterior probabilities of each voxel. Given the posterior probability $p_k(x|\theta)$ of the voxel $x$ that belongs to the $k$th category, where $\theta$ is the model parameters, the hippocampal subfield label of the voxel $x$ is determined by.

$$L(x) = argmax_{k \in \mathbb{C}} p_k(x|\theta), \tag{15}$$

where $\mathbb{C} = \{1, 2, ..., K\}$, and $K$ represents the number of categories. Fig. 12 shows the architecture of the network.

ResDUNet consists of an encoder path, a decoder path, and an embedded dilated dense network. The encoder path contains three ResBlocks (residual blocks) and two $2 \times 2 \times 2$ max-pooling operations with a stride of 2. Correspondingly, the decoder path consists of one $4 \times 4 \times 4$ deconvolution with a stride of 2 and two ResBlocks. Specifically, a $1 \times 1 \times 1$ convolution is added to the last decoder block, which outputs $K$ feature maps (such that $K$ is the number of the label categories including the background). Some padded convolution layers are also used to maintain the spatial dimension.

Similar to ResUNet in Section 3.1.1, the same structure is used here for the ResBlock, and skip connections are also added between the encoder and decoder path, to recover the lost image information and allow an uninterrupted gradient flow from the deeper layer to the shallow one. Specifically, the feature maps before the first and the last pooling layers in the encoder path are

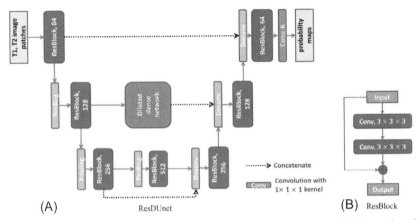

**FIG. 12** Framework of the ResDUNet. The number in each operation rectangle is the number of kernels. All the operations are implemented in a 3D manner. "c" denotes the concatenation (Zhu, Shi, et al., 2019).

concatenated to the corresponding feature maps in the decoder path. Since the level of the features in the encoder block is much lower than that in the decoder block, this direct concatenation of the features may not obtain the optimal results; thus a dilated dense modulator is further incorporated into the network to provide multiscale features for the decoder block while keeping a high spatial resolution. The dilated dense network takes the feature maps of the first encoder block as input, and the output features are then concatenated to the corresponding feature maps in the decoder block.

### The dilated dense network

The dilated dense network is motivated by the dilated convolutions (Yu & Koltun, 2015), which is originally proposed for the semantic image segmentation to arbitrarily enlarge the receptive field. Fig. 13 illustrates the dilated convolutional kernels with different dilation rates. Let $F : \mathbb{Z}^3 \to \mathbb{R}$ be a 3D discrete function, and $h : \Omega_r \to \mathbb{R}$ be a discrete filter with a dilation rate $l$, where $\Omega_r = [-r, r]^3 \cap \mathbb{Z}^3$, the dilated convolution $*_l$ can then be defined as follows (Yu & Koltun, 2015):

$$(F *_l h)(p) = \sum_{s + lt = p} F(s) h(t). \tag{16}$$

Note that when $l = 1$, the dilated convolution becomes the normal convolution.

As shown in Fig. 14, the dilated dense modulator consists of several dilated convolutions with dense connections (Huang, Liu, Weinberger, & van der Maaten, 2016). Specifically, different dilation rates are used for the dilated convolutions to enlarge the receptive field. Dense connections are applied to

Hippocampus segmentation in MR images Chapter | 10 **207**

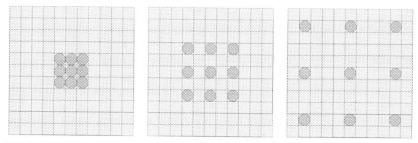

**FIG. 13** Illustration of dilated convolutional kernels: 1—dilated convolutional kernel *(left)*; 2—dilated convolutional kernel *(middle)*; 4—dilated convolutional kernel *(right)* (Zhu, Shi, et al., 2019).

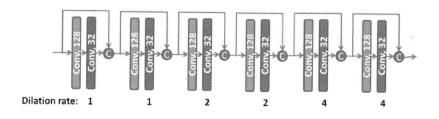

**FIG. 14** The structure of the dilated dense network. The number in each operation rectangle is the number of kernels. All the operations are implemented in a 3D manner, and "c" denotes the concatenation (Zhu, Shi, et al., 2019).

concatenate all the previously generated features to the current feature maps. In order to avoid overfitting, dropout operations are used after each $3 \times 3 \times 3$ convolution with a dropout rate of 0.5 (Srivastava, Hinton, Krizhevsky, Sutskever, & Salakhutdinov, 2014).

The dilated dense network enables ResDUNet to capture the contextual image information while keeping a high spatial resolution and generate multiscale image features. Moreover, two different kinds of features are fused: the features provided by the dilated dense network and those provided by the contracting-expanding path, which provides more abundant image information for the dense prediction.

### 3.2.2 An evaluation of the end-to-end deep learning methods

In this subsection, we present an evaluation of the dilated residual dense U-net method with three state-of-the-art networks: the 3D U-net (Çiçek, Abdulkadir, Lienkamp, Brox, & Ronneberger, 2016), ConvNet (Yu, Yang, Chen, Qin, & Heng, 2017), and hippocampal subfield segmentation method (HIPS).

For the evaluation, we used the publicly available Kulaga-Yoskovitz dataset (https://www.nitrc.org/projects/mni-hisub25), on which the HIPS method

**208** Big data in psychiatry and neurology

obtained the best segmentation results so far (Romero, Coupé, & Manjón, 2017), with the Dice and ASSD metrics. The Kulaga-Yoskovitz dataset contains the data of 25 adult subjects (31 ± 7 years, 12 males). The data of each subject consist of an isotropic 3D-MPRAGE T1-weighted image (TR = 3000 ms; TE = 4.32 ms; TI = 1500 ms; flip angle = 7°; matrix size = 336 × 384; FOV = 201 × 229 mm$^2$; 240 axial slices with 0.6 mm slice thickness resulting in 0.6 × 0.6 × 0.6 mm$^3$ voxels; acquisition time = 16.48 min), an anisotropic 2D T2-weighted TSE image (TR = 10,810 ms; TE = 81 ms; flip angle = 119°; matrix size = 512 × 512; FOV=203 × 203 mm$^2$, 60 coronal slices oriented perpendicular to the hippocampal long axis, slice thickness of 2 mm, resulting in 0.4 × 0.4 × 2.0 mm$^3$ voxels; acquisition time = 5.47 min), and a manually labeled image for the hippocampal subfields including CA1–3, SUB, and CA4/DG (Kulaga-Yoskovitz et al., 2015). All the T1w and T2w images underwent an automated correction of the intensity nonuniformity and standardization. All the images are linearly registered to the MNI152 space and resampled to a resolution of 0.4 × 0.4 × 0.4 mm$^3$.

Fivefold cross-validation was adopted to evaluate the performance. Fifteen subjects were selected for training, 5 subjects for validation, and the remaining 5 subjects for testing. Random cropping was applied to augment the training dataset. Specifically, ∼1300 patches with the size of 32 × 32 × 32 were randomly extracted from each subject. All the extracted patches contain at least one hippocampal voxel. In the test phase, the input image was processed patch by patch (with a stride of 8 × 8 × 8), and a majority voting strategy for the overlapping regions was applied to get the whole image prediction. Since both T1w and T2w images were available, the corresponding T1w and T2w image patches were concatenated and used as input for each network.

The networks were implemented using Caffe (Jia et al., 2014) and optimized using Adam. The batch size was set to 3, the weight decay to 0.0005, the momentum to 0.9, and the learning rate to 0.0001 (decreased by a factor of $\gamma = 0.1$ every 10,000 iterations). A weight decay of 0.0005 and a momentum of 0.9 were used in all the networks. The training process was stopped after 60,000 iterations.

The models are trained using the Softmax loss (Gu et al., 2017):

$$L_{Softmax} = - \sum_{i=1}^{N} \sum_{k=1}^{K} 1\{y_i = k\} \log \frac{e^{z_{k,i}}}{\sum_{j=1}^{K} e^{z_{j,i}}}, \tag{17}$$

where $z_{k,\,i}$ represents the $k$th output of the last network layer for the $i$th voxel, $y_i \in \{1, 2, \ldots, K\}$ is the corresponding ground-truth label, and $K$ and $N$ are the number of categories and the number of voxels, respectively. The term $\frac{e^{z_{k,i}}}{\sum_{j=1}^{K} e^{z_{j,i}}}$ represents the prediction probability for the $k$th class of the $i$th voxel, which is computed using the Softmax function.

For a fair comparison, the 3D U-net used in these experiments consisted of encoder and decoder paths, which are the same as in the ResDUnet. The difference is that the dilated dense network was removed and the ResBlock was replaced with $3 \times 3 \times 3$ convolution. ConvNet (Yu et al., 2017) is a volumetric convolutional neural network (CNN) with mixed residual connections, which also consists of three pooling layers and three deconvolutional layers. In ConvNet, residual connections are used between the successive convolution layers to form the residual blocks, and also between the feature maps of the encoder and decoder paths. Besides, ConvNet (Yu et al., 2017) exploits a deep supervision mechanism to accelerate its convergence speed. The same postprocessing was used in all the experiments to remove the tiny isolated blocks of the segmentation results that appear outside the hippocampal region. The published results of HIPS are used as reported in Romero et al. (2017).

Tables 4 and 5 list the Dice and ASSD coefficients of the segmentation results obtained by the four different networks on the Kulaga-Yoskovitz dataset. The results show that the ResDUnet outperforms the 3D U-net (Çiçek et al., 2016) and ConvNet (Yu et al., 2017) in segmenting all the subfields, according to the Wilcoxon signed rank tests with $P < 0.05$. The ResDUnet also outperforms the HIPS method, especially for segmenting the CA4/DG subfield which is the most difficult task (Dalton, Zeidman, Barry, Williams, & Maguire, 2017). Fig. 15 shows the hippocampal subfield segmentations of a randomly selected subject from the Kulaga-Yoskovitz dataset, obtained by manual segmentation and the four different networks. It can be seen that ResDUnet achieves the most accurate results.

**TABLE 4** Mean (STD) values of Dice for each subfield segmentation by four different methods on the KULAGA-YOSKOVITZ dataset. Higher Dice values indicate a better segmentation performance. The best results are marked in bold (Zhu, Shi, et al., 2019).

| | HIPS (Romero et al., 2017) | 3D U-net (Çiçek et al., 2016) | ConvNet (Yu et al., 2017) | ResDUnet |
|---|---|---|---|---|
| CA1–3 | 0.916(0.015) | 0.916(0.011)[a] | 0.918(0.010)[a] | **0.920 (0.01)** |
| CA4/ DG | 0.862(0.034) | 0.871(0.021)[a] | 0.870(0.016)[a] | **0.879 (0.02)** |
| SUB | 0.886(0.021) | 0.883(0.016)[a] | 0.887(0.018)[a] | **0.888 (0.018)** |
| Average | 0.888 | 0.890 | 0.892 | **0.896** |

[a]Indicates that ResDUnet achieves a significant improvement over the corresponding method in the Wilcoxon signed rank tests with P < 0.05.

**TABLE 5** Mean (STD) values of ASSD for each subfield segmentation by four different networks on the KULAGA-YOSKOVITZ dataset. Smaller ASSD values indicate a better segmentation performance. The best results are marked in bold (Zhu, Shi, et al., 2019).

|  | 3D U-net (Çiçek et al., 2016) | ConvNet (Yu et al., 2017) | ResDUnet |
| --- | --- | --- | --- |
| CA1–3 | 0.065(0.011)[a] | 0.064(0.009)[a] | **0.062 (0.010)** |
| CA4/DG | 0.077(0.014)[a] | 0.079(0.015)[a] | **0.072 (0.014)** |
| SUB | 0.069(0.013)[a] | 0.066(0.013)[a] | **0.065 (0.013)** |
| Average | 0.070 | 0.070 | **0.066** |

[a]Indicates that ResDUnet achieves a significant improvement over the corresponding method in the Wilcoxon signed rank tests with $P < 0.05$.

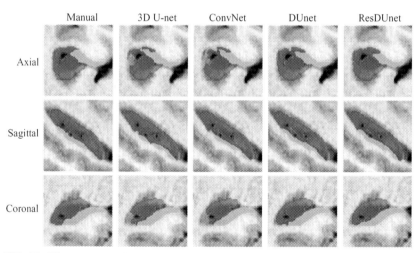

**FIG. 15** Hippocampal subfield segmentations of a randomly selected subject from the Kulaga-Yoskovitz dataset, obtained by manual segmentation and four different networks (Zhu, Shi, et al., 2019).

## 4 Conclusion

In this chapter, we have introduced the state-of-the-art methods for automatically segmenting the hippocampus in the MRI images and presented the recently proposed methods for this task. There are many different techniques applied to the hippocampus segmentation, this chapter mainly included the widely used multiatlas-based methods and deep learning-based methods. The multiatlas-based methods usually consist of three steps: atlas selection, image registration, and label fusion, in which the label fusion is the most important step. Three effective label fusion strategies, weighted voting, local learning, and supervised metric learning have been designed to leverage the different context information to compensate for the anatomical differences between the target image and atlas images. Different from the multiatlas-based methods, the learning-based methods learn the mapping from the extracted features to the class labels, which avoids the time-consuming image registration process in the inference. Deep learning-based methods combine the hierarchical feature learning and classification model into a unified framework for end-to-end optimization. To take advantage of both multiatlas-based methods and deep learning-based methods, a multiatlas-based deep learning method is also presented for hippocampus segmentation.

In future research, the following research directions may lead to an even more efficient and effective performance for the hippocampus segmentation.

- Novel combination of multiatlas and deep learning methods. Even though deep learning has achieved an excellent performance in the task of hippocampus segmentation, it usually cannot deal with some special cases. If useful information is introduced from the most similar atlas images to facilitate the model optimization, the model capability of the trained deep networks may be improved.
- More effective feature representation. The features play an important role in the hippocampus segmentation. While deep networks automatically learn hierarchical features from the data itself, and handcrafted features are extracted based on the expert experience, it is desired to combine them to achieve a more discriminative representation.
- Using fewer or weak annotations. While the manual annotation is time consuming and laborious, semisupervised, unsupervised, or weakly supervised methods with fewer or weak annotations can reduce the cost in terms of the time and human efforts needed for the annotation.
- Domain adaptation between different resources. In real applications, the data from different resources may be with large distribution differences, such that the existing models or atlases can hardly perform on the new dataset. Domain adaptation between the different resources is one way to improve the flexibility of automatic methods.

## Acknowledgments

H. Zhu was supported by National Natural Science Foundation of China [61602307, 61877039] and Natural Science Foundation of Zhejiang Province [LY19F020013].

## 212 Big data in psychiatry and neurology

# References

Anand, K. S., & Dhikav, V. (2012). Hippocampus in health and disease: An overview. *Annals of Indian Academy of Neurology, 15*, 239.

Andersen, P., Morris, R., Amaral, D., Bliss, T., & O'Keefe, J. (2006). *The hippocampus book.* Oxford University Press.

Artaechevarria, X., Munoz-Barrutia, A., & Ortiz-de-Solorzano, C. (2009). Combination strategies in multi-atlas image segmentation: Application to brain MR data. *IEEE Transactions on Medical Imaging, 28*, 1266–1277.

Avants, B. B., Epstein, C. L., Grossman, M., & Gee, J. C. (2008). Symmetric diffeomorphic image registration with cross-correlation: Evaluating automated labeling of elderly and neurodegenerative brain. *Medical Image Analysis, 12*, 26–41.

Bai, W., Shi, W., Ledig, C., & Rueckert, D. (2015). Multi-atlas segmentation with augmented features for cardiac MR images. *Medical Image Analysis, 19*, 98–109.

Bartlett, T. Q., Vannier, M. W., McKeel, D. W., Jr., Gado, M., Hildebolt, C. F., & Walkup, R. (1994). Interactive segmentation of cerebral gray matter, white matter, and CSF: Photographic and MR images. *Computerized Medical Imaging and Graphics, 18*, 449–460.

Bast, T. (2011). The hippocampal learning-behavior translation and the functional significance of hippocampal dysfunction in schizophrenia. *Current Opinion in Neurobiology, 21*, 492–501.

Boccardi, M., Bocchetta, M., Morency, F. C., Collins, D. L., Nishikawa, M., Ganzola, R., et al. (2015). Training labels for hippocampal segmentation based on the EADC-ADNI harmonized hippocampal protocol. *Alzheimer's & Dementia, 11*, 175–183.

Cao, L., Li, L., Zheng, J., Fan, X., Yin, F., Shen, H., et al. (2018). Multi-task neural networks for joint hippocampus segmentation and clinical score regression. *Multimedia Tools and Applications, 77*, 29669–29686.

Carmo, D., Silva, B., Yasuda, C., Rittner, L., & Lotufo, R. (2020). *Hippocampus segmentation on epilepsy and alzheimer's disease studies with multiple convolutional neural networks. arXiv preprint arXiv:2001.05058.*

Chang, C.-C., & Lin, C.-J. (2011). LIBSVM: A library for support vector machines. *ACM Transactions on Intelligent Systems and Technology (TIST), 2*, 27.

Chen, Y., Shi, B., Wang, Z., Sun, T., Smith, C. D., & Liu, J. (2017). Accurate and consistent hippocampus segmentation through convolutional LSTM and view ensemble. In *International workshop on machine learning in medical imaging* (pp. 88–96).

Chen, Y., Shi, B., Wang, Z., Zhang, P., Smith, C. D., & Liu, J. (2017). Hippocampus segmentation through multi-view ensemble ConvNets. In *2017 IEEE 14th international symposium on biomedical imaging (ISBI 2017)* (pp. 192–196).

Çiçek, Ö., Abdulkadir, A., Lienkamp, S. S., Brox, T., & Ronneberger, O. (2016). 3D U-Net: Learning dense volumetric segmentation from sparse annotation. In *International conference on medical image computing and computer-assisted intervention* (pp. 424–432).

Coupé, P., Manjón, J. V., Fonov, V., Pruessner, J., Robles, M., & Collins, D. L. (2011). Patch-based segmentation using expert priors: Application to hippocampus and ventricle segmentation. *NeuroImage, 54*, 940–954.

Cummings, J., Ritter, A., & Zhong, K. (2018). Clinical trials for disease-modifying therapies in Alzheimer's disease: A primer, lessons learned, and a blueprint for the future. *Journal of Alzheimer's Disease, 64*, S3–S22.

Dalton, M. A., Zeidman, P., Barry, D. N., Williams, E., & Maguire, E. A. (2017). Segmenting subregions of the human hippocampus on structural magnetic resonance image scans: An illustrated tutorial. *Brain and Neuroscience Advances, 1*. 2398212817701448.

Dhikav, V., & Anand, K. S. (2007). Glucocorticoids may initiate Alzheimer's disease: A potential therapeutic role for mifepristone (RU-486). *Medical Hypotheses, 68*, 1088–1092.

Dhikav, V., & Anand, K. (2011). Potential predictors of hippocampal atrophy in Alzheimer's disease. *Drugs & Aging, 28*, 1–11.

Dill, V., Franco, A. R., & Pinho, M. S. (2015). Automated methods for hippocampus segmentation: The evolution and a review of the state of the art. *Neuroinformatics, 13*, 133–150.

Dinsdale, N. K., Jenkinson, M., & Namburete, A. I. (2019). Spatial warping network for 3D segmentation of the hippocampus in MR images. In *International conference on medical image computing and computer-assisted intervention* (pp. 284–291).

Dou, Q., Yu, L., Chen, H., Jin, Y., Yang, X., Qin, J., et al. (2017). 3D deeply supervised network for automated segmentation of volumetric medical images. *Medical Image Analysis, 41*, 40–54.

Fang, L., Zhang, L., Nie, D., Cao, X., Rekik, I., Lee, S.-W., et al. (2019). Automatic brain labeling via multi-atlas guided fully convolutional networks. *Medical Image Analysis, 51*, 157–168.

Freeborough, P. A., Fox, N. C., & Kitney, R. I. (1997). Interactive algorithms for the segmentation and quantitation of 3-D MRI brain scans. *Computer Methods and Programs in Biomedicine, 53*, 15–25.

Ghanei, A., Soltanian-Zadeh, H., & Windham, J. P. (1998). Segmentation of the hippocampus from brain MRI using deformable contours. *Computerized Medical Imaging and Graphics, 22*, 203–216.

Gray, H. (1924). *Anatomy of the human body*. Lea & Febiger.

Gu, J., Wang, Z., Kuen, J., Ma, L., Shahroudy, A., Shuai, B., et al. (2017). Recent advances in convolutional neural networks. *Pattern Recognition, 77*, 354–377.

Guillaumin, M., Verbeek, J., & Schmid, C. (2009). Is that you? Metric learning approaches for face identification. In *2009 IEEE 12th international conference on computer vision* (pp. 498–505).

Hao, Y., Wang, T., Zhang, X., Duan, Y., Yu, C., Jiang, T., et al. (2014). Local label learning (LLL) for subcortical structure segmentation: Application to hippocampus segmentation. *Human Brain Mapping, 35*, 2674–2697.

He, K., Zhang, X., Ren, S., & Sun, J. (2016a). Deep residual learning for image recognition. *Proceedings of the IEEE Conference on Computer Vision and Pattern Recognition*, 770–778.

He, K., Zhang, X., Ren, S., & Sun, J. (2016b). Identity mappings in deep residual networks. In *European conference on computer vision* (pp. 630–645).

Heckemann, R. A., Hajnal, J. V., Aljabar, P., Rueckert, D., & Hammers, A. (2006). Automatic anatomical brain MRI segmentation combining label propagation and decision fusion. *NeuroImage, 33*, 115–126.

Huang, G., Liu, Z., Weinberger, K. Q., & van der Maaten, L. (2016). *Densely connected convolutional networks. arXiv preprint arXiv:1608.06993*.

Jafari-Khouzani, K., Elisevich, K. V., Patel, S., & Soltanian-Zadeh, H. (2011). Dataset of magnetic resonance images of nonepileptic subjects and temporal lobe epilepsy patients for validation of hippocampal segmentation techniques. *Neuroinformatics, 9*, 335–346.

Jia, Y., Shelhamer, E., Donahue, J., Karayev, S., Long, J., Girshick, R., et al. (2014). Caffe: Convolutional architecture for fast feature embedding. In *Presented at the proceedings of the 22nd ACM international conference on multimedia*.

Johnston, D., & Amaral, D. G. (2004). Hippocampus. *The Synaptic Organization of the Brain*. Oxford University.

Khlif, M. S., Egorova, N., Werden, E., Redolfi, A., Boccardi, M., DeCarli, C. S., et al. (2019). *A comparison of automated segmentation and manual tracing in estimating hippocampal volume in ischemic stroke and healthy control participants. 21* (p. 101581). NeuroImage: Clinical.

**214** Big data in psychiatry and neurology

Kim, M., Wu, G., & Shen, D. (2013). Unsupervised deep learning for hippocampus segmentation in 7.0 Tesla MR images. In *International workshop on machine learning in medical imaging* (pp. 1–8).

Kulaga-Yoskovitz, J., Bernhardt, B. C., Hong, S.-J., Mansi, T., Liang, K. E., Van Der Kouwe, A. J., et al. (2015). Multi-contrast submillimetric 3 tesla hippocampal subfield segmentation protocol and dataset. *Scientific Data, 2*, 150059.

Landman, B. A., Lyu, I., Huo, Y., & Asman, A. J. (2020). Multiatlas segmentation. In *Handbook of medical image computing and computer assisted intervention* (pp. 137–164). Elsevier.

Lanke, V., Moolamalla, S., Roy, D., & Vinod, P. (2018). Integrative analysis of hippocampus gene expression profiles identifies network alterations in aging and Alzheimer's disease. *Frontiers in Aging Neuroscience, 10*, 153.

Liao, S., Gao, Y., Lian, J., & Shen, D. (2013). Sparse patch-based label propagation for accurate prostate localization in CT images. *IEEE Transactions on Medical Imaging, 32*, 419–434.

Long, J., Shelhamer, E., & Darrell, T. (2015). Fully convolutional networks for semantic segmentation. In *Proceedings of the IEEE conference on computer vision and pattern recognition* (pp. 3431–3440).

Pizer, S. M., Fletcher, P. T., Joshi, S., Thall, A., Chen, J. Z., Fridman, Y., et al. (2003). Deformable m-reps for 3D medical image segmentation. *International Journal of Computer Vision, 55*, 85–106.

Rohlfing, T., Brandt, R., Menzel, R., & Maurer, C. R., Jr. (2004). Evaluation of atlas selection strategies for atlas-based image segmentation with application to confocal microscopy images of bee brains. *NeuroImage, 21*, 1428–1442.

Romero, J. E., Coupé, P., & Manjón, J. V. (2017). HIPS: A new hippocampus subfield segmentation method. *NeuroImage, 163*, 286–295.

Ronneberger, O., Fischer, P., & Brox, T. (2015). U-net: Convolutional networks for biomedical image segmentation. In *International conference on medical image computing and computer-assisted intervention* (pp. 234–241).

Rousseau, F., Habas, P. A., & Studholme, C. (2011). A supervised patch-based approach for human brain labeling. *IEEE Transactions on Medical Imaging, 30*, 1852–1862.

Roy, A. G., Conjeti, S., Navab, N., Wachinger, C., & Initiative, A. S. D. N. (2019). QuickNAT: A fully convolutional network for quick and accurate segmentation of neuroanatomy. *NeuroImage, 186*, 713–727.

Sabuncu, M. R., Yeo, B. T. T., Van Leemput, K., Fischl, B., & Golland, P. (2010). A generative model for image segmentation based on label fusion. *IEEE Transactions on Medical Imaging, 29*, 1714–1729.

Shen, D., Moffat, S., Resnick, S. M., & Davatzikos, C. (2002). Measuring size and shape of the hippocampus in MR images using a deformable shape model. *NeuroImage, 15*, 422–434.

Shen, D., Wu, G., & Suk, H.-I. (2017). Deep learning in medical image analysis. *Annual Review of Biomedical Engineering, 19*, 221–248.

Srivastava, N., Hinton, G., Krizhevsky, A., Sutskever, I., & Salakhutdinov, R. (2014). Dropout: A simple way to prevent neural networks from overfitting. *The Journal of Machine Learning Research, 15*, 1929–1958.

Thyreau, B., Sato, K., Fukuda, H., & Taki, Y. (2018). Segmentation of the hippocampus by transferring algorithmic knowledge for large cohort processing. *Medical Image Analysis, 43*, 214–228.

Toriwaki, J., & Yoshida, H. (2009). *Fundamentals of three-dimensional digital image processing*. London: Springer.

Wang, S., He, K., Nie, D., Zhou, S., Gao, Y., & Shen, D. (2019). CT male pelvic organ segmentation using fully convolutional networks with boundary sensitive representation. *Medical Image Analysis, 54*, 168–178.

Wang, H., Suh, J. W., Das, S. R., Pluta, J. B., Craige, C., & Yushkevich, P. A. (2013). Multi-atlas segmentation with joint label fusion. *IEEE Transactions on Pattern Analysis and Machine Intelligence, 35*, 611–623.

Wang, S., Wang, Q., Shao, Y., Qu, L., Lian, C., Lian, J., et al. (2020). Iterative label Denoising network: Segmenting male pelvic organs in CT from 3D bounding box annotations. *IEEE Transactions on Biomedical Engineering*.

Wang, F., Zuo, W., Zhang, L., Meng, D., & Zhang, D. (2015). A kernel classification framework for metric learning. *IEEE Transactions on Neural Networks and Learning Systems, 26*, 1950–1962.

Wu, Z., Gao, Y., Shi, F., Ma, G., Jewells, V., & Shen, D. (2018). Segmenting hippocampal subfields from 3T MRI with multi-modality images. *Medical Image Analysis, 43*, 10–22.

Xing, E. P., Jordan, M. I., Russell, S., & Ng, A. Y. (2002). Distance metric learning with application to clustering with side-information. *Advances in Neural Information Processing Systems*, 505–512.

Yang, H., Sun, J., Li, H., Wang, L., & Xu, Z. (2018). Neural multi-atlas label fusion: Application to cardiac MR images. *Medical Image Analysis, 49*, 60–75.

Yu, F., & Koltun, V. (2015). *Multi-scale context aggregation by dilated convolutions. arXiv preprint arXiv:1511.07122*.

Yu, L., Yang, X., Chen, H., Qin, J., & Heng, P.-A. (2017). Volumetric ConvNets with mixed residual connections for automated prostate segmentation from 3D MR images. *AAAI*, 66–72.

Zhao, K., Ding, Y., Han, Y., Fan, Y., Alexander-Bloch, A. F., Han, T., et al. (2020). Independent and reproducible hippocampal radiomic biomarkers for multisite Alzheimer's disease: Diagnosis, longitudinal progress and biological basis. *Science Bulletin*.

Zhu, H., Adeli, E., Shi, F., & Shen, D. (2020). FCN based label correction for multi-atlas guided organ segmentation. *Neuroinformatics*, 1–13.

Zhu, H., Cheng, H., Yang, X., Fan, Y., & Initiative, A. S. D. N. (2017). Metric learning for multi-atlas based segmentation of hippocampus. *Neuroinformatics, 15*, 41–50.

Zhu, H., & He, G. (2020). Joint neighboring coding with a low-rank constraint for multi-atlas based image segmentation. *Journal of Medical Imaging and Health Informatics, 10*, 310–315.

Zhu, H., Shi, F., Wang, L., Hung, S.-C., Chen, M.-H., Wang, S., et al. (2019). Dilated dense U-net for infant Hippocampus subfield segmentation. *Frontiers in Neuroinformatics, 13*, 30.

Zhu, H., Tang, Z., Cheng, H., Wu, Y., & Fan, Y. (2019). Multi-atlas label fusion with random local binary pattern features: Application to hippocampus segmentation. *Scientific Reports, 9*, 1–14.

# Chapter 11

# A scalable medication intake monitoring system

Diane Myung-Kyung Woodbridge and Kevin Bengtson Wong

*Data Science, University of San Francisco, San Francisco, CA, United States*

## 1 Introduction

Life expectancy has increased rapidly since the Age of Enlightenment as scientific discoveries have brought advances in medical understanding and treatment. However, in 1900, even in the richest countries, the average life expectancy was under 50 years, with a global average life expectancy of 31 years (Prentice, 2008). Since then, the global average life expectancy has doubled to over 70 years. This drastic improvement in lifespan is largely due to advances in medicine and pharmacology that have reduced infection.

Pharmacology has historically been the basis of most medical treatment and remains a preferred intervention method in both preventing and fighting disease. Advances in research and development in this field continue to provide and discover new medicines and improve the effectiveness of existing medicines. The problem is that these benefits are not often realized, as it is estimated by the World Health Organization (WHO) that 50% of patients with chronic illness do not take their medications as prescribed (Brown et al., 2016).

Factors contributing to poor medication adherence include those that are related to patients (e.g., suboptimal health literacy and lack of involvement in the treatment decision-making process), those that are related to physicians (e.g., prescription of complex drug regimens, communication barriers, ineffective communication of information about adverse effects, and provision of care by multiple physicians), and those that are related to health-care systems (e.g., office visit time limitations, limited access to care, and lack of health information technology) (Brown & Bussell, 2011). As a result, poor adherence to medication puts a significant strain on health-care systems and results in increased morbidity and death. The problem with medication adherence is so significant that in its 2003 report (Sabaté et al., 2003), the WHO estimated that poor adherence to medication incurs costs of more than $100 billion per year. The same report included a quote from Haynes et al. stating that "increasing the

Big Data in Psychiatry and Neurology. https://doi.org/10.1016/B978-0-12-822884-5.00020-9
Copyright © 2021 Elsevier Inc. All rights reserved.

effectiveness of adherence interventions may have a far greater impact on the health of the population than any improvement in specific medical treatments." It is also proposed that solutions to medication adherence are nontrivial and require easy adaptation to real-life scenarios in terms of practicality. Medical workers cannot constantly watch over patients and remind them to take their medication properly. However, technology can be leveraged to improve adherence.

We believe that a smartwatch could be an incredible tool for improving medication adherence. Not only are smartwatches relatively accessible to patients and unobtrusive compared to other solutions, but are also paired with distributed computing and advances in cloud systems that allow processing of high-frequency data. Smartwatches could be used to track automatically when patients take medication using physical sensors on the device. In light of this, we developed an application that collects sensor data from the device and streams this data to Amazon Web Services (AWS; Amazon Web Services, 2020 a), Simple Storage Service (S3; Amazon Web Services, 2020 c), and a cloud storage service. With data collected from research participants performing various activities while wearing the watch, we tested various processing methods and machine learning models to determine if an automatic detection of medication intake is possible.

Formulated as a classification task, the goal of the models was to automatically classify and differentiate pill and liquid medication taking from other common activities such as texting, drinking water, or walking. Machine learning models use statistics and computation to find patterns in massive amounts of data and are proving to be effective solutions for data abundant problems across industries (West & Allen, 2018). High-frequency smartwatch sensor data is an area seemingly ripe for the application of machine learning. The results for the task at hand and implications for broader medication adherence are promising.

The rest of this chapter is organized as follows: Section 2 covers existing medication intake monitoring procedures and systems. Sections 3 and 4 contain system architecture and algorithm details. Section 5 describes experiment design, specifications of different computing settings, and experiment results under various machine learning algorithms. Section 6 provides a conclusion and future work suggestions.

## 2 Related work

Medication adherence can be defined as a patient taking medication at a prescribed dosage, timing, and frequency (Mrosek, Dehling, & Sunyaev, 2015). A high medication adherence rate helps reduce health complications, hospital readmissions, and health-care costs (Osterberg & Blaschke, 2005). Therefore, it is critical to monitor medication adherence for improving and enhancing clinical outcomes.

Medication adherence monitoring can be approached in two ways: direct and indirect monitoring. Direct monitoring includes direct observation of the patient's medication intake behaviors and laboratory detection of the medication from the patient's blood or urine samples. While direct monitoring is accurate, it is difficult to be a continuous and long-term alternative as it involves more time, effort, and cost. Indirect monitoring tracks medication adherence via medication intake logs, pharmacy refill information, and technology-based monitoring systems (Aldeer, Javanmard, & Martin, 2018). As medication intake logs and pharmacy refill information cannot provide precise medication intake time along with dosage, technology-based monitoring systems become more reliable, accurate, and cost-efficient options.

Most technology-based systems utilize one or a combination of computer vision, proximity sensors, and inertial sensors. Computer vision-based medication adherence monitoring systems require cameras that can capture a patient and his/her motions. Furthermore, they apply advanced computer vision algorithms for object segmentation, classification, detection, and tracking (Huynh, Meunier, Sequeira, & Daniel, 2009; Valin, Meunier, St-Arnaud, & Rousseau, 2006). Batz, Batz, da Vitoria Lobo, and Shah (2005) proposed a computer vision system for medication intake monitoring. The developed system utilizes a single camera and applies skin segmentation, face detection, hand and bottle tracking, and occlusion handling for improving object detection. Tucker et al. (2015) developed a data mining-driven methodology, which utilizes Microsoft Kinect vision sensors, to model and predict patients' adherence to medication protocols based on variations in their motions.

Proximity tags use contactless communication protocols including radio-frequency identification (RFID) and near-field communication (NFC). Proximity sensors establish communication and exchange data between an unpowered tag and a powered reader within relatively close proximity (about 10 cm for NFC and 1.5 m for RFID). A proximity tag is capable of storing and transferring information such as medication names and regimens. Agarawala, Greenberg, and Ho (2004) implemented a medication intake monitoring system using an RFID tag attached to a medication bottle and a stand that embeds an RFID reader and LED lights. The medication regimen from the tag is transmitted to the reader and the system sends a reminder by changing the color of LED lights. Once the medication bottle is not within the proximity of the reader, it assumes medication consumption.

Recent advances in inertial sensors have reduced sensor size and weight, while increasing reliability. An accelerometer measures acceleration in its own instantaneous rest frame while a gyroscope measures orientation and angular velocity. Many studies have shown that using sensors or devices that embed inertial sensors provides a high degree of accuracy for detecting motions during medication intake. Kalantarian, Alshurafa, and Sarrafzadeh (2016) utilize smartwatches attached to both of a patient's wrists for collecting and processing accelerometer and gyroscope data in order to detect a series of activities

220  Big data in psychiatry and neurology

including opening a bottle and twisting a cap by using the distribution of the sensor readings. Kalantarian extended his study to offer a system and algorithms based on data collected from a smart necklace. The system offers opportunities to detect whether the medication has been ingested based on the skin movement in the lower part of the neck during a swallow using a piezoelectric sensor (Kalantarian, Motamed, Alshurafa, & Sarrafzadeh, 2016).

In order to provide long-term continuous medication adherence monitoring, indirect monitoring which can monitor a patient's everyday medication intake behavior is better suited than direct monitoring. While vision-based systems do not require a user to wear a sensor device, they suffer from low accuracy, high cost, usability, and privacy issues. In addition, the proximity tag suffers from low accuracy and requires locating a medication container near a powered reader for data collection. The inertial sensor-based approach requires that sensors be attached to a patient's wrist(s) or neck, causing acceptability and comfort issues as significant barriers to adoption. However, remarkable improvements to smartwatches have been made in recent years—smartwatches embed high-frequency sensors, provide applications, are light to wear, support seamless data transfer, have long battery life, and are of durable quality. Furthermore, the recent growth of the smartwatch market has helped improve the social acceptance of smartwatch adoption (i.e., a recent report shows that 20.1 million smartwatches were sold in the United States in 2019; Statista Research Department, 2019). A survey with 221 people from Kalantarian's work shows that 72% of participants responded positively to wearing a smartwatch (Kalantarian, Alshurafa, & Sarrafzadeh, 2016).

In order to collect and monitor a massive volume of high-frequency data, it is required to have a system that can store and process large volumes efficiently. While using a high-performance computing (HPC) system provides high processing speed, this option is very expensive and not reliable in case of system or network failures (Woodbridge, Wilson, Rintoul, & Goldstein, 2015). Therefore, a distributed computing system that utilizes a cluster of many commodity computers to store and process data will be a cost-efficient and reliable solution (Fan & Bifet, 2013; Rathore, Paul, Ahmad, Anisetti, & Jeon, 2017).

As cost, accuracy, ease of use, and acceptability determine successful medication adherence (Fozoonmayeh et al., 2020), this study focuses on designing and developing a single wrist-worn smartwatch-based medication intake monitoring system connected to a scalable data storage, processing, and machine learning pipeline.

## 3 System architecture

The main focus of this study is the development of a data science pipeline which collects, stores, and processes a large volume of high-frequency sensor data (Fig. 1). At the start of the pipeline is a smartwatch application developed for collecting inertial and sound data, in light of the goal of providing a

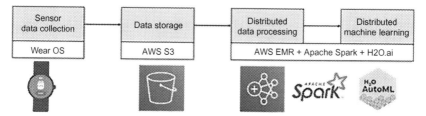

**FIG. 1** System workflow.

user-friendly, accurate, and socially acceptable sensor system. A smartwatch is able to transmit data via a cellular or WiFi connection to cloud-based scalable data storage. In order to preprocess data to transform sensor data into a feature vector amenable to machine learning, we utilize distributed computing, which combines a cluster of commodity computers connected in a network. With the intention of improving productivity and reducing human errors of domain experts, we attempt to apply automated machine learning pipeline tools in this study.

## 3.1 Smartwatch application

Smartwatches are effective activity monitoring devices because they already contain embedded sensors that can capture a wide range of movements. For example, smartwatches contain a three-axis accelerometer, gyroscope, NFC, and heart rate monitor. These seamlessly integrated sensors provide a much less obtrusive monitoring experience in comparison to smartphones or other wearable devices such as a heart rate monitor chest strap. Sensor data collected from a smartwatch application plays a critical role in providing contextual information which can be used for analyzing user behavior and generating relevant feedback for patients. In addition, in this context, the information provided from a smartwatch is more accessible and of better quality than information provided from other devices including a laptop, tablet, or smartphone, due to compactness and adjacency to the user (Ho, 2015).

In this study, we utilized an LG Watch Sport—the first Android watch running Android Wear OS 2.0, which provides an improved user interface and cellular connectivity (Berzati, Ippisch, & Graffi, 2018). The list of available biosensors that LG Watch Sport supports is listed in Table 1. As the LG Watch Sport supports cellular connectivity, collected sensor data can be directly transmitted to cloud storage without being synchronized to a smartphone or without WiFi connectivity.

In this study, we collected three-axis accelerometer and gyroscope data (with a sensor delay of up to 5 ms) along with sound data. The two inertial sensors play a critical role in detecting and distinguishing physical activities. Sound data contributes to identifying unique noises during medication intake such as

## TABLE 1 A list of biosensors embedded in LG Watch Sport and monitored attributes.

| Sensor | Monitored attributes |
| --- | --- |
| Gyroscope | Rotation |
| Accelerometer | Acceleration |
| Photoplethysmogram (PPG) | Heart rate |
| Barometer | Atmosphere pressure |

the rattling sound of pills in a pill bottle. In order to save storage space on the device and reduce the amount of data transferred over the network, the system collects activity sensor data only when there is a change in sensor readings.

## 3.2 Cloud services

Accelerometer and gyroscope sensors embedded in the smartwatch collect three-dimensional data with a frequency of 200 Hz. For audio data, the sampling rate is 48,000 Hz. This multidimensional high-frequency time-series data lends itself naturally to scalable solutions for data storage, data preprocessing, and machine learning model development.

Cloud computing utilizes storage and computing resources located in remote data centers connected via a network, and provides services on demand. Cloud computing is highly scalable, reacting to user needs dynamically by scaling resources, and providing IT infrastructure and maintenance services. Allowing resources and services to be shared by multiple users, cloud computing minimizes costs and has become an economical and powerful tool (Chaczko, Mahadevan, Aslanzadeh, & Mcdermid, 2011; Chieu, Mohindra, Karve, & Segal, 2009; Furht & Escalante, 2010; Zissis & Lekkas, 2012). Therefore, a cloud service which is scalable and accessible could be the best solution for storing and processing the high-frequency sensor data in the multiuser setting. Acknowledging these constraints, we identified AWS as a platform that provides cost-effective storage and computing frameworks (Amazon Web Services, 2020 a).

## 3.3 Data storage

For storing raw sensor data collected from a smartwatch, we utilized networked data stores which support high data availability. With AWS Simple Storage Service (S3), data is accessible from anywhere with an option to replicate data in multiple servers across different regions. S3 additionally offers a secure infrastructure via access policy options which allow only authorized users to access the data. AWS S3 also ensures scalability and flexibility by parallelizing requests and allowing any size and type of object, while minimizing time and cost for server maintenance (Amazon Web Services, 2020 c).

## 3.4 Distributed data processing

Hadoop's MapReduce, introduced in 2004, implemented efficient distributed techniques in an attempt to process and analyze vasts amounts of data (Dean & Ghemawat, 2008). MapReduce splits data into smaller chunks across multiple nodes in a cluster and processes a task, for example, filtering and sorting, completely in parallel. The outputs of the map processes become the input of a reduce operation, which performs a summary operation, for example, counting and collecting. This highly effective model allows users to design programs with successive Map and Reduce operations, and is a popular and powerful programming paradigm (Hadoop, 2020).

Apache Spark adopts the MapReduce model but executes a task closer to 100 times faster than Hadoop MapReduce by processing data in memory. Also, Spark uses efficient job scheduling and adds an additional recovery model using a directed acyclic graph (DAG) representation while still running 10 times faster in disk than Hadoop MapReduce (Apache Spark, 2019; Gu & Li, 2013; Zaharia, Chowdhury, Franklin, Shenker, & Stoica, 2010).

For processing sensor data and applying machine learning algorithms using Spark, we utilized AWS Elastic MapReduce (EMR; Fig. 2; Amazon Web Services, 2020 b) which uses Hadoop's YARN (Yet Another Resource Negotiator) for provisioning the cluster's hardware resources (Elastic Compute Cloud [EC2] instances) and installs the required software for running Apache Spark.

## 4 Algorithms

With the goal of processing high-frequency sensor data and classifying medication intake activities, we designed and developed a distributed preprocessing

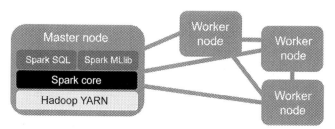

**FIG. 2** AWS EMR cluster architecture.

algorithm to impute missing data and extract descriptive statistical features from both inertial and audio data. Using the preprocessed data, we applied distributed AutoML to find the best performing algorithms for activity type classification.

## 4.1 Distributed preprocessing

As mentioned in Section 3.1, a smartwatch application only records readings to a file when there are new events detected by sensors, saving battery, and storage. In order to capture relevant and meaningful signatures of high-frequency data streams and pass this into machine learning algorithms, it is first necessary to select sensor data of interest from a log and apply missing data imputation.

As the length of time series data varies, the developed algorithm discretizes outputs of missing data imputation into a standard number of windows and extracts statistical features of data from each window in addition to the entire (global) recording. We reduced the time series data length of $n$ to the length of $f$ ($f \leq n$) and calculated statistics for the entire data and over each sliding window. When the original time series after imputing missing data is $C = c_1, \ldots, c_n$, the mean over the sliding window ($\overline{C}$) is calculated by Eq. (1). In addition to the mean, we also calculated other aggregate measures including minimum, maximum, standard deviation, 5th, 25th, 50th, 75th, and 95th percentiles for the entire time frame and each sliding window. Including and calculating statistical values other than the mean as features of the data help estimate the data distribution and add resilience to outliers. For example, percentile values provide more detail about the distribution of the data (Bruce & Bruce, 2017).

$$\overline{\mu}_i = \frac{f}{n} \sum_{j=\frac{n}{f}(i-1)+1}^{\frac{n}{f}i} c_j \qquad (1)$$

In order to process high-frequency data which was collected every 5 ms, the preprocessing step (shown in Fig. 3) was developed in a distributed manner. The pseudocode of the designed algorithm is defined in Algorithm 1. The

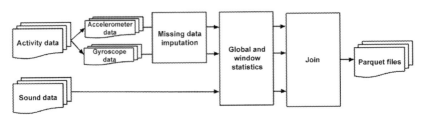

FIG. 3 Cloud-based distributed preprocessing pipeline.

A scalable medication intake monitoring system **Chapter | 11** **225**

---

**ALGORITHM 1  Preprocessing algorithm pseudocode**

```
 1: function MISSINGDATAIMPUTATION(x)
 2:     array_size = int((max_timestamp(x)−min_timestamp(x))/time_interval)+1
 3:     reading[i] = [None] * array_size
 4:     for i in range(0, len(x)) do
 5:         if first_reading == None then
 6:             first_reading = x[i]
 7:         end if
 8:         reading[int((time(x[i])−min_timestamp(x))/time_interval)] = x[i]
 9:     end for
10:     for i in range(0, array_size) do
11:         if readings[i] == None then
12:             if prev_reading ! = None then
13:                 readings[i] = prev_reading
14:             else
15:                 readings[i] = first_reading
16:             end if
17:         else
18:             prev_reading = readings[i]
19:         end if
20:     end for
21: end function
22:
23: function GLOBALDATASTATS(data)
24:     return mean(data), std(data), min(data), percentile(data, 5),
25:            percentile(data, 25), median(data), percentile(data, 75),
26:            percentile(data, 95), max(data)
27: end function
28:
29: function WINDOWDATASTATS(data, window_size)
30:     window_stats = [ ]
31:     windows = array_split(data, window_size)
32:     for i in range(0, window_size) do
33:         window_stats.append(mean(windows[i]))
34:         window_stats.append(std(windows[i]))
35:         window_stats.append(min(windows[i]))
36:         window_stats.append(percentile(windows[i], 5))
37:         window_stats.append(percentile(windows[i], 25))
38:         window_stats.append(median(windows[i]))
39:         window_stats.append(percentile(windows[i], 75))
40:         window_stats.append(percentile(windows[i], 95))
41:         window_stats.append(max(windows[i]))
42:     end for
43:     return window_stats
44: end function
45:
```

*Continued*

**226** Big data in psychiatry and neurology

---

**ALGORITHM 1 Preprocessing algorithm pseudocode—cont'd**

46: **function** PREPROECESSINGDATASTREAM(*x, window_size*)
47:    **if** *x* is *activity_data* **then**
48:        *complete_reading* = MissingDataImputation(*x*)
49:    **else if** *x* is *sound_data* **then**
50:        *complete_reading* = *x*
51:    **end if**
52:    *global_stats* = GLOBALDATASTATS(*complete_reading*)
53:    *window_stats* = WINDOWDATASTATS(*complete_reading, window_size*)
54:    **return** *concatenate*(*global_stats, window_stats*)
55: **end function**
56:
57: **preprocessed_data** = PreproecessingDataStream(*activity, window_size*)
58:                            .leftOuterJoin(PreproecessingDataStream(*audio,*
*window_size*))

---

preprocessing step yields $f+1$ features for a three-axis accelerometer, three-axis gyroscope, and single-channel audio sensors per activity from each subject.

The output of this step was stored in parquet which is a columnar storage formatted for the Hadoop distributed file system (HDFS). HDFS is a distributed file system which provides high-throughput access to large data sets. HDFS coordinates storage and replication across nodes for ensuring high system availability and fault tolerance (Borthakur et al., 2008).

## 4.2 Distributed AutoML and machine learning

In order to accurately classify the medication intake activity, we grouped the activity labels into binary classes—medication intake activities and nonmedication intake activities (i.e., 1 or 0). Using these labels, we applied automated machine learning along with traditional machine learning techniques and compared their predictive performance using metrics such as AUC scores, as well as execution time.

Automated machine learning (AutoML) enables automation of time and cost-intensive model building processes including model and feature selection and hyperparameter optimization. Therefore, AutoML helps optimize machine learning tasks with less of a need and investment of human experts in machine learning and statistics (Gijsbers et al., 2019; Zöller & Huber, 2019).

A recent benchmark study that compares various AutoML platforms showed that H2O works the best in various classification tasks (Gijsbers et al., 2019). H2O AutoML algorithms include random forests, gradient boosted models, XGBoost, generalized linear models (GLMs), and artificial neural networks (ANNs) with varying specifications—including hyperparameters. H2O AutoML also trains Stacked Ensemble models using the best-trained model from each family of models with the best performance to produce a highly predictive ensemble model (H2O.ai, 2020).

### 4.2.1 Random forest

A random forest classifier is an ensemble-based supervised learning algorithm that aggregates multiple decision trees (Safavian & Landgrebe, 1991). The algorithm uses random sampling of training data when building trees, and a random subset of features when splitting the nodes. Each decision tree in a random forest learns from random samples which are drawn using bootstrapping methods. This inherent randomness within the trees avoids overfitting issues complicit with overly deterministic decision trees, which allows random forest to perform well without much hyperparameter tuning. Predictions for testing are calculated by averaging the predictions of each decision tree (Breiman, 2001).

### 4.2.2 Gradient boosting

Gradient boosting is another ensemble method similar to its random forest counterpart. They both benefit from a collection of decision trees, subsequently making a prediction based on the weighted scoring from each of those trees (Mason, Baxter, Bartlett, & Frean, 2000). The primary difference in gradient boosting is that the first tree is used to make a prediction, and, once evaluated, an additional tree is added such that it minimizes the error of the previous trees. Trees are added, one at a time, until a robust model is developed (Friedman, 2001).

### 4.2.3 XGBoost

Extreme gradient boosting (XGBoost) is a specific open-source implementation of gradient boosting. XGBoost uses second-order gradients to minimize the loss function and advanced regularization, such as $L1$ and $L2$ regularization to control overfitting (Ng, 2004). XGBoost is a sparsity-aware algorithm for sparse data and achieves high scalability and accuracy, gaining attention as an algorithm of choice for many winning teams of machine learning competitions (Chen & Guestrin, 2016).

### 4.2.4 Generalized linear models

GLM estimate regression models that predict the probability that an input value belongs to a particular category by fitting the data to a linear model (McCullagh, 2018). The main strength of the regression models is the interpretability of the model output and speed of model training. The algorithm can also be regularized to avoid overfitting and is frequently used as a base model for classification problems.

### 4.2.5 Artificial neural networks

The ANN model used in H2O AutoML is based on a multilayer feedforward ANN trained with stochastic gradient descent (Candel, Parmar, LeDell, & Arora, 2016). ANN is the first and simplest type of ANN and consists of input, hidden, and output layers. The hidden layer includes multiple layers, where each neuron in one layer has a directed connection to the neurons of the subsequent layer. ANN is capable of learning weights that map any input to the output in the connection using gradient descent. ANN works well with tabular data as input, while not as well with sequential data (Fan, Qian, Xie, & Soong, 2014).

### 4.2.6 Stacked ensemble

Stacked ensemble applies ensemble methods to obtain the best predictive model using multiple base machine learning algorithms. A stacked ensemble combines diverse and strong machine learning algorithms and creates a weighted combination model. To create this weighted combination, called as a super learner, the stacked ensemble algorithm applies cross-validation to select weights to combine candidate base learners. Studies show that stacked ensemble yields high performance (Van der Laan, Polley, & Hubbard, 2007).

## 5 Experiment results

For validating the designed data science pipeline, we deployed cloud-based distributed systems for storing and processing smartwatches sensor data. Section 5.1 describes the details of the hardware and human subjects, along with performed activities. Section 5.2 demonstrates the accuracy and time efficiency of the developed system compared to nondistributed computing and within various cluster settings.

## 5.1 Experiment setting

In this study, the system was designed to store a large volume of high-frequency sensor data, extract features, and apply AutoML to choose the best models with scalability and time efficiency using cloud-based frameworks. For validating the developed pipeline, we recruited 44 subjects (Table 2) to collect sensor data while performing various activities listed in Table 3. For liquid and pill medication intake activities, subjects used standard child-proof medication bottles with twist caps. At the time of recording, subjects would twist the cap off to open the medication bottle, take the medication then close the cap. Texting, writing, and walking activities were recorded for approximately 30–60 s. Visualizations in Figs. 4 and 5 show that sensor and sound readings have distinct features for different activities. All procedures performed in studies involving human participants were in accordance with the University of San Francisco, Institutional Review Board (IRB) for the Protection of Human Subjects.

For evaluating the efficiency of data processing and machine learning algorithms on different cluster settings, we used various cluster specifications including the number of nodes, CPUs, and memory sizes listed in Table 4. We also compared distributed computing with nondistributed computing, and all nondistributed computing was performed on a single instance with 4 CPUs and 16 GB memory.

A scalable medication intake monitoring system **Chapter | 11** **229**

**TABLE 2** Recruited subject information.

| Criteria | | Number of subjects |
|---|---|---|
| Dominant hand | Right | 38 |
| | Left | 6 |
| Watch wrist | Right | 20 |
| | Left | 24 |
| Sex | Female | 19 |
| | Male | 25 |
| Age group | 18–20 | 6 |
| | 20–24 | 20 |
| | 25–29 | 6 |
| | 30–34 | 3 |
| | 35–39 | 1 |
| | 45–49 | 3 |
| | 55–60 | 5 |

**TABLE 3** Activity types and watch wrists (each subject repeated each activity five times).

| Activity | Activity class | Watch wrist |
|---|---|---|
| Pill medication intake | Medication intake | Left |
| | | Right |
| Liquid medication intake | | Left |
| | | Right |
| Texting | Nonmedication intake | Preferred wrist by the subject |
| Writing | | |
| Walking | | |
| Bottled water intake | | |

**230** Big data in psychiatry and neurology

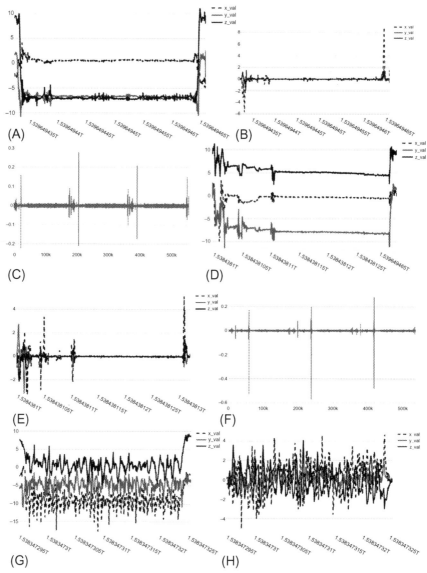

**FIG. 4** Example sensor readings from accelerometer, gyroscope, and audio readings of nonmedication intake activities. (A) Texting (accelerometer); (B) texting (gyroscope); (C) texting (audio); (D) writing (accelerometer); (E) writing (gyroscope); (F) writing (audio); (G) walking (accelerometer); (H) walking (gyroscope);

*(Continued)*

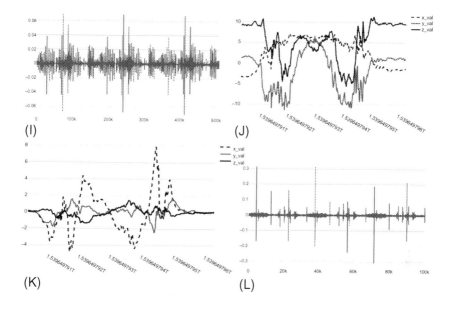

**FIG. 4, CONT'D** (I) walking (audio); (J) bottled water intake (accelerometer); (K) bottled water intake (gyroscope); and (L) bottled water intake (audio).

### 5.2 Results

To evaluate the performance of our models, we compared the preprocessing time, model fitting time, and the AUC score. AUC is the area under the receiver operating characteristic (ROC) curve (Eq. 4). The ROC curve plots the true positive rate (TPR) against the false positive rate (FPR), where TPR is on the $y$-axis and FPR is on the $x$-axis. AUC varies between 0 and 1, where 1 is the best and 0 is the worst.

$$TPR = \frac{TP}{TP + FN} \quad (2)$$

$$FPR = \frac{FP}{TN + FP} \quad (3)$$

$$AUC = \int_0^1 TPR(FPR^{-1}(x))\,dx \quad (4)$$

In order to determine the optimal window size for predicting medication intake activities, we tuned the number of the windows—between 5 and 100, in increments of 5. We found that naturally, the bigger the window count is, the longer the preprocessing takes (Fig. 6). We also compared the execution

## 232  Big data in psychiatry and neurology

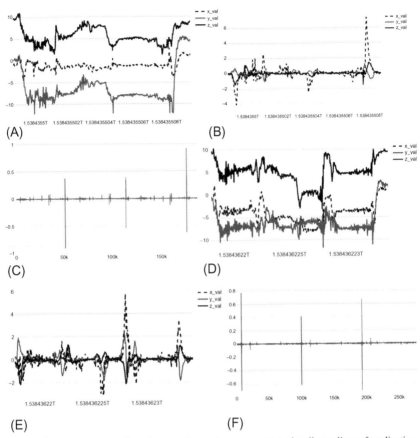

FIG. 5 Example sensor readings from accelerometer, gyroscope, and audio readings of medication intake activities. (A) Pill medicine intake (accelerometer); (B) pill medicine intake (gyroscope); (C) pill medicine intake (audio); (D) liquid medicine intake (accelerometer); (E) liquid medicine intake (gyroscope); and (F) Liquid medicine intake (audio).

**TABLE 4** EMR cluster configurations used for launching Apache Spark (given CPU and memory information are for each node).

| Cluster ID | Number of nodes | Number of CPUs | Memory |
|---|---|---|---|
| Cluster 1 | 3 | 4 | 16 GB |
| Cluster 2 | 5 | 4 | 16 GB |
| Cluster 3 | 7 | 4 | 16 GB |
| Cluster 4 | 3 | 8 | 32 GB |
| Cluster 5 | 3 | 16 | 64 GB |

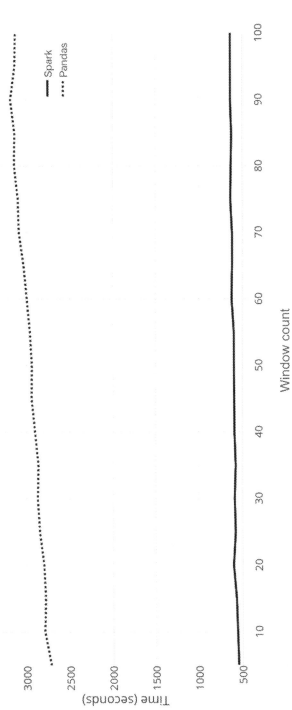

**FIG. 6** Execution time comparison between a three-node Spark cluster (Cluster 1 in Table 4) and Pandas—both executed on machines with the same specifications.

time of distributed preprocessing with a nondistributed algorithm using Pandas, the library written for the Python programming language for data manipulation and analysis. The result shows that the developed preprocessing algorithm takes between 2722.14 and 3125.80 s on Pandas, whereas it only takes between 546.63 and 633.73 s using Apache Spark.

In order to find the optimal window count that yields the best model with the highest AUC, H2O AutoML was performed with 10-fold validation. In $n$-fold validation, the original sample is randomly partitioned into $n$-equal-sized subsets, and a single subset is held as a validation set, with the remaining subsets used as a training dataset. This cross-validation process is then repeated $n$ times, using a new subset as a validation set for each repetition. Finally, evaluation metrics are averaged for a better approximation of the true performance on unseen data. Fig. 7 shows the highest AUC value along with the execution time among 10 returned models for each window count. The experiment result shows that the window count of five returns the best model with the AUC value of 0.975 within 805.70 s on *Cluster 1*. In this case, the best-returned model was the stacked ensemble of gradient boosting, XGBoost, and random forest

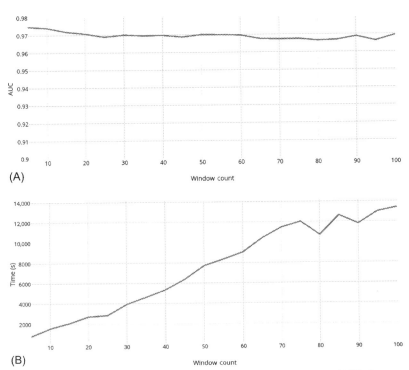

**FIG. 7** AutoML experiment results with varying window count on Cluster 1. (A) Window count and AUC of the best model; (B) window count and execution time for returning 10 best models using AutoML.

models. As Fig. 7 shows, the execution time grows linearly as the number of the features increases while the AUC is only between 0.966 and 0.974. In addition, random forest, gradient boosting, and XGBoost and their stacked ensemble models yield the best results for the given varying window counts of 5–100.

Furthermore, we compared execution time of various cluster settings given in Table 4 to find the optimal cluster settings that yield fastest execution time for preprocessing and AutoML (Fig. 8). The result shows that running a cluster with bigger memory and more CPUs improves execution time more than increasing the number of nodes with the given data set. Although adding more nodes speeds up in many cases, cross-node connections and network traffic affected speed degradation in some cases. Through this experiment, we could find that window count of five yields the highest time efficiency in addition to the highest AUC in all tested cluster settings. A related study predicting remaining pill count using inertial sensors and sound also showed a window count of five as optimal (Cheon et al., 2020).

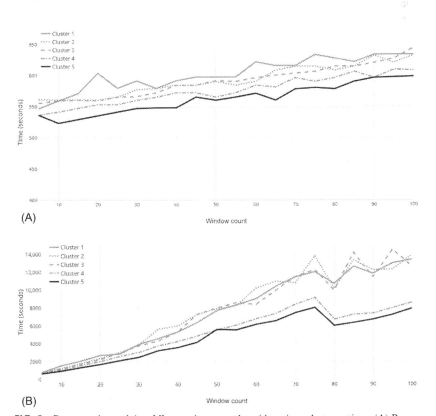

**FIG. 8** Preprocessing and AutoML experiment results with various cluster settings. (A) Preprocessing execution time comparison of various cluster settings; (B) AutoML execution time comparison of various cluster settings.

As the random forest, gradient boosting, and XGBoost models yield high accuracies using H2O AutoML, we implement and compare the same algorithms in Scikit-learn, a machine learning library for the Python programming language. Using the same parameters to the extent possible, H2O-based distributed machine learning algorithms yield higher AUC values than their counterpart algorithms implemented in Scikit-learn (Fig. 9). This could be due to the randomness of the algorithms or due to differences in algorithm implementation details including default parameters and functions.

## 6 Conclusion

It is critical to improve medication adherence rates, as poor adherence causes increased levels of hospital readmission, emergency department and physician visits, morbidity, and other associated health-care costs. To improve medication adherence, we developed a user-friendly, affordable, and scalable medication adherence monitoring system.

We attempted to improve usability for monitoring a patient's medication intake activities by utilizing a smartwatch, an increasingly popular and powerful wearable device with various embedded biosensors and convenient, efficient user interfaces. For storing and processing transmitted data from smartwatches in the multiuser setting, we store and process data using cloud-based scalable storage and computing resources. For improving the efficiency of data preprocessing and machine learning, we designed and implemented our processing framework using Apache Spark, a MapReduce paradigm-based distributed computing framework, and H2O AutoML, a distributed machine learning framework. AutoML helps discover best performing machine learning models and hyperparameters, while minimizing human experts' time. Our best model was a stacked ensemble model achieving the AUC value of 0.975 within 647.07 s on a three-node cluster with 16 CPUs and 64 GB memory. Resulting in a highly predictive system for classifying medication intake, this is promising for the intersection of wearable technology, distributed data processing, and improved medication adherence.

Adding extra features using other biosensors embedded in a smartwatch may further enhance accuracy and reliability, although it would require more preprocessing and training time. In addition to the biosensors and microphone utilized in this study, many smartwatches are equipped with NFC which establishes communication and exchanges data between two electronic devices within close proximity. Developing and incorporating existing components and data sources like the NFC reader into a smartwatch-based medication adherence system in novel ways could yield even higher accuracy rates and reliability.

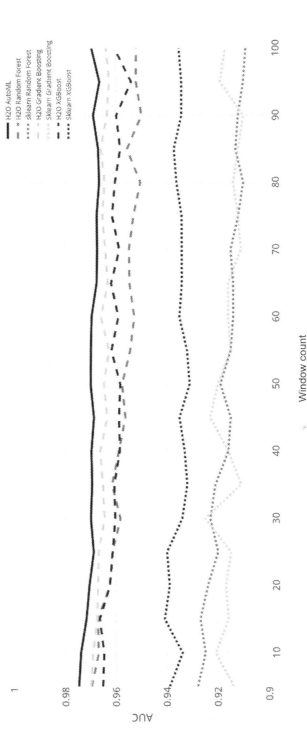

**FIG. 9** AUC comparison of H2O AutoML and distributed machine learning algorithms on a three-node Spark cluster (Cluster 1 in Table 4) and corresponding machine learning algorithms using Scikit-learn on a single computer.

238 Big data in psychiatry and neurology

## Acknowledgments

This work was supported by Jesuit Foundation Grant, University of San Francisco Faculty Development Fund, and Systers Pass-it-on Award by Anita Borg Institute for Women and Technology. Any opinions, findings, conclusions, or recommendations expressed in this material are those of the authors and do not necessarily reflect the views of the funding organizations.

## References

Agarawala, A., Greenberg, S., & Ho, G. (2004). *The context-aware pill bottle and medication monitor*. University of Calgary.

Aldeer, M., Javanmard, M., & Martin, R. P. (2018). A review of medication adherence monitoring technologies. *Applied System Innovation, 1*(2), 14.

Amazon Web Services. (2020a). *Amazon*. https://aws.amazon.com.

Amazon Web Services. (2020b). *Amazon EMR*. https://aws.amazon.com/emr.

Amazon Web Services. (2020c). *Amazon S3*. https://aws.amazon.com/s3/.

Spark, Apache. (2019). *Apache spark: Lightning-fast cluster computing*. http://spark.apache.org.

Batz, D., Batz, M., da Vitoria Lobo, N., & Shah, M. (2005). A computer vision system for monitoring medication intake. In *The 2nd Canadian conference on computer and robot vision (CRV'05)* (pp. 362–369).

Berzati, B., Ippisch, A., & Graffi, K. (2018). An android wear OS framework for sensor data and network interfaces. In *2018 IEEE 43rd Conference on local computer networks workshops (LCN Workshops)* (pp. 98–104).

Borthakur, D. (2008). HDFS architecture guide. *Hadoop Apache Project, 53*(1–13), 2.

Breiman, L. (2001). Random forests. *Machine Learning, 45*(1), 5–32.

Brown, M. T., Bussell, J., Dutta, S., Davis, K., Strong, S., & Mathew, S. (2016). Medication adherence: Truth and consequences. *The American Journal of the Medical Sciences, 351*(4), 387–399.

Brown, M. T., & Bussell, J. K. (2011). Medication adherence: WHO cares? *Mayo Clinic Proceedings, 86*(4), 304–314.

Bruce, P., & Bruce, A. (2017). *Practical statistics for data scientists: 50 essential concepts*. O'Reilly Media, Inc.

Candel, A., Parmar, V., LeDell, E., & Arora, A. (2016). *Deep learning with H2O*. H2O.ai Inc.

Chaczko, Z., Mahadevan, V., Aslanzadeh, S., & Mcdermid, C. (2011). Availability and load balancing in cloud computing. In *International conference on computer and software modeling, Singapore: Vol. 14*.

Chen, T., & Guestrin, C. (2016). XGBoost: A scalable tree boosting system. In *Proceedings of the 22nd ACM SIGKDD international conference on knowledge discovery and data mining* (pp. 785–794).

Cheon, A., Jung, S. Y., Prather, C., Sarmiento, M., Wong, K., & Woodbridge, D. M.-K. (2020). A machine learning approach to detecting low medication state with wearable technologies. In *International conference of the IEEE engineering in medicine and biology society (EMBC)*..

Chieu, T. C., Mohindra, A., Karve, A. A., & Segal, A. (2009). Dynamic scaling of web applications in a virtualized cloud computing environment. In *2009 IEEE international conference on e-business engineering* (pp. 281–286).

Dean, J., & Ghemawat, S. (2008). MapReduce: Simplified data processing on large clusters. *Communications of the ACM, 51*(1), 107–113.

Fan, W., & Bifet, A. (2013). Mining big data: Current status, and forecast to the future. *ACM SIGKDD Explorations Newsletter, 14*(2), 1–5.

Fan, Y., Qian, Y., Xie, F.-L., & Soong, F. K. (2014). TTS synthesis with bidirectional LSTM based recurrent neural networks. In *Fifteenth annual conference of the international speech communication association..*

Fozoonmayeh, D., Le, H. V., Wittfoth, E., Geng, C., Ha, N., Wang, J., … Woodbridge, D. M.-K. (2020). A scalable smartwatch-based medication intake detection system using distributed machine learning. *Journal of Medical Systems, 44*(4), 1–14.

Friedman, J. H. (2001). Greedy function approximation: a gradient boosting machine. *Annals of Statistics, 29*(5), 1189–1232.

Furht, B., & Escalante, A. (2010). *Vol. 3. Handbook of cloud computing.* Springer.

Gijsbers, P., LeDell, E., Thomas, J., Poirier, S., Bischl, B., & Vanschoren, J. (2019). An open source AutoML benchmark. *arXiv preprint arXiv:1907.00909.*

Gu, L., & Li, H. (2013). Memory or time: Performance evaluation for iterative operation on Hadoop and Spark. In *2013 IEEE 10th international conference on high performance computing and communications & 2013 IEEE international conference on embedded and ubiquitous computing (HPCC_EUC)* (pp. 721–727).

H2O.ai. (2020). *AutoML: Automatic machine learning.* https://docs.h2o.ai/h2o/latest-stable/h2o-docs/automl.html.

Hadoop, A. (2020). *Apache Hadoop.* http://hadoop.apache.org.

Ho, A. (2015). Step-by-step android wear application development. *Amazon Digital Services.*

Huynh, H. H., Meunier, J., Sequeira, J., & Daniel, M. (2009). Real time detection, tracking and recognition of medication intake. *World Academy of Science, Engineering and Technology, 60,* 280–287.

Kalantarian, H., Alshurafa, N., & Sarrafzadeh, M. (2016). Detection of gestures associated with medication adherence using smartwatch-based inertial sensors. *IEEE Sensors Journal, 16*(4), 1054–1061.

Kalantarian, H., Motamed, B., Alshurafa, N., & Sarrafzadeh, M. (2016). A wearable sensor system for medication adherence prediction. *Artificial Intelligence in Medicine, 69,* 43–52.

Mason, L., Baxter, J., Bartlett, P. L., & Frean, M. R. (2000). Boosting algorithms as gradient descent. In *Advances in neural information processing systems* (pp. 512–518).

McCullagh, P. (2018). *Generalized linear models.* Routledge.

Mrosek, R., Dehling, T., & Sunyaev, A. (2015). Taxonomy of health IT and medication adherence. *Health Policy and Technology, 4*(3), 215–224.

Ng, A. Y. (2004). Feature selection, L 1 vs. L 2 regularization, and rotational invariance. In *Proceedings of the twenty-first international conference on machine learning* (p. p. 78).

Osterberg, L., & Blaschke, T. (2005). Adherence to medication. *New England Journal of Medicine, 353*(5), 487–497.

Prentice, T. (2008). Health, history, and hard choices: Funding dilemmas in a fast-changing world. *Nonprofit and Voluntary Sector Quarterly, 37*(Suppl. 1), 63S–75S.

Rathore, M. M., Paul, A., Ahmad, A., Anisetti, M., & Jeon, G. (2017). Hadoop-based intelligent care system (HICS) analytical approach for big data in IoT. *ACM Transactions on Internet Technology (TOIT), 18*(1), 1–24.

Sabaté, E., Rand, C., Hotz, S., De Castro, S., Karkashian, C., Chesney, M., ... Dick, J. et al. (2003). *Adherence to long-term therapies: evidence for action* (pp. 19–25). World Health Organization.

Safavian, S. R., & Landgrebe, D. (1991). A survey of decision tree classifier methodology. *IEEE Transactions on Systems, Man, and Cybernetics, 21*(3), 660–674.

## 240 Big data in psychiatry and neurology

Zöller, M.-A., & Huber, M. F. (2019). Benchmark and survey of automated machine learning frameworks. *arXiv preprint arXiv:1904.12054*.

Statista Research Department. (2019). *Smartwatch devices unit sales in the United States from 2016 to 2020*. https://www.statista.com/statistics/381696/wearables-unit-sales-forecast-united-states-by-category/.

Tucker, C. S., Behoora, I., Nembhard, H. B., Lewis, M., Sterling, N. W., & Huang, X. (2015). Machine learning classification of medication adherence in patients with movement disorders using non-wearable sensors. *Computers in Biology and Medicine, 66*, 120–134.

Valin, M., Meunier, J., St-Arnaud, A., & Rousseau, J. (2006). Video surveillance of medication intake. In *2006 international conference of the IEEE engineering in medicine and biology society* (pp. 6396–6399).

Van der Laan, M. J., Polley, E. C., & Hubbard, A. E. (2007). Super learner. *Statistical Applications in Genetics and Molecular Biology, 6*(1), 25.

West, D., & Allen, J. (2018). *Brookings*.

Woodbridge, D. M.-K., Wilson, A. T., Rintoul, M. D., & Goldstein, R. H. (2015). Time series discord detection in medical data using a parallel relational database. In *2015 IEEE international conference on bioinformatics and biomedicine (BIBM)* (pp. 1420–1426).

Zaharia, M., Chowdhury, M., Franklin, M. J., Shenker, S., & Stoica, I. (2010). Spark: cluster computing with working sets. *HotCloud, 10*(10–10), 95.

Zissis, D., & Lekkas, D. (2012). Addressing cloud computing security issues. *Future Generation Computer Systems, 28*(3), 583–592.

Chapter 12

# Evaluating cascade prediction via different embedding techniques for disease mitigation

Abhinav Choudhury[a], Shubham Shakya[b], Shruti Kaushik[a], and Varun Dutt[a]

[a]*Indian Institute of Technology Mandi, Kamand, India,* [b]*National Institute of Technology Kurukshetra, Kurukshetra, India*

## 1 Introduction

Understanding how the diffusion of innovation occurs inside a social network can help create strategies for better and faster dissemination of information. This understanding is highly relevant in the healthcare domain, where it is paramount that medical innovations and guidelines are disseminated appropriately and swiftly (Kasthuri, 2018). The adoption of innovations (e.g., medications) inside a physician's social network may differ from physician to physician. While some physicians may adopt an innovation early, others may embrace it late or not at all. This innovation adoption by a physician is highly influenced by interpersonal communication with the members of her ego network (Choudhury, Kaushik, & Dutt, 2017). Most diffusion studies conclude that active users (who adopted a medication) exert an influence on their inactive neighbors to perform an action (i.e., adopting the medication). The diffusion process that takes place inside a social network assumes that people tend to perform an action (adopting a medication) if they see their social contacts (friends, family, acquaintances) performing that action in the network (Domingos & Richardson, 2001). As such, understanding the underlying dynamics of the social network may lead to better dissemination of medical information and guidelines.

Information cascades, which describe the flow of information inside a social network, play a pivotal role in the process of information diffusion. The modeling of information cascades has many applications, such as identifying opinion

Big Data in Psychiatry and Neurology. https://doi.org/10.1016/B978-0-12-822884-5.00006-4
Copyright © 2021 Elsevier Inc. All rights reserved.

leaders; growth of a cascade; detecting diffusion sources of a particular story, post, or a tweet (Bao, Cheung, & Liu, 2016; Cheng, Dow, Kleinberg, & Leskovec, 2014; Kempe, Kleinberg, & Tardos, 2003; Zhu, Chen, & Ying, 2017). If information cascades can be predicted, one can make informed decisions in multiple scenarios. For example, by predicting the next user that will get infected, healthcare professionals may help stop the spread of diseases. Also, the prediction of a potential influence of a rumor may enable administrators to take safeguards against the spread of misinformation (Li, Ma, Guo, & Mei, 2017).

Most prior works have investigated the prediction of diffusion cascades using generative models based on epidemiology (Cheng et al., 2014) or using a mixture of structural, content, and temporal features (Matsubara, Sakurai, Prakash, Li, & Faloutsos, 2012). Some recent works have also used machine learning-based models (Bourigault, Lamprier, & Gallinari, 2016; Li et al., 2017; Wang, Zheng, Liu, & Chang, 2017). In most social networks, information is usually diffused over the underlying social graph representing the users' social structure. For predicting cascade, researchers have assumed strong correlations between the spread of content and structural properties of the underlying social graph (Cheng et al., 2014). Therefore researchers mostly relied on manually extracted features from the underlying network and the cascade itself (Kefato et al., 2018). Unfortunately, manually extracting structural features from a social graph needs extensive domain knowledge, and it is a time-consuming process that may not lead to good feature extraction.

In recent years, certain graph embedding techniques (Cai, Zheng, & Chang, 2018) have been developed that aim to represent a graph as a low-dimensional (embedding's dimensions) vector, where the graph structures are preserved (i.e., connected nodes are closer to each other). This representational learning method helps obtain data representations that make it easier to extract useful structural information from graphs (Cai et al., 2018). Prior researchers have used graph embeddings to assist in information diffusion prediction on graphs (Bourigault et al., 2016; Li et al., 2017; Wang et al., 2017). For example, Bourigault et al. (2016) generated node embeddings of users based on the closeness of these embeddings and the predicted diffusion between users. Similarly, DeepCas (Li et al., 2017) generated embeddings of the diffusion subgraph using the Gated Recurrent Unit (GRU) (Cho, Van Merriënboer, Bahdanau, & Bengio, 2014) to predict the future cascade size of Twitter tweets. Topo-LSTM (Wang et al., 2017) used a recurrent neural network (RNN) (Rumelhart, Hinton, & Williams, 1986), which took dynamic directed acyclic graphs (DAGs) as input to generate embeddings for each node for the diffusion prediction. However, all the models used in (Bourigault et al., 2016; Li et al., 2017; Wang et al., 2017) have used a variant of RNN, which are memory-based networks, which may not be the appropriate choice when there is a scarcity of data (Feng, Zhou, & Dong, 2019).

Furthermore, most graph embedding techniques may share different design choices and hyperparameters. One important hyperparameter is the number of embedding dimensions (Patel & Bhattacharyya, 2017). The embedding dimensions are usually decided via a rule of thumb, mostly between 16 and 256 based on graph size, or by trial and error (Patel & Bhattacharyya, 2017). However, there have not been rigorous studies on the number of embedding dimensions used during training different graph embeddings.

The primary objective of this research is to bridge the literature gaps highlighted before and evaluate different graph embedding techniques with different embedding dimensions via two machine learning (ML) models in the healthcare domain. Specifically, in this research, different ML models rely upon different embedding techniques and dimensions for predicting the next set of physicians that will be activated in an information cascade inside a physician's social network.

In what follows, we first provide a brief review of related work involving cascade prediction and graph embeddings. Then, we explain the data used in this research and describe the diffusion cascade prediction problem. In the next section, we explain the methodology of the three embedding techniques and implement machine learning models to predict the next set of users that become activated in an information cascade. Furthermore, we present our experimental results and discuss the implications of this research and its possible extensions.

## 2 Background

In social network analysis (SNA) (Scott, 1991), the structure of social interactions is viewed as a network composed of nodes (people) interconnected by edges (social relations, friendship, or advice). SNA is an ideal approach for describing interaction patterns to study how social influence is transmitted among physicians and how it affects their contingent behavior (Rumelhart et al., 1986). Social networks are important channels for the diffusion of innovations and social influence (Scott, 1991). Most diffusion processes can be viewed as a collection of interaction traces (e.g., prescription logs of physicians) indicating when people post, tweet a piece of information, or buy a particular product (Bourigault et al., 2016). Most often, one only observes the when and where the actions had taken place rather than why it happened (Bourigault et al., 2016). Therefore modeling the diffusion process mostly boils down to approximating this mechanism based on the interaction traces. As such, the diffusion cascades of particular types of information like Tweets/microblogs, photos, videos, and academic papers have been empirically proven to be predictable to some extent (Li et al., 2017).

In literature, cascade prediction is formulated either as a classification problem (popularity of a piece of information) or as a regression problem (future size of a cascade) (Li et al., 2017). But prior work has not attempted to predict the next set of people to whom the information is going to be diffused. Moreover,

**244** Big data in psychiatry and neurology

prior research toward predicting information cascades has relied on handcrafted features extracted from the underlying network structure and the cascade itself (Kefato et al., 2018).

Manually extracting structural features from a social graph is a time-consuming exercise and it may not lead to good feature extraction. Additionally, handcrafted features are not able to fully represent both the local and the global structure of a graph and the complex interaction between local and global properties (Kefato et al., 2018). However, with the recent development of graph embedding techniques for representing graph structures, we can preserve the underlying structure of the social network and get a better representation of the local and global structures. Graph embedding represents a graph as low-dimensional vectors such that the graph structure is preserved. Researchers have used as graph embedding techniques like Node2Vec (Grover & Leskovec, 2016) and DeepWalk (Perozzi, Al-Rfou, & Skiena, 2014) for predicting interests of users in social networks as well as link prediction tasks like predicting future relationships of users in social networks. Node2Vec and DeepWalk are deep learning methods, where the embeddings are generated from random walks sampled from the graph. Wang, Cui, and Zhu (2016) used Structural Deep Network Embedding (SDNE), which uses autoencoders (AE) (Salakhutdinov & Hinton, 2009) to generate graph embeddings to perform multilabel classification of nodes. While there are multiple graph embedding techniques, not much research has been done on the effect of size of embedding dimensions on these embeddings. The embedding dimensions are usually decided via a rule of thumb (Patel & Bhattacharyya, 2017).

Graph embeddings have also been successfully implemented in cascade prediction tasks. For example, the Embedded-IC algorithm (Bourigault et al., 2016) takes a cascade-based modeling approach, which assumes that an active node can activate an inactive node. It differentiates its activate nodes as senders and inactive nodes as receivers. The receivers (inactive nodes) receive information from the sender (active nodes) in a diffusion cascade. The algorithm learns an embedding vector for each role. Then, it models an activation taking place based on the closeness of the receivers' embedding vector and the senders' embedding vectors.

Similarly, another approach named DeepCas (Li et al., 2017) is designed to predict the future cascade size. It creates a subgraph of the activated nodes based on the cascade at each time. Then, it decomposes the subgraph into random walk paths and uses a Gated Recurrent Unit (GRU) to learn an embedding vector of the subgraph. Based on this subgraph embedding vector, it predicts the future size of the cascade.

Moreover, prior studies used only memory-based networks like LSTM or GRU to make cascade predictions. However, memory-based networks may not perform well when there is a scarcity of data (Feng et al., 2019). Furthermore, to the best of author's knowledge, there is no research on performing cascade predictions of diffusion of medical information (e.g., new medications) inside a physician's social network using different embedding techniques. Furthermore, for cascade prediction tasks, memory-based networks (e.g., LSTMs or GRUs) have not been compared against memoryless networks (e.g., MLPs).

In this research, we try to overcome these limitations. For learning the representation of the underlying network structure, we evaluate three different graph embedding techniques: DeepWalk (Perozzi et al., 2014), Node2Vec (Grover & Leskovec, 2016), and Structured Deep Network Embedding (SDNE) (Wang et al., 2016). We also vary the embedding dimensions between 16, 32, and 64 to investigate the effect of higher dimensions on the prediction task. Lastly, we generate embeddings that are feed into a memoryless Multilayer Perceptron (MLP) model (Hastie, Tibshirani, & Friedm, 2009) and a memory-based Long Short-Term Memory (LSTM) model (Hochreiter & Schmidhuber, 1997). Both these models predict the next set of physicians that will adopt a new medication.

# 3 Method

## 3.1 Dataset

We used a real-world physician dataset PhysicianSN (Choudhury, Kaushik, & Dutt, 2018) to predict the diffusion cascade of certain medications prescribed in a social network. The PhysicianSN is a physician social network dataset, where the physicians are the nodes and edges represent the relationship between them. We used a medical-prescriptions dataset and the Healthcare Organization Services (HCOS) (IMS Health, 2012) physician-affiliation dataset to extract physicians' attributes. The graph had 288 nodes and 38,795 edges. The relationship between the physicians was created based on the similarity between physicians' attributes. We used the prescription record of 45 medicines (prescribed between April 1996 and April 2017) as our cascade data, of which 35 (80%) were used for training and 10 (20%) were used for testing. Then, we converted the diffusion cascade of each medicine in the training and test sets into smaller supervised sets (mini-batches) containing the chosen look-back period. After the supervised training set had been created, we used 10% of the training data as a validation set. The first prescription of the medicine by a physician was considered as the time of adoption. The average length of a cascade was 95 physicians, i.e., on an average, a new medication was diffused to 95 physicians in the social network.

## 3.2 Generating graph embeddings

### 3.2.1 Graph embedding

Given the input graph $G = (V, E)$ and the embedding dimension d ($d " \ |V|$), graph embedding aims to represent $G$ as a vector in the $d$-dimensional space,

**246** Big data in psychiatry and neurology

in such a way that graph properties are preserved. In graph embedding, either the whole graph is represented as a $d$-dimensional vector or parts of the graph like nodes, and edges are represented as a set of $d$-dimensional vectors. Most graph embedding techniques try to preserve the following properties:

First-order proximity (Cai et al., 2018): The first-order proximity, $s_{i,\,j}^{(1)}$, between node $v_i$ and node $v_j$ (where $v_i$ and $v_j$ are vertices of the graph G) is the weight of the edge $e_{ij}$, i.e., $A_{i,\,j}$ (where $A$ is the adjacency matrix detailing the connection between nodes).

Second-order proximity (Cai et al., 2018): The second-order proximity $s_{i,\,j}^{(2)}$ between node $v_i$ and $v_j$ is the similarity between $v_i$'s neighborhood and $v_j$'s neighborhood.

We evaluated DeepWalk (Perozzi et al., 2014), Node2Vec (Grover & Leskovec, 2016) and Structured Deep Network Embedding (SDNE) (Wang et al., 2016) algorithms to generate the $d$-dimensional node embeddings of each node in the PhysicianSN graph. Each of these techniques relies upon deep learning approaches to preserve the first-order proximity and the second-order proximity of the embeddings (Cai et al., 2018). We also varied the embedding dimension in the range 16, 32, and 64 in each of the graph embeddings. Once the embeddings were generated, we used the t-Distributed Stochastic Neighbor Embedding (t-SNE) (Maaten & Hinton, 2008) to plot the embeddings in a 2-dimensional space. The t-SNE algorithm calculates a similarity measure between pairs of instances in both the high- and low-dimensional space and then optimizes the two similarity measures using a cost function (Maaten & Hinton, 2008).

## DeepWalk

DeepWalk (Perozzi et al., 2014) is a graph embedding technique that adopts the SkipGram language model (McCormick, 2019) for embedding the nodes. Skip-Gram maximizes the probability of co-occurrence among words that appear within a window $w$. DeepWalk takes a graph G and samples uniformly a random vertex $v_i$ as the root of the random walk. It then samples a set of paths from the node $v_i$ of length $l$. Each path is analogous to a sentence, and each node is analogous to a word. We generate $r$ walks per each node. Then, SkipGram is applied on the paths to maximize the probability of observing a node's neighborhood conditioned on its embedding. In this way, nodes with similar neighborhoods shared similar embeddings. DeepWalk preserves the second-order proximity between nodes.

## Node2Vec

Node2Vec (Grover & Leskovec, 2016) is similar to DeepWalk in that it also adopts the SkipGram model for graph embeddings. However, the random walk generation process is different. In Node2Vec, a random walk of fixed length $l$ is

Evaluating cascade prediction via different embedding techniques **Chapter | 12 247**

simulated from a source node ($v$). Let $n_i$ denote the $i$th node in the walk, starting with $n_0 = v$. Nodes $n_i$ are generated based on Eq. (1):

$$P(n_i = x \mid n_{i-1} = v) = \begin{cases} \dfrac{\pi_{vx}}{Z}, & \text{if } (v,x) \in E \\ 0, & \text{otherwise} \end{cases} \tag{1}$$

where $\pi_{vx}$ is the unnormalized transition probability between nodes $v$ and $x$, and $Z$ is the normalizing constant.

In Node2Vec, $\pi_{xv} = \alpha_{p,q}(t,x) \cdot w_{v,x}$, where $w_{v,x}$ is the static edge weight and $\alpha_{p,q}(t,x)$, is a bias term defined as follows:

$$\alpha_{p,q}(t,x) = \begin{cases} \dfrac{1}{p}, & \text{if } d_{tx} = 0 \\ 1, & \text{if } d_{tx} = 1 \\ \dfrac{1}{q}, & \text{if } d_{tx} = 2 \end{cases} \tag{2}$$

where $d_{tx}$ denotes the shortest path distance between nodes $t$ and $x$ (see Fig. 1). Fig. 1 illustrates how Node2Vec performs a random walk. The random walk starts for a source node and transitions to each of its neighbor node-based transition probability $\pi_{xv} = \alpha_{p,q}(t,x) \cdot w_{v,x}$ where $\alpha$ is the search bias and $w_{v,x}$ is the static edge weight. Moreover, $\alpha$ is calculated based on the shortest path distance between nodes denoted by $d_{tx}$. As can be seen from Fig. 1, currently, the walk has transitioned from $t$ to $v$, and it is evaluating its next node to select from node $v$. The edge labels indicate the search biases $\alpha$ and the distance between nodes $d_{tx}$. Since the walk has already transitioned from node $t$ to reach $v$, the shortest distance from $t$ to $v$ ($d_{tv}$) is equal to zero. As per Eq. (2), if $d_{tv} = 0$, $\alpha_{p,q}(t,v) = \frac{1}{p}$, i.e., the walk has $\frac{1}{p} \cdot w_{v,x}$ probability of transitioning back to $t$ from $v$. Next, $X_1$ is a neighbor to both $t$ and $v$, as such the shortest distance from $t$ to $X_1$ ($d_{tx_1}$) is equal to 1, thereby $\alpha_{p,q}(t,x_1) = 1$, i.e., the walk has $1 \cdot w_{v,x}$, probability of transitioning

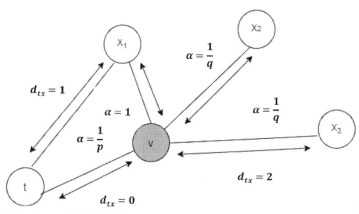

**FIG. 1** Random walk in Node2Vec.

**248** Big data in psychiatry and neurology

to $X_1$. Lastly, both $X_2$ and $X_3$ are neighbors of $v$ but not of $t$. So, the shortest path from $t$ to $X_2$ and $X_3$ is through $v$, as such $d_{tx_2} = 2$, which according to Eq. (2), implies that $\alpha_{ap,q}(t, x_2) = \frac{1}{q}$, i.e., the walk has $\frac{1}{q}.w_{v,x}$ probability of transitioning from $v$ to $X_2$ or $X_3$.

The in-out parameter $q$ determines how far or close does the walk traverse from the source node $t$. If q > 1, the random walk is more inclined toward nodes closer to $t$, whereas if $q < 1$, the walk is more prone toward node further away from $t$.

The return parameter $p$ controls the likelihood of backtracking to a visited node. A high value of $p$ (>max $(q, 1)$) ensures that one is less inclined to backtrack to an already visited node in the following two steps. While a low value of $p$(< min $(q, 1)$) indicates that walk is inclined to backtrack a step, thus keeping the walk local. Node2Vec also preserves the second-order proximity between nodes.

## Structured deep network embedding (SDNE)

SDNE (Wang et al., 2016) uses a semisupervised model comprising of two autoencoders (Salakhutdinov & Hinton, 2009) that preserves the first-order proximity and the second-order proximity of the node embeddings by minimizing the following objective function:

$$L_{sdne} = L_{2nd} + \alpha L_{1st} + \nu L_{reg} \qquad (3)$$

where $L_{2nd}$ is the reconstruction loss of the autoencoder, $L_{1st}$ is the first-order proximity loss, and $L_{reg}$ is an L2-norm regularizer term to prevent overfitting. The parameter $\alpha$ balances the weight of the first-order proximity, and second-order proximity between vertices and $\nu$ is the regularization constant. The $L_{2nd}$ is calculated as follows:

$$L_{2nd} = \sum_{i=1}^{n} ||(\hat{x}_i - x_i) \odot b_i||_2^2 \qquad (4)$$

where $x$ is the input and $\hat{x}_i$ is the reconstructed output, $\odot$ is the Hadamard operator, and $b_i = \{b_{i,j}\}_{j=1}^{n}$ is a penalty term. If $s_{i,j} = 0, b_{i,j} = 1 \; else \; b_{i,j} = \beta, \beta > 1$, where $s_{i,j}$ is an entry in the adjacency matrix $S$. An autoencoder (AE) aims to minimize the reconstruction error of the output and input samples by using the encoder and decoder networks. The encoder maps input data to a representation space (embedding dimension), and the decoder maps the representation space (embedding dimension) to a reconstruction space. The reconstruction loss $L_{2nd}$ helps preserve the second-order proximity. Due to the sparsity of graphs (e.g., social networks), the number of nonzero elements in the adjacency matrix $S$ is far less than that of zero elements. Thus if $S$ is used as the input to the traditional AE, it is more prone to reconstruct the zero elements in $S$. To address this problem, a penalty term is imposed on the reconstruction error of the nonzero elements than that of zero elements.

Evaluating cascade prediction via different embedding techniques Chapter | 12 **249**

The $L_{1st}$ loss is calculated as follows:

$$L_{1st} = \sum_{i,j=1}^{n} s_{i,j} \big|\big| (y_i - y_j) \big|\big|_2^2 \tag{5}$$

where $y_i$ and $y_j$ are the embeddings of vertex $v_i$ and $v_j$, respectively. The SDNE consists of two AEs (left and right networks), where each AE receives as input the node adjacency vectors $s_i$ of all pairs of nodes connected by edges on the input graph. In an Adjacency vector (a row from adjacency matrix, S), positive values indicate a connection between the node and the selected node. For each node, the adjacency vector is defined as follows:

$s_i = \{s_{i,j}\}_{j=1}^{n}$, $s_{i,j} > 0$ if and only if there exists a link between $v_i$ and $v_j$.

Since vertices $v_i$ and $v_j$ are neighbors, their embeddings should be similar (i.e., first-order proximity). Thus a distance loss is added to the objective function, which computed the distance between embeddings of $v_i$ and $v_j$.

The $L_{reg}$ loss is defined as follows:

$$L_{reg} = \frac{1}{2} \sum_{k=1}^{K} \left( \big|\big| W^{(k)} \big|\big|_F^2 + \big|\big| \hat{W}^{(k)} \big|\big|_F^2 \right) \tag{6}$$

where $W^{(k)}$ and $\hat{W}^{(k)}$ are the $k$th layer encoder and decoder weight matrices.

We used the backpropagation algorithm for training the AE, and we minimized the $L_{sdne}$ loss using stochastic gradient descent.

### 3.2.2 Model calibration

Each embedding technique generated $d$-dimensional node embeddings for the 288 nodes' social network. We generated node embedding of three different dimensions: 16, 32, and 64. This was done to understand the effect of embedding dimensions on the prediction result. The parameter settings used for Node2Vec and DeepWalk were similar. Specifically, we set window size $w = 10$, walks per node $r = 500$, and maximum length $l = 95$ (same as the average cascade length). The $p$ and $q$ parameters in Node2Vec were set to 1 and 0.5, respectively, based on (Grover & Leskovec, 2016). In the case of SDNE, the encoder and decoder in the AE model consisted of two fully connected layers with 256 and 128 neurons, and $d$ neurons in the latent layer, where $d$ is the embedding dimension. The AE was trained for 1000 epochs with a batch size of 1. We ran each model 5 times and calculated the mean and standard deviation of the training and test evaluation metric for each of the models over these five runs.

### 3.3 Cascade prediction

We used both memoryless and memory-based models to predict the diffusion cascades inside the social network. In this paper, we used a Multilayer

Perceptron model as the memoryless model and a Long Short-Term Memory model as the memory-based model. All models forecasted one-step ahead with walk-forward validation. In one-step-ahead walk-forward validation, a model uses training data to predict the next time-step. This prediction is then evaluated against the actual value. Next, the actual value corresponding to the prediction is added to the training data, and the process is repeated by predicting the value for the next time-step.

### 3.3.1 Multilayer perceptron (MLP)

An MLP is a variant of the original perceptron model (see Fig. 2A) proposed by Rosenblatt (Rosenblatt, 1962).

A neuron (represented as $\sum$ in Fig. 2A) computes a weighted sum of the inputs, followed by a nonlinear activation $\varphi$ of the calculated sum, as shown

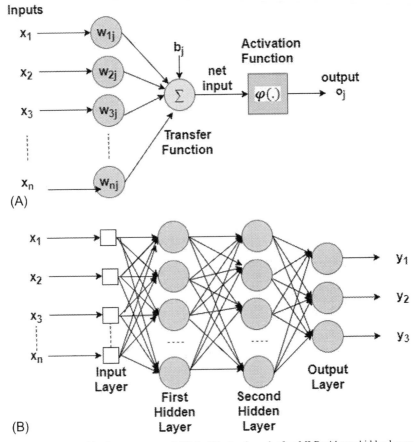

**FIG. 2** (A) Rosenblatt's perceptron and (B) Architectural graph of an MLP with two hidden layers.

Evaluating cascade prediction via different embedding techniques **Chapter | 12 251**

in Eq. (3). Neural network architectures are considered as the universal function approximators because of the presence of activation functions. An activation function helps generate mappings from inputs to outputs, and it provides the neural network model the ability to learn complex data representations (Chung, Lee, & Park, 2016). There are a number of activation functions proposed in the literature, which include the sigmoid, tanh, and ReLU (Chung et al., 2016; Krizhevsky, Sutskever, & Hinton, 2012). Krizhevsky et al. (2012) have shown that the sigmoid and tanh activation functions suffer from the problem of vanishing gradient, while the ReLU activation function overcomes the vanishing gradient problem and provides faster convergence. Also, the function is computationally efficient to calculate. Thus, based on the literature, we used the ReLU activation function in the MLP model. The output $o_i$ of a neuron in the MLP was defined as per the following equation:

$$o_i = \varphi \left( \sum_{j=1}^{d} (x_j w_{ij} + b_j) \right) \tag{7}$$

where $d$ is the length of the input vector, $x_i$ is a single instance of the input vector, and $b_j$ and $w_{ij}$ are the bias and weights associated with each $x_j$. Fig. 2B shows the architecture of an MLP with two hidden layers. A typical MLP is composed of multiple hidden layers, with multiple neurons in each layer where every neuron in a layer (say $i$) is fully connected to every other neuron in the next layer (i.e., $i + 1$).

### 3.3.2 Long short-term memory (LSTM)

An LSTM (Hochreiter & Schmidhuber, 1997) model is a recurrent neural network (RNN) model with the capacity of remembering the values from earlier stages in the network (i.e., the model possesses memory). The architecture of an LSTM consists of units called memory cells. Fig. 3 shows an LSTM memory cell containing self-connections and special multiplicative units called gates. These connections remember the temporal state of the memory cells, and the gates control the flow of information. Each memory cell contains an input gate, an output gate, and a forget gate. The flow of input activations is controlled by the input gate, while the flow of cell activations into the remaining network is controlled by the output gate. Furthermore, the internal state of the cell is scaled by the forget gate and is added back to the cell as input through a self-recurrent connection. In Fig. 3, $c_{t-1}$ is the previous cell state, $h_{t-1}$ is the previous cell output, $x_t$ represents the input to the memory cell, $c_t$ represents the new cell state, and $h_t$ represents the output of the hidden layer at time t.

### 3.3.3 Model calibration

Both the MLP and LSTM were trained on the training set and validated on the validation set. The LSTM model comprised of a single LSTM layer followed by

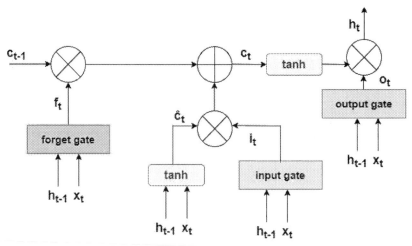

FIG. 3  An illustration of an LSTM memory cell. ((Source: (Kaushik et al., 2020)))

a dropout layer and a fully connected output layer with 288 nodes with the softmax activation function. We also used a 20% recurrent dropout in the LSTM layer. The MLP model consisted of 2 fully connected layers, followed by a fully connected output layer with 288 nodes with a softmax activation function. Both dropout and recurrent dropout were used only in the LSTM model, while no dropout was used in the MLP model. This arrangement of dropouts in LSTM and MLP models was made because the LSTM model was more likely to overfit due to a higher number of parameters (# parameter: 189744) compared to the MLP model (# parameter: 87216). All hidden layers in both models had ReLU activation. We used the cross-entropy loss as the loss function. The number of layers was kept small to reduce overfitting. We also used early stopping (Chollet, 2015) (i.e., training will stop when the chosen performance measure stops improving). In this paper, we have used validation loss as the performance measure for early stopping, i.e., if the validation loss stops decreasing, then the model's training is stopped. Furthermore, we used a genetic algorithm (Whitley, 1994) with a 5% mutation and an 80% crossover rate to tune the following hyperparameters using training data: number of neurons in a layer, batch size, epochs, and dropout rate. The hyperparameters were varied in the following ranges: the number of neurons in a layer (16, 32, 64, 128, 256, and 512), batch size (16, 32, 64, 128, 256, 512, and 1024), and the number of epochs (5 to 20) in increments of 1. For both the LSTM and MLP model, we set the look-back period to be 5, i.e., for predicting $t$, we look at 5 ($t-1$, ..., $t-5$) prior time-steps. After training, all models were tested on the test dataset. During training, we did not want to give an advantage to the memory-based LSTM model over the memoryless MLP model. Therefore we varied the same range of hyperparameters for both the MLP and LSTM models.

Evaluating cascade prediction via different embedding techniques Chapter | 12 **253**

## 3.4 Evaluation metrics

To evaluate our results, we used mAP@k[a] (Wang et al., 2016) to compare our algorithms. The map@k computes the mean average precision at $k$ for a set of predictions. So, if the algorithm predicts $n$ nodes, only the $k$ topmost nodes out of $n$ predicted nodes are compared with the actual result. mAP@k is calculated as follows:

$$mAP@k = \frac{1}{|U|} \sum_{u=1}^{U} \frac{1}{m} \sum_{i=1}^{k} P(i).rel(i) \tag{8}$$

where $|U|$ is the total number of cascades, m is the number of nodes in the cascade, P is the precision, $P(k)$ is the precision at $k$ calculated by considering only the first $k$ predictions from 1 to $k$, and $rel(k)$ is just an indicator that says whether that $k_{th}$ item was relevant ($rel(k) = 1$) or not ($rel(k) = 0$).

$$precision\,(P) = \frac{\#correctpositive}{\#predictedpositive} \tag{9}$$

For this research, we have selected the $k$ value to be 10.

## 4 Results

### 4.1 Cascade prediction using MLP

Fig. 4 shows the MLP model's training and test mAP@10 score for different graph embedding algorithms. As shown in Fig. 4, the best training map@10 ($= 0.74$) was obtained for SDNE-16 and SDNE-32 embeddings while the best test mAP@10 score ($= 0.66$) was obtained on the SDNE-32 embeddings (i.e., an embedding dimension of 32) by the MLP model. The MLP model comprised of 2 layers with 128 neurons in the first layer and 256 neurons in the second layer and was trained for 12 epochs with 128 batch size. We also found that Node2Vec gave the worst training and test mAP@10 scores among all the embeddings irrespective of the embedding dimensions.

### 4.2 Cascade prediction using LSTM

Fig. 5 shows the LSTM model's training and test mAP@10 score for different graph embedding algorithms. As shown in Fig. 5, the best training mAP@10 score ($= 0.63$) was obtained on SDNE-16 embeddings, while the best test mAP@10 score ($= 0.58$) was obtained by the LSTM model for the SDNE-64

---

a. The map@k score was calculated using the ml_metrics python library. Available at https://github.com/benhamner/Metrics

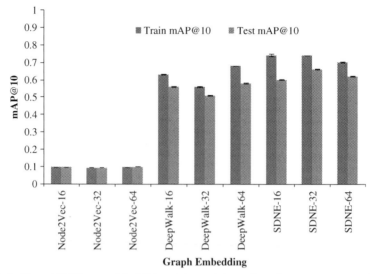

**FIG. 4** The mAP@10 score of the MLP model on training and test data.

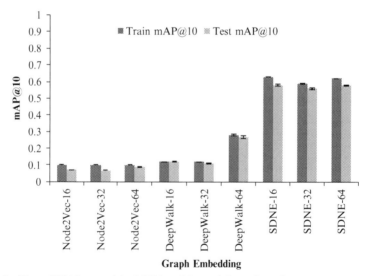

**FIG. 5** The mAP@10 score of the LSTM model on training and test data.

embeddings. The LSTM model comprised of a 1-LSTM layer with 512 neurons and was trained for 16 epochs with a batch size of 64 iterations. Similar to the MLP model, the worst test mAP@10 was given by the Node2Vec embedding. Moreover, even the DeepWalk embedding did not give a high-test mAP@10 score on the LSTM model.

Evaluating cascade prediction via different embedding techniques **Chapter | 12** **255**

## 4.3 Model comparison

Table 1 presents the mean and the standard deviation of the training and test mAP@10 scores obtained over the 5 runs from both MLP and LSTM models. Overall the best mAP@10 score was given by the MLP model on the SDNE-32 embeddings.

Fig. 6 shows the t-SNE visualization of the different embedding dimensions for each of the three embedding techniques for which the highest test mAP@10

**TABLE 1** Model Comparison of different embedding algorithm and embedding dimensions.

| Model | Embedding algorithm | Embedding dimension | Training map@10 | Test map@10 |
|---|---|---|---|---|
| MLP | SDNE | 32 | $0.74 \pm 0.019^a$ | $0.66 \pm 0.02$ |
| LSTM | SDNE | 64 | $0.62 \pm 0.02$ | $0.57 \pm 0.05$ |

[a] The standard deviation in the map@10 score.

was obtained. T-SNE visualizes the high-dimensional node embeddings (32-dimension SDNE embedding, 64-dimension DeepWalk embedding, and 64-dimension Node2Vec embedding) in a 2-dimensional space. Each node in the social network is represented with a different color, and similar nodes are clustered together. The denser the node clusters, more similar are the embeddings of the nodes in that cluster. As can been seen from the t-SNE visualization of DeepWalk (Fig. 6B) and Node2Vec (Fig. 6C) embeddings, nodes are segregated into dense clusters while in SDNE embeddings there are no such clusters. These differences in clustering could explain the poor performances of DeepWalk and Node2Vec algorithms compared to the SDNE algorithm.

## 5 Discussion and conclusions

Information cascades play a pivotal role in the process of information diffusion, as it describes the flow of information inside a social network. Most diffusion processes can be viewed as a collection of interaction traces (e.g., prescription logs of physicians) indicating when people post/tweet a piece of information or buy a particular product. Most often, one only observes the when and where the actions had taken place rather than why it happened (Bourigault et al., 2016). Therefore modeling the diffusion process mostly boils down to approximating this mechanism based on the interaction traces. Prior research had modeled the diffusion process as to how a piece of information like a tweet or meme diffused inside an online social network (e.g., Twitter or Facebook) using the action of

**256** Big data in psychiatry and neurology

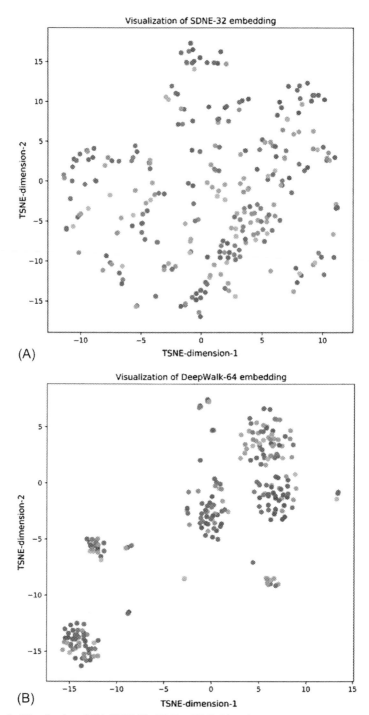

**FIG. 6** Visualization of (A) SDNE-32, (B) DeepWalk-64, and

*(Continued)*

Evaluating cascade prediction via different embedding techniques Chapter | 12  **257**

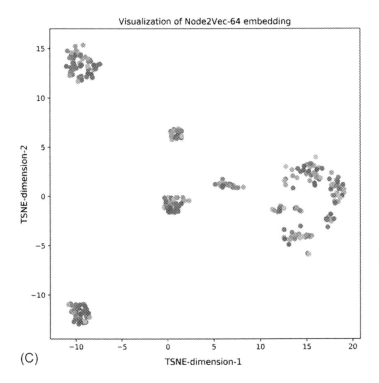

**FIG.6, CONT'D** (C) Node2Vec-64 embeddings using t-SNE.

liking or retweeting of tweets/memes/posts as an indicator of diffusion (Bourigault et al., 2016; Li et al., 2017; Wang et al., 2017). Bourigault et al. (2016) and Wang et al. (2017) generated graph embedding of the underlying social network and the cascade itself to predict future diffusion. However, no such analysis was done in the field of healthcare, where it is of utmost importance that accurate information is swiftly disseminated. In this research, we investigate how medication diffusion occurs inside a physician's social network. We performed a comprehensive evaluation of different graph embedding techniques and studied the effect of different embedding dimensions on future cascade prediction using both memory and memoryless networks in the healthcare domain. We tried SDNE, DeepWalk, and Node2Vec algorithms to extract low-dimensional embeddings of the underlying network and used prescription logs of 45 medications as our cascade data. We then fed the embeddings into MLP and LSTM models to predict the next set of physicians who will adopt the medication in a cascade.

Firstly, we found that both MLP and LSTM gave a high-test mAP@10 score on the SDNE embeddings. This result was as per our expectations

**258** Big data in psychiatry and neurology

as the SDNE method preserves both the first-order proximity and the second-order proximity of the graph structure (Cai et al., 2018). We also found that Node2Vec gave the worst test mAP@10 score, followed by Deep-Walk. One likely reason for this finding could be due to the presence of dense clusters in the Node2Vec and DeepWalk representation of the underlying social network. Clusters are obstacles to diffusion cascades as people are hesitant to adopt new information if they belong to a cluster (Easley & Klienberg, 2010). Furthermore, we also found that generally higher dimensional embeddings performed better than its lower dimensional counterparts, specifically in the case of Node2Vec and DeepWalk, while for SDNE, the highest test mAP@10 was obtained for embedding dimension of 32 followed by 64 and 16. Although not much research has been done into the evaluation of different embedding dimensions in graphs, a rule of thumb from word embedding literature is that the embedding dimension may be the 4th root of the size of the vocabulary (i.e., the number of the distinct word in the corpus) (Google Developers, 2017). Our results disagree with this rule as we found that higher dimensional embeddings performed better in general in predicting diffusion inside social networks.

Next, we found that the MLP model performed better than the LSTM model on predicting the next set of users that will be activated in an information cascade inside a network (Table 1). This result is not as per our expectation as we expected memory-based models like LSTM to perform better compared to memoryless models. A likely reason for this finding could be that mostly memory-based networks like LSTM and GRU need a considerable amount of data for training. Also, even though we used dropout, recurrent dropout, and early stopping, LSTMs were still overfitting and were unable to generalize on the test data due to having a significantly larger number of parameters compared to MLP.

Understanding how diffusion takes place inside a physician's social network is paramount as it helps in better dissemination of medical information and guidelines. However, the availability of cascade data like prescription logs is most scarce, and the available data is mostly small. In this paper, we have tried to tackle this issue by evaluating three different graph embedding techniques (DeepWalk, Node2Vec, and SDNE) with three different embedding dimensions (16, 32, and 64) using two machine learning models (MLP and LSTM) for predicting the next set of users that will be activated in an information cascade inside a physician social network. As per our findings, we believe that by preserving both first- and second-order proximity of the underlying social network, we were able to perform better in cascade prediction tasks specifically in predicting the next set of users in the cascade. We also found that for smaller datasets, MLPs performed better than LSTMs. As information regarding the social structure of physicians as well as diffusion information relating to healthcare innovation (specifically a medication) is likely sparse, one may generate high-dimensional embeddings of the social network and use MLPs to make future predictions.

Evaluating cascade prediction via different embedding techniques Chapter | 12 **259**

One particular limitation of our work is that we only generated the embeddings of the underlying structure and not the cascade itself. This process adapted well in prediction as the underlying link structure was representative of the diffusion channels; however, in the case of online social networks, the diffusion may not always propagate through links between users. In the future, we plan to generate embeddings of both the underlying social network as well as the cascade itself to predict the next set of users as well as the size of future cascades.

## Acknowledgment

The project was supported by grant (award: # IITM/CONS/RxDSI/VD/33) to Dr. Varun Dutt.

## References

Bao, Q., Cheung, W. K., & Liu, J. (2016). Inferring motif-based diffusion models for social networks. In *Proceedings of the twenty-fifth international joint conference on artificial intelligence* (pp. 3677–3683). New York, New York, United States: AAAI Press.

Bourigault, S., Lamprier, S., & Gallinari, P. (2016). Representation learning for information diffusion through social networks: an embedded cascade model. In *Ninth ACM international conference on web search and data mining* (pp. 573–582). San Francisco, California: Association for Computing Machinery.

Cai, H., Zheng, V. W., & Chang, K. C.-C. (2018). A comprehensive survey of graph embedding: Problems, techniques, and applications. *IEEE Transactions on Knowledge and Data Engineering, 1616–1637.*

Cheng, J., Dow, P. A., Kleinberg, J. M., & Leskovec, J. (2014). Can cascades be predicted? In *Proceedings of the 23rd international conference on World wide web* (pp. 925–936). Seoul: Association for Computing Machinery.

Cho, K., Van Merriënboer, B., Bahdanau, D., & Bengio, Y. (2014). *On the properties of neural machine translation: Encoder-decoder approaches.* arXiv preprint arXiv:1409.1259.

Chollet, F. (2015). *Deep learning library for theano and tensorflow.* Retrieved March 12, 2020, from Keras.io https://www.tensorflow.org/api_docs/python/tf/keras/callbacks/EarlyStopping.

Choudhury, A., Kaushik, S., & Dutt, V. (2017). Social-network analysis for pain medications: Influential physicians may not be high-volume prescribers. In *IEEE/ACM International Conference on Advances in Social Networks Analysis and Mining (ASONAM)* (pp. 881–885). Sydney: IEEE.

Choudhury, A., Kaushik, S., & Dutt, V. (2018). Social-network analysis in healthcare: Analysing the effect of weighted influence in physician networks. In R. Alhajji, & U. K. Wiil (Eds.), *Vol. 7 (17). Network modeling analysis in health informatics and bioinformatics* Springer.

Chung, H., Lee, S. J., & Park, J. G. (2016). Deep neural network using trainable. In *International joint conference on neural networks (IJCNN)* (pp. 348–353). IEEE.

Domingos, P., & Richardson, M. (2001). Mining the network value of customers. In *International conference on knowledge discovery and data mining* (pp. 57–66). San Francisco: Association for Computing Machinery.

Easley, D., & Klienberg, J. (2010). Modeling network traffic using game theory. *Networks, crowds, and markets.* New York: Cambridge University Press.

Feng, S., Zhou, H., & Dong, H. (2019). Using deep neural network with small dataset to predict material defects. *Materials & Design, 162,* 300–310.

**260** Big data in psychiatry and neurology

Google Developers. (2017, November 20). *Introducing TensorFlow feature columns.* Retrieved April 4, 2020, from https://developers.googleblog.com/2017/11/introducing-tensorflow-featurecolumns.html.

Grover, A., & Leskovec, J. (2016). node2vec: Scalable feature learning for networks. In *The 22nd ACM SIGKDD international conference on knowledge discovery and data mining* (pp. 855–864). San Francisco,California, USA: Association for Computing Machinery.

Hastie, T., Tibshirani, R., & Friedm, J. H. (2009). *The elements of statistical learning.* New York: Springer.

Hochreiter, S., & Schmidhuber, J. (1997). Long short-term memory. *Neural Computation,* 1735–1780.

IMS Health. (2012). *Healthcare organization services: Professional and organization affiliations maintenance process.* Retrieved from http://us.imshealth.com/legal/ServicePlanDetails-HCOS.pdf.

Kasthuri, A. (2018). Challenges to healthcare in India—The five A's. *Indian Journal of Community Medicine, 43,* 141–143.

Kaushik, S., Choudhury, A., Sheron, P. K., Dasgupta, N., Natarajan, S., Pickett, L. A., et al. (2020). AI in healthcare: Time-series forecasting using statistical, neural, and ensemble architectures. *Frontiers in Big Data, 3,* 4. https://doi.org/10.3389/fdata.2020.00004.

Kefato, Z. T., Sheikh, N., Bahri, L., Soliman, A., Montresor, A., & Girdzijauskas, S. (2018). CAS2-VEC: Network-agnostic cascade prediction in online social networks. In *Fifth international conference on social networks analysis, management and security* (pp. 72–79). IEEE.

Kempe, D., Kleinberg, J., & Tardos, É. (2003). Maximizing the spread of influence through a social network. In *The ninth ACM SIGKDD international conference on knowledge discovery and data mining* (pp. 137–146). Washington, D.C.: Association for Computing Machinery.

Hinton, G. E., Krizhevsky, A., & Sutskever, I. (2012). ImageNet classification with deep convolutional neural networks. *Advances in Neural Information Processing, 25,* 1097–1105. Nevada.

Li, C., Ma, J., Guo, X., & Mei, Q. (2017). DeepCas: An end-to-end predictor of information cascades. In *Proceedings of the 26th international conference on world wide web* (pp. 577–586). Perth: International World Wide Web Conferences Steering Committee.

Maaten, L. V., & Hinton, G. (2008). Visualizing data using t-SNE. *Journal of Machine Learning Research, 9,* 2579–2605.

Matsubara, Y., Sakurai, Y., Prakash, B. A., Li, L., & Faloutsos, C. (2012). Rise and fall patterns of information diffusion: Model and implications. In *The 18th ACM SIGKDD international conference on knowledge discovery and data mining* (pp. 6–14). Beijing: Association for Computing Machinery.

McCormick, C. (2019). *The Skip-Gram Model.* 19, April. Retrieved April 12, 2020, from Word2Vec Tutorial http://www.mccormickml.com.

Patel, K., & Bhattacharyya, P. (2017). Towards lower bounds on number of dimensions for word embeddings. In *Proceedings of the eighth international joint conference on natural language processing* (pp. 31–36). Taipei: Asian Federation of Natural Language Processing.

Perozzi, B., Al-Rfou, R., & Skiena, S. (2014). DeepWalk: Online learning of social representations. In *The 20th ACM SIGKDD international conference on knowledge discovery and data mining* (pp. 701–710). New York USA: Association for Computing Machinery.

Rosenblatt, F. (1962). *Principles of neurodynamics: Perceptrons and the theory of brain mechanisms.* New York: Spartan Books.

Rumelhart, D. E., Hinton, G. E., & Williams, R. J. (1986). Learning internal representations by error propagation. In D. E. Rumelhart, & J. L. McClelland (Eds.), *Parallel distributed processing: Explorations in the microstructure of cognition* (pp. 318–362). Cambridge, MA: MIT Press.

Salakhutdinov, R., & Hinton, G. (2009). Semantic hashing. *International Journal of Approximate Reasoning, 50*(7), 969–978.

Scott, J. (1991). *Social network analysis: A handbook.* SAGE Publications.

Wang, D., Cui, P., & Zhu, W. (2016). Structural deep network embedding. In *The 22nd ACM SIGKDD international conference on knowledge discovery and data mining* (pp. 1225–1234). San Francisco, California, USA: Association for Computing Machinery.

Wang, J., Zheng, V. W., Liu, Z., & Chang, K. C.-C. (2017). Topological recurrent neural network for diffusion prediction. In *IEEE international conference on data mining (ICDM)* (pp. 475–484). New Orleans, LA: IEEE.

Whitley, D. (1994). A genetic algorithm tutorial. *Statistics and Computing, 4*(2), 65–85.

Zhu, K., Chen, Z., & Ying, L. (2017). Catch'em all: Locating multiple diffusion sources in networks with partial observations. In *The thirty-first AAAI conference on artificial intelligence* (pp. 1676–1682). San Francisco: AAAI Press.

# Chapter 13

# A two-stage classification framework for epileptic seizure prediction using EEG wavelet-based features

Sahar Elgohary[a], Mahmoud I. Khalil[a], and Seif Eldawlatly[a,b]
[a]*Computer and Systems Engineering Department, Faculty of Engineering, Ain Shams University, Cairo, Egypt, *[b]*Faculty of Media Engineering and Technology, German University in Cairo, Cairo, Egypt*

## 1 Introduction

Epilepsy is a neurological condition that is distinguished by the disruption of nerve cell activity in the brain causing seizures, which can be defined as periods of unusual behavior and sometimes loss of consciousness (Fisher et al., 2005). Other symptoms may include staring blankly for a few seconds or twitching of arms and legs (Shorvon, 2010). In some cases, psychological complications such as anxiety and depression have been reported to occur as a result of epilepsy (Cianchetti et al., 2018). Approximately, 50 million people suffer from epilepsy all over the world; the number of people having active epilepsy at a given time is roughly between 4 and 10 per 1000 people (World Health Organization, 2019). Seizures can cause serious complications and accidents for patients due to their unexpected nature. Moreover, they can happen while patients are performing dangerous activities such as driving or swimming (Panayiotopoulos, 2007).

One way to monitor and diagnose epilepsy is using Electroencephalography (EEG) activity (Khalil & Misulis, 2006). Epileptic EEG signals have four main categories: (1) The ictal stage that represents the seizure occurrence episode, (2) the preictal stage which is the window of time before the seizure onset, (3) the postictal stage which is the window following the seizure, and (4) the interictal stage which can be defined as the normal stage when it is far from the start or the end of seizures (Chiang, Chang, Chen, Chen, & Chen, 2011). According to the aforementioned EEG classification, the preictal stage is the transitional interval

**264** Big data in psychiatry and neurology

that happens between the interictal stage and the seizure (Litt & Echauz, 2002). Consequently, recognizing the preictal stage, through performing a classification between interictal and preictal stages, can be considered as an alarm for patients of a potential seizure in order to stop any potentially dangerous activity.

Wavelet transform is a powerful tool to study nonstationary signals, thus it is appropriate for EEG analysis and it has been used extensively in seizure prediction and detection applications (Alickovic, Kevric, & Subasi, 2018; Carney, Myers, & Geyer, 2011). The main factor to be considered when utilizing wavelet transform is determining the discriminating and relevant features for classification to be extracted from different subbands of wavelet coefficients. For instance, Nasehi et al. (Nasehi & Pourghassem, 2011) proposed a method that is based on extracting wavelet coefficients features such as mean, maximum, minimum, and variance of coefficients, and using Fisher's discriminant analysis in classification. Additionally, Holla et al. (Holla & Aparna, 2012) proposed a method that is based on extracting nonlinear features from wavelet coefficients as approximate entropy and sample entropy, and used K-Nearest Neighbors in classification (Altman, 1992). Acharya et al. (Acharya, Sree, Ang, Yanti, & Suri, 2012) examined the use of nonlinear features in seizure prediction such as approximate entropy and sample entropy and achieved a classification accuracy of 99.7%. Moreover, Park et al. (Park, Luo, Parhi, & Netoff, 2011) proposed a method that utilizes spectral power analysis of EEG of different subbands and used support vector machine (SVM) classification (Cortes & Vapnik, 1995). Wavelet energy has also been used in seizure prediction by Gigola et al. where they could forecast the seizure onset time by 70 mins before the onset (Gigola, Ortiz, D'attellis, Silva, & Kochen, 2004). Discrete wavelet transform and SVM have been used as well by Tzimourta et al. in seizure detection which achieved a sensitivity and specificity of 93% and 99%, respectively (Tzimourta et al., 2017).

More recently, Convolutional Neural Networks (CNNs) and deep neural networks have been utilized in the task of seizure prediction (Khan, Marcuse, Fields, Swann, & Yener, 2018). Khan et al. have proposed using wavelets with CNNs. Wavelet output is constructed by performing convolution on each channel of the EEG signal with wavelet functions at different scales as convolution filters. The reported sensitivity is 87.8% and a low false prediction rate of 0.142 false positives/hour is achieved. Despite the promising results achieved using deep learning in this task, a significantly long training data is required to train the developed network.

The aforementioned methods achieved significantly high accuracies; however, there are other challenges and enhancements that need to be tackled to have more practically feasible methods. One of the major challenges is the differences between seizure patterns that a single patient can have (Chiang et al., 2011). As a result, a model that is trained and tested by the same set of seizures, by randomly splitting the data to training and testing sets, could produce overly optimistic performance and may fail to reliably generalize to unseen future

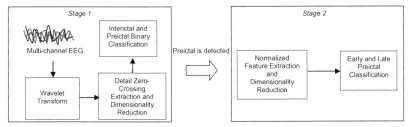

**FIG. 1** Overview of the proposed two-stage wavelet-based framework. In the first stage, interictal and preictal classification is performed. EEG signals from different channels were divided to 5 s windows, next, wavelet transform is computed per window. This is followed by feature extraction and reduction. Finally, classification is performed between preictal and interictal classes. If the preictal class is detected, stage 2 is then performed to further classify the preictal class to early and late preictal classes after feature extraction and dimensionality reduction.

seizures. Another challenge is developing practical tools that could run on relatively low computational resources setups. This includes developing simple feature extraction methods in addition to selecting appropriate recording channels in order to have a feasible real-time system with few channels to avoid difficulties in usage and setup. Additionally, the need to have long durations of recording training data as well as including multiple seizures, as used by the aforementioned methods, could be burdensome for patients during the setup procedure of the real-time system. Finally, there is a need to provide more accurate estimation for the onset time rather than only detecting the preictal state.

In this chapter, we propose a two-stage wavelet-based method, shown in Fig. 1, that addresses such limitations using a novel feature extraction method, which has not been used in seizure prediction before, that identifies the number of zero-crossings in wavelet detail coefficients. The first stage of the proposed framework involves a method for seizure occurrences prediction through binary classification between preictal and interictal classes. In the second stage, we further classify the preictal data—after being detected in the first stage—to early and late preictal states to have more accurate measure for seizure onset time.

## 2 Materials and methods

### 2.1 Dataset

In this work, we examined the performance of the introduced framework using the publicly available CHB-MIT scalp EEG dataset which was recorded at the Children's Hospital Boston (Goldberger et al., 2000). EEG data were recorded from pediatric patients with intractable seizures. The sampling rate of the dataset is 256 Hz and the recording consists of long-term scalp EEG recordings of 24 patients. The number of channels in most patients is 22. Subjects whose data fit the seizure prediction problem are supposed to include long and uninterrupted interictal recording periods between two seizure onsets. In addition,

**266** Big data in psychiatry and neurology

the data labeled as interictal should be continuous without gaps in recording to guarantee that no seizure has happened in that gap time. Finally, the recording should include several seizures for each patient to be able to take enough preictal data and perform reliable testing on unseen seizures. Since the dataset was developed mainly for seizure detection and not prediction, not all recordings are suited for our problem, therefore, 8 patients were selected out of the 24 patients who fulfill the aforementioned conditions. The preictal data were selected to be the 1-h recordings preceding seizure onsets, while the interictal data were selected such that there are at least 4 h separation before or after any seizure onset to avoid intersection with the preictal or postictal states. The selection of the data is based on previous studies and the American Society of Epilepsy seizure prediction challenge guidelines (American Epilepsy Society Seizure Prediction Challenge, 2014; Chang, Chen, Chiang, & Chen, 2012). When comparing the proposed method to other methods that did not use the same dataset as will be detailed later, the other methods were reimplemented on the used dataset to have a common reference for comparison. In addition, optimum hyperparameter selection search is performed as well for the implemented methods that are compared to our method to ensure fair comparison.

## 2.2 Two-stage zero-crossings wavelet-based framework

In this section, we introduce the proposed wavelet-based framework that utilizes wavelet detail coefficients zero-crossings count as the proposed feature. The framework is composed of two stages: the first stage focuses on the binary classification between preictal and interictal classes, and the second stage focuses on further classification of the detected preictal class into early and late preictal classes to have a more accurate seizure onset estimation.

In this framework, continuous wavelet transform is employed which can be defined as (Burrus, Gopinath, & Guo, 1997)

$$W(a,b) = \frac{1}{\sqrt{a}} \int_{-\infty}^{\infty} x(t)\psi^* \left( \frac{t-b}{a} \right) dt \tag{1}$$

where $W$ is the continuous wavelet transform, $\psi(t)$ is the analyzing wavelet function, $a$ is the scalar parameter, $b$ is the position parameter, and $(\cdot)^*$ refers to the complex conjugate. Continuous wavelet is split to discrete wavelet by using binary scales. In this case, the scaled and shifted wavelet function takes the form

$$\psi(j,b) = \frac{1}{\sqrt{2^j}} \psi \left( \frac{t-b}{2^j} \right) \tag{2}$$

where $j$ denotes the scale, $b$ denotes the shift, and $\psi$ is the unscaled unshifted mother wavelet. The two parts of the wavelet transform are approximation coefficients ($A_n$, where $n$ is wavelet level index) that represent the lower frequencies, and detail coefficients ($D_n$, where $n$ is wavelet level index) that represent the higher frequencies.

### 2.2.1 Stage 1: Interictal and preictal binary classifications method

In this section, we introduce the first stage of the proposed method which is the binary classification stage between preictal and interictal classes. In this stage, first, input EEG signals are divided into successive time windows. Wavelet transform is computed per time window, and features are extracted afterward from wavelet levels 1 and 2 detail coefficients. This is followed by classifier training and automatically selecting the highest performing channels for each subject using cross-validation. The feature utilized in our method is the count of zero-crossings events in the wavelet detail coefficients per channel, where zero-crossings can be defined as a change of sign across two successive coefficients.

To illustrate the basic idea of the proposed feature, we demonstrate in this section using Haar wavelet transform for feature extraction. However, different wavelet functions were also examined as demonstrated later. The zero-crossings in detail coefficients in case of using Haar wavelet function indicate a slope direction change in the time-domain signal (when calculating detail coefficients of the first level), or in the upper wavelet level (when calculating detail coefficients of higher levels). As demonstrated in Fig. 2A, a change in signal direction occurs between the points labeled as 1, 2 and the points labeled as 3, 4. This corresponds to a change of sign in wavelet detail coefficients as demonstrated in Fig. 2B. Therefore counting the zero-crossings occurrences in the wavelet detail coefficients is equivalent to counting the occurrences of slope change.

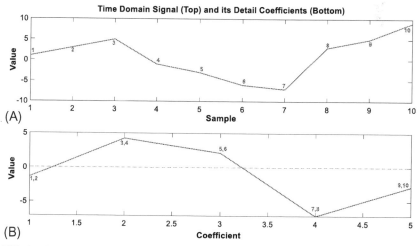

**FIG. 2** A sample signal demonstrating zero-crossings in wavelet details coefficients. A signal is shown in (A) time domain along with (B) its corresponding detail coefficients. In (B), labels in red indicate which two samples were considered in the calculations.

After extracting features, dimensionality reduction is performed to select significant features only. The introduced approach is illustrated in Fig. 3, where, first, each channel is used separately as a feature vector to the classifier. This is followed by sorting channels based on the computed classification accuracy of the first step. An incremental search for the highest performing channels is then implemented, where in each iteration the next highest performing channel is added to the set and the accuracy of the model is again computed with the accumulated channels. The search ends when the accuracy of the model with the current channel set saturates or declines. This step improves the accuracy considerably by eliminating irrelevant features. Additionally, it has a potential to significantly increase the practicality of potential real-time applications by reducing the needed computation time.

We compared the aforementioned dimensionality reduction approach to using all channels in classification (i.e. without dimensionality reduction), using Principal Component Analysis (PCA), and using autoencoders (Baldi, 2012; Jolliffe, 2002) for dimensionality reduction. In case of PCA, the feature vector that is fed to the classifier is the projection of the extracted features (number of zero-crossings) on the selected principal components (PCs), where the number of PCs is a hyperparameter that is determined using cross-validation during

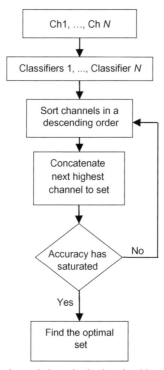

**FIG. 3** Dimensionality reduction and channel selection algorithm.

training. Additionally, autoencoders were used as a channel selection and reduction method after the feature extraction phase, where the output of the encoder function, represented as the hidden layer of the network, is used as the feature vector that is fed to the classifier. We also examined the effect of changing the number of hidden nodes and changing the encoder and decoder functions of the autoencoder.

The selection of preictal training and testing data was done such that seizures included in training data are not included in testing data to ensure the capability of the proposed method to generalize to future unseen preictal data. Additionally, most seizure prediction methods need long hours of recording in training which imposes the need for several recording sessions from patients to be included in the training data. To tackle this challenge, we included only 10 min of data in the training from each class. To assess the generality of the results, for each subject, a $k$-fold approach is used in which one seizure is used for testing and the rest are used for training (where $k$ is set to 2 or 3 according to the number of seizures available for each subject). The final accuracy is the average of all trials where in each trial the testing seizure is changed. Furthermore, for each of the previous trials, 10 min are selected randomly from each class for training (which corresponds to 240 data points where each data point is a 5-s window) and 2 min for testing (which corresponds to 48 data points). Additionally, the last setup is repeated 10 times and averaged.

The main classifier that we use in the proposed model is Support Vector Machine (SVM) which obtains a discriminating hyperplane, that could be either linear or nonlinear, that has a maximum-margin in the higher feature space induced by kernel function $q$ (Cortes & Vapnik, 1995). In our method, we used the Radial Basis Function (RBF) kernel that has the form: $q(x, z) = e^{-\gamma \|x-z\|^2}, \gamma > 0$, where $\gamma = \frac{1}{2\sigma^2}$ (Cao, Naito, & Ninomiya, 2008). The RBF sigma value is a hyperparameter that is determined for each subject using cross-validation. Other classifiers were examined including K-Nearest Neighbors (KNN), Softmax, Neural Networks, and Classification Trees (Gold & Rangarajan, 1996; Loh, 2011).

The evaluation measures used in our work are sensitivity (percentage of correctly classified preictal points to the sum of preictal and interictal points), specificity (percentage of correctly classified interictal points to the sum of preictal and interictal points), and accuracy (percentages of correctly classified points, both preictal and interictal, to the total number of points). These measures are used to ensure that the results are not biased to one class over the other.

### 2.2.2 Stage 2: Preictal classification into early and late classes

In the previous section, we discussed the first stage of the proposed method where we classify EEG segments to either preictal or interictal states. We also discussed some enhancements that can be done to seizure prediction methods to make them more practical including channel selection and using shorter training data duration. Despite these enhancements, the first stage cannot accurately

**270** Big data in psychiatry and neurology

predict when the seizure onset might occur. The first stage only predicts the seizure occurrences possibility, but it gives no information about the onset time. In this section, we introduce the second stage of the proposed method to address this challenge. In addition, we examine different variations of selecting and labeling the two classes: early and late preictal.

### 2.2.2.1 Different selections of early and late preictal classes: Twelve 5 min consecutive classes

In this part, the classification of the preictal hour data is performed as follows: the preictal hour is divided to 12 classes. Each class represents a 5-min block. Class 1 is the first block in the preictal hour, while class 12 is the last block in the preictal hour. Multiclass classification is then performed to classify any given data segment as belonging to one of the 12 classes. We have examined the model using both modes when training and testing were done using the same seizures, and when different seizures were used in training and testing. We examined Linear Discriminant Analysis (LDA) and KNN as linear and non-linear classifiers, respectively, in the multiclass classification using wavelet zero-crossings features. In addition, we studied different approaches in selecting the classes. For instance, we tried dividing the preictal hour to blocks of consecutive 10 min instead of 5 min.

### 2.2.2.2 Different selections of early and late preictal classes: Binary classification between early and late classes

In addition to multiclass classification between consecutive blocks of the preictal hour, we have also examined using only two classes where we divided the preictal hour to early and late preictal classes. We examined setting the early and late preictal classes such that the early class is the first 30 min of the preictal hour, while the late class is the last 30 min of the preictal hour. In this experiment, the training and testing data were selected to be from different seizures. The second setting we examined was to have a gap between the early and late preictal classes such that the differences between the recorded signals are significant. We selected the early class to be the first 10 min of the preictal hour and the late class to be the last 10 min of the preictal hour.

Moreover, to add more robustness to the model against differences in the preictal class of different seizures, we add a normalization step in the feature extraction block. Instead of directly computing the zero-crossings in each window, we calculate the number of consecutive zero-crossings relative to the total number of zero-crossings occurrences. In this way, the feature is not the absolute values of zero-crossings occurrences. We compared the performance of the model using the two different approaches in feature extraction. The modified normalized feature can be represented as

$$Feature = \frac{\#Consecutive\_zero\_crossings, \, (i = j + 1)}{\#zero\_crossings} \tag{3}$$

where indicates the count, $i$ is the parsed index of the current occurrence of zero-crossings in wavelet detail coefficient, and $j$ is the index of the previous zero-crossing occurrence. When $i = j + 1$, we have two consecutive occurrences of zeros crossings.

## 2.3 Comparative analysis methods

In this section, we briefly summarize the methods we compared our approach to.

### 2.3.1 Wavelet features-based method

This method is based on extracting the following features from wavelet coefficients (Nasehi & Pourghassem, 2011): $Mean = \frac{1}{N}\sum_{i=1}^{N} x_i$, $Variance = \frac{1}{N}\sum_{i=1}^{N} (x_i - Mean)^2$, $Minimum\ value$, $Maximum\ value$, and $Number\ of\ zero\ coefficients$. For this method, we have examined and compared calculating those features for detail coefficients and approximation coefficients, and we have examined different wavelet functions such as db8, db4, and Haar functions as well as different classifiers including SVM, LDA, and KNN.

### 2.3.2 Wavelet entropy-based method

We examined another wavelet-based method, proposed by Holla et al. (Holla & Aparna, 2012), that is based on Approximation Entropy $ApEn(m,r,N)$ (where $m$ represents the length of the compared segment of data, $r$ is the similarity threshold, and $N$ is the number of sample points) (Pincus, 1991). $ApEn$ is a method used for measuring the amount of predictability and regularity in a time series and its fluctuations. $ApEn$ is defined as follows

$$ApEn = \ln\left(\frac{C_m(r)}{C_{m+1}(r)}\right) \tag{4}$$

where

$$C_m(r) = \frac{\sum_{i=1}^{i=N-m+1} C_{im}(r)}{N - m + 1} \tag{5}$$

and

$$C_{im}(r) = \frac{n_{i,m}(r)}{N - m + 1} \tag{6}$$

where $C_{im}(r)$ represents the percentages of patterns of length $m$ that are similar to the pattern of length $m$ starting from index $i$ (i.e.: $p_m(i)$), relative to the total number of patterns $N - m + 1$. $C_m(r)$ is the mean of all values of $C_{im}(r)$ calculated in the previous step. $ApEn$ is the natural logarithmic of the frequency of repetitive subsequences of length $m$ compared to subsequences of length $m + 1$. This implies that the larger the value of $ApEn$, the more the similarity of the

**272** Big data in psychiatry and neurology

patterns is coincidental as similar patterns are not followed by other similar patterns. Similarly, smaller values of *ApEn* imply high predictability in the time series.

### 2.3.3 Wavelet energy-based method

Wavelet energy has been utilized as a univariate feature in seizure prediction literature (Direito, Dourado, Vieira, & Sales, 2008; Gigola et al., 2004; Rasekhi, Mollaei, Bandarabadi, Teixeira, & Dourado, 2013; Sridevi et al., 2019). In our study, we compared the proposed zero-crossings model against wavelet energy-based models. First, the signal energy $E_j$ of each level $j$ is computed as follows:

$$E_j = \sum_{i=1}^{N_j} d_j^2(i), \tag{7}$$

where $d_j(i)(i = 1, \ldots, N_j)$ are the wavelet coefficients in level $j$, and $N_j$ is the number of coefficients in level $j$. The energy of each level is then normalized by the total energy in the signal.

### 2.3.4 Spectral power-based method

In this method, the features are based on spectral power analysis of EEG signals (Park et al., 2011). The power of different subband is calculated and then normalized by the total power. The subbands studied are delta (0.5–4 Hz), theta (4–8 Hz), alpha (8–13 Hz), and beta (13–30) Hz, and the gamma band which is decomposed to 4 other subbands (30–47 Hz, 53–75 Hz, 75–97 Hz, and 103–128 Hz), and finally the last feature is the total power.

## 3  Results

In this section, we discuss the performance of both stages of the proposed method. We also compare the proposed wavelet feature to other previously introduced methods and we discuss other parameters such as wavelet function selection, dimensionality reduction approaches, and classification methods.

### 3.1  Stage 1: Interictal and preictal binary classification

#### 3.1.1  Comparison with other methods

We first examined the performance of the Stage 1 of the proposed framework to classify the recorded EEG signals as belonging to either the preictal or the interictal classes. The proposed method was compared to the wavelet entropy and wavelet features methods with two cases of comparison: the first case is when the training and testing data were selected by randomly splitting the same set of seizures, and the second case when the training and testing seizures are different. Fig. 4 demonstrates the results in both cases of the proposed method compared to other methods. The proposed method achieves significantly higher accuracy in both cases with mean accuracy of 98% $\pm$ 2% in the first case,

# A two-stage classification framework for seizure prediction

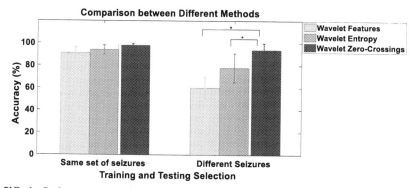

**FIG. 4** Performance comparison between the proposed zero-crossings method and the wavelet features and wavelet entropy methods. The figure shows the accuracies in two cases: the first case is when the training and testing data are selected by randomly splitting the same set of seizures, and the second case is when the testing seizures are different from the training seizures. (*$P < 0.01$, two-sample $t$-test).

and, more importantly, 94% ± 5% in the second case ($P < 0.01$, two-sample $t$-test). This indicates that the proposed method generalizes better to future unseen seizures compared to other examined methods whose performance declines significantly in the second case when the testing is based on seizures different from those used in training. The comparison was done using SVM classifier and combining the highest performance channels using the proposed dimensionality reduction approach, which are the factors that maximize the performance as will be seen in the next comparisons.

### 3.1.2 Detail versus approximation coefficients for feature extraction

In order to identify whether the detail or approximation coefficients provide better separation between the preictal and interictal classes, we compare the performance of the approach when each of the two types of coefficients is used. Fig. 5A shows the zero-crossings count of both detail and approximation coefficients. The figure shows a subset of feature points plotted with the two highest performing channels. As demonstrated in the figure, detail coefficients (left) have higher separability between preictal and interictal states when compared to approximation coefficients (right). However, we note that the detail coefficients zero-crossings count in the preictal and interictal states does not have the same pattern across subjects. For instance, for some subjects, the count of preictal zero-crossings was observed to be higher than the count of interictal zero-crossings, while for other subjects the opposite was observed. However, in both cases, the zero-crossings count can be highly separating among the two classes of interictal and preictal data. Fig. 5A shows an example of one subject where the count of preictal detail coefficients zero-crossings is less than the interictal detail coefficients zero-crossings count, while Fig. 5B shows another subject

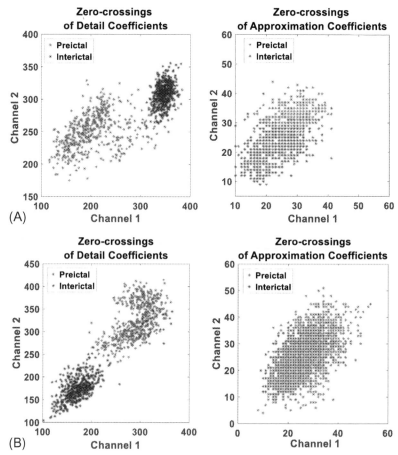

FIG. 5 A subset of interictal and preictal feature space of zero-crossings count for detail and approximation coefficients for two subjects. (A) Subject 1, zero-crossings of detail coefficients (top left), and zero-crossings of approximation coefficients (top right). For this subject, the number of zero-crossings occurrences in the preictal detail coefficients is less than those in interictal detail coefficients. Additionally, detail coefficients features have a higher separation between the classes. (B) Subject 2, Zero-crossings of detail coefficients (bottom left), Zero-crossings of approximation coefficients (bottom right). For this subject, the number of zero-crossings occurrences in the preictal detail coefficients is higher than those in interictal detail coefficients.

where the opposite happens. However, in both cases, the figures show that using detail coefficients in feature extraction achieves significant separation between the two classes as opposed to using approximation coefficients.

### 3.1.3 Dimensionality reduction and selection

We compared the performance of our proposed approach to the performance achieved when channel selection is not used and when either PCA or autoencoder approaches are used for dimensionality reduction. For autoencoders, the output of the hidden layer is used as the feature vector of the classifier. We studied the effect of changing the encoding and decoding functions as well as changing the number of hidden nodes of the autoencoder on accuracy.

Fig. 6A demonstrates a comparison between the four approaches where the accuracy is 70% ± 20% when using all channels, 78% ± 11% using PCA, 82% ± 14% using autoencoders, and 94% ± 5% using the proposed channel selection method. SVM classifier and Haar wavelet function were used in this comparison. Additionally, we examined the specificity and sensitivity of all methods as demonstrated in Fig. 6B. When using all channels, the sensitivity (i.e.: predicting preictal class) drops significantly, while PCA and autoencoders achieve comparable performance in detecting both classes. More importantly, the proposed channel selection method achieves high sensitivity and specificity in detecting both classes.

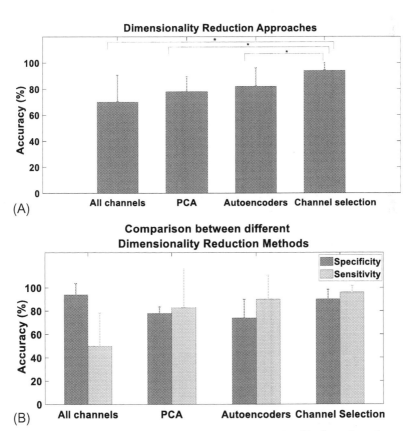

FIG. 6 (A) Accuracy of different dimensionality reduction approaches. The figure shows the accuracy of using all channels, using PCA, using autoencoders, and using the proposed channel selection approach (last bar) which achieves the highest accuracy (*$P < 0.01$, two-sample $t$-test). (B) Performance of different dimensionality reduction approaches showing the specificity and sensitivity of different methods.

We also examined the performance of the autoencoder approach when the activation function of the encoder and decoder is varied to examine if there is a specific combination of activation functions that can achieve a significant higher accuracy. Fig. 7A demonstrates the performance of autoencoders where we calculated the average accuracy for all eight subjects while unifying the hidden size layer but with changing all the combinations of encoder and decoder functions (linear, log-sigmoid, saturating linear). The six combinations have no significant effect on accuracy. In all of them, the average is approximately 80% while the standard deviation across subjects is relatively high ($\pm 14\%$). This analysis was performed while unifying the hidden layer size to be 5 nodes. However, other values of the hidden layer size were examined, and the results were also comparable.

Moreover, we examined different classifiers including Softmax, KNN, SVM-RBF, Neural nets, and Classification trees at the output of the

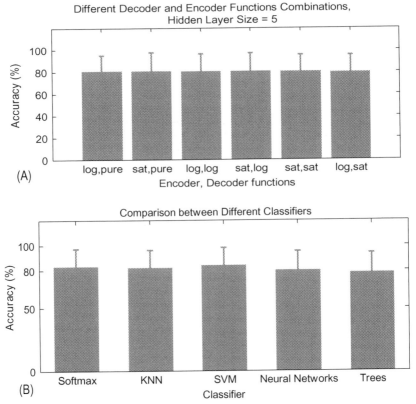

FIG. 7 Autoencoder results analysis. (A) Average accuracy vs. encoder, decoder transfer functions for all subjects while unifying the hidden layer size. (B) Performance comparison of different classifiers showing the output of SoftMax, KNN, SVM, neural networks and trees.

A two-stage classification framework for seizure prediction Chapter | 13   **277**

autoencoder (Gold & Rangarajan, 1996; Loh, 2011). As demonstrated in Fig. 7B, the results are similar where the maximum mean accuracy is obtained using SVM or Softmax (83% ± 14%), while the minimum mean accuracy is obtained using trees (76% ± 15%). These results indicate that the proposed channel selection approach outperforms other examined dimensionality reduction approaches.

## 3.2   Stage 2: Preictal classification into early and late stages

In this section, we demonstrate the results of the second stage of the approach in which we further classify the preictal interval identified by the first stage to either early or late preictal stage. We show the performance with respect to different selections of early and late preictal intervals and show the performance of different parameters such as the features and classifiers used.

### 3.2.1   Different selections of early and late preictal classes

#### 3.2.1.1   Twelve 5 min consecutive classes

In this setup, the performance of the classification is evaluated as follows:

- A percentage of how many points (where each point represents the features computed from a 5-s window), from all classes, were correctly classified.
- A confusion matrix that includes the percentages of how many points (with respect to the testing set) were misclassified by up to 10 min, 10–20 min, 20–30, and more than 30 min, before or after the actual seizure onset time.

We first examined the performance when the training and testing data were selected by randomly splitting from the same seizures. We examined linear and nonlinear classifiers in the multiclass classification using wavelet zero-crossings features. Fig. 8A shows the classification results using KNN and LDA. The figure shows the percentage of correctly classified points, and the percentages of each case in the confusion matrix metrics that were mentioned earlier. The figure shows that LDA can achieve higher accuracy than KNN. In case of LDA, 81% of the points were classified correctly or with a difference of maximum 10 min from their actual timing (49% of points were classified correctly in addition to 32% of points classified with a difference of 5–10 min).

Additionally, using LDA classifier, we compared using entropy and the proposed zero-crossings feature extraction methods in the same multiclass classification of different preictal classes. Results are illustrated in Fig. 8B showing that the proposed approach achieves higher accuracy compared to the entropy-based method, where the correct percentage is 49% ± 8% in case of using the zero-crossings method, and 34% ± 13% in case of using the entropy method.

We next examined the performance when the training and testing data were taken from different seizures. Similar to the previous case, the proposed approach performs better than the entropy feature extraction method. However,

**278** Big data in psychiatry and neurology

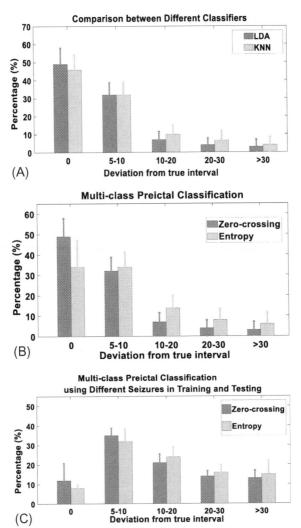

**FIG. 8** (A) Comparison between LDA and KNN classifiers in multiclass classification of preictal states when using same seizures in training and testing. The figure shows the percentage of samples for which their timing was perfectly classified (Correct), classified within 5–10 min, 10–20 min, 20–30 min, or more than 30 min, from their actual seizure onset time. (B) Comparison between entropy and zero-crossings based methods in multiclass classification of preictal states for the same data and time intervals in (A). (C) Performance of the proposed zero-crossings based methods versus the entropy method in multiclass classification of preictal states when using different seizures in training and testing for the same time intervals in (A).

the percentage of correctly classified points drops significantly to 13% ± 4% in case of using the zero-crossings based method, and 8% ± 2% in case of using entropy-based method, as shown in Fig. 8C. This could be attributed to the lack of significant differences across successive 5-min intervals.

### 3.2.1.2 Binary classification between early and late classes

Given that using a large number of classes (12 classes) did not achieve high accuracy as demonstrated in the previous section, we examined the performance of the model when dividing the preictal hour to only two classes (early versus late preictal stage). We first set the early and late classes to the first and last 30 min of the preictal hour, respectively. Using zero-crossings, the proposed channel selection approach, and SVM classifier, an accuracy of 70% ±9% is achieved between the early and late classes, which is higher than the output of classification in the 12-class setting. We then examined the performance after increasing the gap between the early and late preictal classes such that the differences between the recorded signals are significant. We selected the early class to be the first 10 min of the preictal hour and the late class to be the last 10 min of the preictal hour. This setting achieved a higher accuracy (as expected) of 74% ± 10%.

In the mentioned comparisons, there were some subjects that achieved a higher accuracy than others (93% as opposed to 68%) as shown in the first column of Table 1. Therefore we analyzed the differences between the features in the two subjects corresponding to the aforementioned accuracies. Fig. 9A shows a sample of two channels for the higher accuracy subjects, where each point represents the zero-crossing count for a window, and the points of preictal data for two different seizures are shown, along with showing the early and late preictal data for the first and last 30 min of each seizure. Fig. 9B shows the same analysis for the other subject which achieved the 68% accuracy. In both figures, Early 1 and Early 2 points are the first 30 min of the preictal hour for seizures 1 and 2, respectively, while Late 1 and Late 2 points are the last 30 min of the preictal hour for seizures 1 and 2, respectively. When comparing the two cases, it can be seen that there is a major challenge in classifying preictal data as the two preictal hours data points do not lie in the same numerical range. Therefore training using data of one seizure and testing with data of another might fail.

To resolve this issue, we proposed adding a normalization step in the feature extraction phase to make the preictal data of different seizures within the same range. Therefore instead of directly computing the zero-crossings in each window, we calculated the number of consecutive zero-crossings relative to the total number of zero-crossings occurrences. In this way, the feature is not the absolute values of zero-crossings occurrences. Table 1 demonstrates a comparison between the two setups when using the feature vector as zero-crossings, as opposed to the normalized zero-crossings (second column). As shown, when relying on the normalized zero-crossings features, the mean accuracy of

**TABLE 1** A comparison of the accuracy obtained in the second stage of the framework when using zero-crossings and normalized zero-crossings features.

| Patient number | Zero-crossings | Normalized zero-crossings |
|---|---|---|
| Patient 22 | 68% | 77% |
| Patient 1 | 93% | 95% |
| Patient 9 | 60% | 72% |
| Patient 7 | 82% | 86% |
| Patient 18 | 71% | 79% |
| Patient 5 | 67% | 73% |
| Patient 13 | 77% | 78% |
| Patient 6 | 76% | 84% |
| Mean% ± std% | 74% ± 10% | 80.5% ± 7.5% |

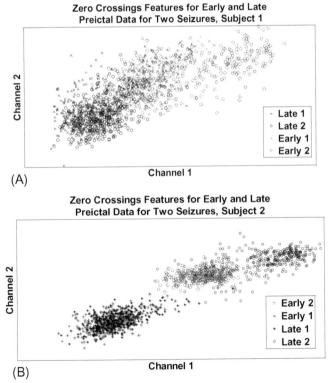

**FIG. 9** Early and late preictal data for two seizures for two subjects. (A) A subject with relatively high separation between the early and late classes (patient 1). (B) Another subject with low separation between the early and late classes (patient 22).

classifying the first and last 10 min of preictal data increased from 74% ± 10% to 80.5% ± 7.5%, and the standard deviation between the subjects has decreased as well, which indicates more consistency in the classification.

### 3.2.2 Performance of different wavelet functions and levels

Moreover, we also examined the feature of normalized zero-crossings from several aspects including the selection of the wavelet function in wavelet transform calculation. We examined db8 from Daubechie wavelets, coif1 from Coiflet family (Chakrabarti, Swetapadma, Ranjan, & Pattnaik, 2020; Chui, 2014), bior 1.1 from biorthogonal wavelet (Dahmen & Micchelli, 1997), and Haar wavelet. Haar function achieves the highest accuracy of 80.5% ± 7.5%, while 'db8' achieves the least accuracy of 70% ± 6% (when setting the early and late classes to the first and last 10 min of the preictal hour). The comparison of these functions is demonstrated in Fig. 10. This is also consistent with the results obtained from the similar experiment done on stage 1 of the method, where the Haar wavelet function achieved a higher accuracy of 94% ± 5%, while the 'db4' and 'db8' functions achieved an accuracy of 90 ± 8% and 87 ± 9%, respectively.

Additionally, in stage 1 of the method, we studied the performance of higher wavelet levels when using a 4-level wavelet decomposition along with using Haar wavelet function. An accuracy of 93% ± 8% is achieved, which is strongly similar to the 2-level wavelet decomposition results. These results are consistent with the results reported in (Holla & Aparna, 2012) where increasing the number of decomposition levels does not necessarily result in higher accuracies. Thus we limited the number of levels in the wavelet analysis to 2. In this analysis, the classification is done between the preictal and interictal classes using the proposed channel selection method and SVM classifier.

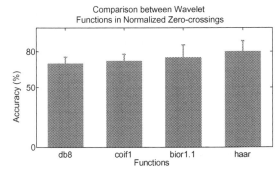

**FIG. 10** Performance of the second stage of the proposed framework when varying the wavelet functions.

### 3.2.3 Performance of different classifiers and feature extraction methods

In addition, we studied the performance of other classifiers along with SVM including linear classifiers such as LDA and nonlinear classifiers such as KNN and classification trees. The performance of these classifiers is demonstrated in Fig. 11A, where SVM achieves the highest accuracy of 80.5% ± 7.5%, LDA achieves an accuracy of 75% ± 12%, classification trees achieve an accuracy of 72% ±12%, while KNN achieves the least accuracy of 70%±11% (when setting the early and late classes to the first and last 10 min of the preictal hour).

Finally, the performance of the normalized zero-crossings method in detecting late and preictal classes was compared against the examined two methods of wavelet energy and spectral power. The reason behind the selection of those two features to compare the proposed feature against is that they both use normalization in feature calculation which we found to be more effective than calculating the absolute value and achieved a higher performance in early and late

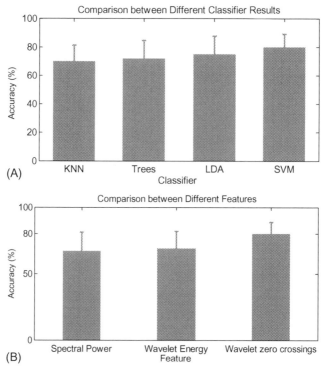

**FIG. 11** (A) The performance of different classifiers including KNN, trees, LDA, and SVM. (B) Comparison between wavelet normalized zero-crossings and other features including normalized spectral power and wavelet energy.

preictal classification. The results of the comparison are demonstrated in Fig. 11B. Results demonstrate that wavelet normalized zero-crossings feature can achieve higher accuracy of 80.5% $\pm$ 7.5% as opposed to the other two studied methods having accuracy of 69% $\pm$ 14% and 67% $\pm$ 12%, respectively (when setting the early and late classes to the first and last 10 min of the preictal hour). In the three cases, the results are obtained using SVM, the proposed channel selection method, and Haar wavelet function which are the model parameters that achieve the highest performance.

## 4  Discussion

In this chapter, we introduced a novel two-stage framework to analyze epileptic EEG data. The goal of the first stage of the approach is to accurately classify an EEG segment to either preictal or interictal, while the goal of the second stage is to classify preictal data into early or late stages. We investigated several feature extraction approaches that were previously introduced. These methods can achieve high performance when applied to preictal patterns that are already included in the training data. However, there are different variations across seizures as the number of different seizure types can be at least 40 and patients may suffer from one or several different seizure types (Council, J. E, 2005). Therefore, to test the aptitude of the proposed approach to generalize to future unseen preictal data, the method testing was done with seizures that were not included in the training data. When examining the previously introduced methods, we found that they fail to achieve such high performance when the preictal training and testing data are taken from different seizures. Thus we proposed a feature extraction method that has not been used in seizure prediction before to face such challenge. The proposed method utilizes the zero-crossings count of wavelet transform. We found that zero-crossings of detail coefficients achieve higher separation than approximation coefficients. A possible interpretation for this observation is that detail coefficients when using Haar function represent changes in the signal pattern or ordering and they are independent of the values of the signal themselves, unlike approximation coefficients which represent an averaging of signal samples. Therefore detail coefficients could be more robust in detecting preictal data of different seizures and overcoming possible differences in them. Additionally, since zero-crossings of wavelet transform provide information about the locations of sharp variation points (Mallat, 1991), in our interpretation, this could provide important information from seizure prediction perspective to capture transient variations.

As part of the proposed two-stage framework, the proposed channel selection method achieves high sensitivity and specificity in detecting both classes compared to using all channels. This demonstrates that the changes that distinguish the two classes persistently across different seizures could be in a local set of channels rather than being global across all channels. In our analysis, we

**284** Big data in psychiatry and neurology

studied shallow autoencoders which performed poorly than the proposed channel selection method; however, deep autoencoders could be further studied.

It is worth mentioning here that in our analysis the input to the second stage in our framework are the ground truth preictal points not the points predicted by the first stage. This is done to assess the performance of the second stage independently; however, in real usage scenarios, the input to the second stage should be the points classified as preictal from the first stage.

As an extension to this work, there could be more profound study on the underlying reasons of differences of preictal data across different seizures that were observed in our work, and how that can be addressed whether by signal denoising, artifacts removal, or different feature extraction methods to have a more reliable seizure prediction method. In addition, our work was based on offline data segmented in a way that matches the definition of preictal and interictal classes. Despite the statistically significant improved results achieved by our method, the performance is likely to need further improvements when tested on online or continuous data. Therefore the framework should be tested on continuous data recordings that do not necessarily meet the conditions of preictal and interictal data selection to assess the false alarms rate and the sensitivity, and to measure when the framework will start producing alarms before seizures. In addition, even though seizure prediction systems are patient specific due to EEG differences between patients, cross-subject training could be further studied as a way to reduce training and data recording time.

## 5 Conclusions

In this chapter, we proposed a two-stage framework using zero-crossings count of wavelet detail coefficients that could be used for seizure prediction. The first stage of the method is preictal and interictal classification in which we demonstrated the efficacy of the framework in achieving reliable classification between preictal and interictal classes. In addition, the proposed method was further evaluated based on sensitivity and accuracy, and was found to achieve higher performance compared to other studied methods. Our results indicated that the performance is maximized when using the proposed channel selection approach in addition to using Haar wavelet in feature extraction and support vector machine for classification. The method was demonstrated to be invariant to differences between seizures and it achieves high accuracy using a short interval of training recording, where only 10 min of training data were used to facilitate the setup process of potential real-time applications. In addition, in stage 2 of the method, we studied the further classification of preictal data to early and late classes, where we examined different schemes for selecting the classes. Our results indicated that when using the normalized zero-crossings feature along with using SVM and the proposed channel selection method, an accuracy of 80.5% could be achieved between the early and late classes which outperforms other examined methods.

## Disclosure statement

No potential conflict of interest was reported by the authors.

## References

Acharya, U. R., Sree, S. V., Ang, P. C. A., Yanti, R., & Suri, J. S. (2012). Application of non-linear and wavelet based features for the automated identification of epileptic EEG signals. *International Journal of Neural Systems, 22*(02), 1250002.

Alickovic, E., Kevric, J., & Subasi, A. (2018). Performance evaluation of empirical mode decomposition, discrete wavelet transform, and wavelet packed decomposition for automated epileptic seizure detection and prediction. *Biomedical Signal Processing and Control, 39*, 94–102.

Altman, N. S. (1992). An introduction to kernel and nearest-neighbor nonparametric regression. *The American Statistician, 46*(3), 175–185.

American Epilepsy Society Seizure Prediction Challenge. (2014). Retrieved March 2017, from https://www.kaggle.com/c/seizure-prediction/data.

Baldi, P. (2012). Autoencoders, unsupervised learning, and deep architectures. In *Paper presented at the proceedings of ICML workshop on unsupervised and transfer learning*.

Burrus, C. S., Gopinath, R. A., & Guo, H. (1997). *Introduction to wavelets and wavelet transforms*.

Cao, H., Naito, T., & Ninomiya, Y. (2008). Approximate RBF kernel SVM and its applications in pedestrian classification. In *Paper presented at the the 1st international workshop on machine learning for vision-based motion analysis-MLVMA'08*.

Carney, P. R., Myers, S., & Geyer, J. D. (2011). Seizure prediction: Methods. *Epilepsy & Behavior, 22*, S94–S101.

Chakrabarti, S., Swetapadma, A., Ranjan, A., & Pattnaik, P. K. (2020). Time domain implementation of pediatric epileptic seizure detection system for enhancing the performance of detection and easy monitoring of pediatric patients. *Biomedical Signal Processing and Control, 59*, 101930.

Chang, N.-F., Chen, T.-C., Chiang, C.-Y., & Chen, L.-G. (2012). Channel selection for epilepsy seizure prediction method based on machine learning. In *Paper presented at the engineering in medicine and biology society (EMBC), 2012 annual international conference of the IEEE*.

Chiang, C.-Y., Chang, N.-F., Chen, T.-C., Chen, H.-H., & Chen, L.-G. (2011). Seizure prediction based on classification of EEG synchronization patterns with on-line retraining and post-processing scheme. In *Paper presented at the engineering in medicine and biology society, EMBC, 2011 annual international conference of the IEEE*.

Chui, C. K. (2014). *An introduction to wavelets. Vol. 1*. Academic Press.

Cianchetti, C., Bianchi, E., Guerrini, R., Baglietto, M. G., Briguglio, M., Cappelletti, S., et al. (2018). Symptoms of anxiety and depression and family's quality of life in children and adolescents with epilepsy. *Epilepsy & Behavior, 79*, 146–153.

Cortes, C., & Vapnik, V. (1995). Support-vector networks. *Machine Learning, 20*(3), 273–297.

Council, J. E. (2005). *Epilepsy prevalence, incidence and other statistics*. Leeds, UK: Joint Epilepsy Council.

Dahmen, W., & Micchelli, C. A. (1997). Biorthogonal wavelet expansions. *Constructive Approximation, 13*(3), 293–328.

Direito, B., Dourado, A., Vieira, M., & Sales, F. (2008). Combining energy and wavelet transform for epileptic seizure prediction in an advanced computational system. In *Paper presented at the biomedical engineering and informatics, 2008. BMEI 2008. International Conference on*.

Fisher, R. S., Boas, W.v. E., Blume, W., Elger, C., Genton, P., Lee, P., et al. (2005). Epileptic seizures and epilepsy: Definitions proposed by the international league against epilepsy (ILAE) and the International Bureau for Epilepsy (IBE). *Epilepsia, 46*(4), 470–472.

**286** Big data in psychiatry and neurology

Gigola, S., Ortiz, F., D'attellis, C., Silva, W., & Kochen, S. (2004). Prediction of epileptic seizures using accumulated energy in a multiresolution framework. *Journal of Neuroscience Methods*, *138*(1), 107–111.

Gold, S., & Rangarajan, A. (1996). Softmax to softassign: Neural network algorithms for combinatorial optimization. *Journal of Artificial Neural Networks*, *2*(4), 381–399.

Goldberger, A. L., Amaral, L. A., Glass, L., Hausdorff, J. M., Ivanov, P. C., Mark, R. G., et al. (2000). Physiobank, physiotoolkit, and physionet components of a new research resource for complex physiologic signals. *Circulation*, *101*(23), e215–e220.

Holla, A. V., & Aparna, P. (2012). A nearest neighbor based approach for classifying epileptiform EEG using nonlinear DWT features. In *Paper presented at the signal processing and communications (SPCOM), 2012 International conference on*.

Jolliffe, I. (2002). *Principal component analysis*. Wiley Online Library.

Khalil, B. A., & Misulis, K. (2006). *Atlas of EEG & seizure semiology*. Philadelphia: Elsevier, Butterworth Heinehmann Edition.

Khan, H., Marcuse, L., Fields, M., Swann, K., & Yener, B. (2018). Focal onset seizure prediction using convolutional networks. *IEEE Transactions on Biomedical Engineering*, *65*(9), 2109–2118.

Litt, B., & Echauz, J. (2002). Prediction of epileptic seizures. *The Lancet Neurology*, *1*(1), 22–30.

Loh, W. Y. (2011). Classification and regression trees. *Wiley Interdisciplinary Reviews: Data Mining and Knowledge Discovery*, *1*(1), 14–23.

Mallat, S. (1991). Zero-crossings of a wavelet transform. *IEEE Transactions on Information Theory*, *37*(4), 1019–1033.

Nasehi, S., & Pourghassem, H. (2011). Automatic prediction of epileptic seizure using kernel fisher discriminant classifiers. In *Paper presented at the intelligent computation and bio-medical instrumentation (ICBMI), 2011 International conference on*.

Panayiotopoulos, C. (2007). Epileptic syndromes and their treatment. In *Neonatal seizures* (2nd ed., pp. 185–206). London: Springer.

Park, Y., Luo, L., Parhi, K. K., & Netoff, T. (2011). Seizure prediction with spectral power of EEG using cost-sensitive support vector machines. *Epilepsia*, *52*(10), 1761–1770.

Pincus, S. M. (1991). Approximate entropy as a measure of system complexity. *Proceedings of the National Academy of Sciences*, *88*(6), 2297–2301.

Rasekhi, J., Mollaei, M. R. K., Bandarabadi, M., Teixeira, C. A., & Dourado, A. (2013). Preprocessing effects of 22 linear univariate features on the performance of seizure prediction methods. *Journal of Neuroscience Methods*, *217*(1), 9–16.

Shorvon, S. D. (2010). *Handbook of epilepsy treatment*. John Wiley & Sons.

Sridevi, V., Reddy, M. R., Srinivasan, K., Radhakrishnan, K., Rathore, C., & Nayak, D. S. (2019). Improved patient-independent system for detection of electrical onset of seizures. *Journal of Clinical Neurophysiology*, *36*(1), 14.

Tzimourta, K. D., Astrakas, L. G., Tsipouras, M. G., Giannakeas, N., Tzallas, A. T., & Konitsiotis, S. (2017). Wavelet based classification of epileptic seizures in EEG signals. In *Paper presented at the 2017 IEEE 30th international symposium on computer-based medical systems (CBMS)*.

World Health Organization. (2019). *Epilepsy fact sheet*. Retrieved April 2020, from https://www.who.int/news-room/fact-sheets/detail/epilepsy.

# Chapter 14

# Visual neuroscience in the age of big data and artificial intelligence

Kohitij Kar[a,b]

[a]*McGovern Institute for Brain Research and Department of Brain and Cognitive Sciences, Massachusetts Institute of Technology, Cambridge, MA, United States,* [b]*Center for Brains, Minds, and Machines, Massachusetts Institute of Technology, Cambridge, MA, United States*

## 1 Confining the problem space

Human vision is a very large scientific domain. It begins as the photons of light reflected from objects around us hit our retina. Vision scientists have dealt with a range of problems, studying the luminance-dependent processing in the retinal ganglion cells to our seamless recognition of visual objects. To keep things tractable, this chapter mainly focuses on recent developments in computational modeling of vision as it applies to a systems neuroscience investigation of "core object recognition"—primates' ability to rapidly identify visual objects in their central visual field (within the central 10 degrees of visual angle) in a single, natural fixation (lasting approximately 200 ms) despite various identity-preserving image transformations (like a variation in pose, size, shape, and color of objects). Understanding the brain mechanisms that seamlessly solve this challenging computational problem has been a key goal of visual neuroscience (Riesenhuber & Poggio, 2000).

Oftentimes, before asking a question in science, it might be important to reflect upon the type of answer that might satisfy one's curiosity. The choice of scientific methods of investigation is often guided by the level of detail we seek in the desired answer. For instance, lesioning a brain area and observing certain behavioral deficits might implicate that area in the behavior, but might not tell us the exact computational role of the area in the behavior. For the latter, we need to rigorously test the predictions of neurally mechanistic models of the area (and most likely a series of brain areas that come together to perform the behavioral task).

Big Data in Psychiatry and Neurology. https://doi.org/10.1016/B978-0-12-822884-5.00015-5
Copyright © 2021 Elsevier Inc. All rights reserved.

**288** Big data in psychiatry and neurology

## 2  Chapter roadmap

In the following sections, I first explain the type of answers one could seek in visual neuroscience. The chapter then focuses on a brief review of how the developments in artificial intelligence (specifically computer vision; AI as applied to vision) and big data along with diverse efforts in neuroscience have shaped the field's recent progress. To complement this section, I further discuss the concepts of "prediction" and "control" as ways to test the current models of vision. Finally, I provide three examples to demonstrate how the vision community is coming together to combine open-access large-scale data collection and model building to develop and promote a more data-driven and quantifiable *modus operandi.*

## 3  Understanding vision—What do we seek to reveal?

Let us imagine that we have entered a board meeting of an organization. We find that all the board members are communicating with each other by modulating the sound of their voice (measured in decibels) and producing words in some language. Now let us assume that we do not understand the language or the role that each member plays in the organization. How shall we go about understanding what's going on? How much information can we gain then by trying to come up with events (e.g., asking them questions in our own language) that will make each individual increase their voice (i.e., shout) to the highest decibel levels? Let us compare this scenario to that of neurons in the brain. Presumably, individual neurons (similar to board members in the analogy) produce dynamic responses (words in some language) and these patterns of responses across the population of neurons (aka distributed population code) lead to different behaviors in an organism (the organization). Neuroscientists, typically quantify the magnitude of the response of each neuron by measuring the number of spikes emitted (similar to sound decibels in the analogy) to estimate the contribution of each neuron to the system's output. Understanding this neural system is equivalent to understanding the language in which the board members are communicating to run the organization. In the classical neuroscience approach, an experimenter tries to modulate input stimuli (based on their own intuition) to best modulate the responses of individual or population of neurons and guesses the "principle" that might underlie the activity of each neuron. An underlying assumption is that a circuit-level theory can be constructed by combining the operation of these individual neurons. The inefficiency of this method becomes very clear if we compare this to deciphering the board meeting scenario. At best, this can be the first step to gain some insight, but it soon becomes close to shooting arrows in the dark. What is the operational goal or objective of the system? Without formulating the investigations in light of this question, it is hard to interpret or build upon most experimental results (for further discussions, see Richards et al., 2019).

One can broadly categorize "understanding" the visual system as solving either of the following two specific problems (separately or combined): the *encoding* problem (how is the visual world represented in the neural activity in the brain), and the *decoding* problem (how does the neural activity of brain areas give rise to different behaviors). Studies that evaluate the tuning properties of neurons (Albright, 1984; Nauhaus, Benucci, Carandini, & Ringach, 2008; Tsao, Freiwald, Knutsen, Mandeville, & Tootell, 2003) are essentially solving the encoding problem. On the other hand, studies that often fall into the domain of decision-making (Britten, Newsome, Shadlen, Celebrini, & Movshon, 1996; Majaj, Hong, Solomon, & DiCarlo, 2015) are essentially solving the decoding problem. It is often very informative to determine which of these two problems is being addressed to assess the strength of the inference presented in a study. It is not hard to find studies where the inferences are articulated in terms of intuitive explanations of a phenomenon rather than a falsifiable model that makes very specific predictions. Throughout the rest of the chapter, I have focused mainly on how to leverage on falsifiable vision models as a measure of success and less on qualitative data descriptor models. So instead of explaining neural representation in terms of qualitative terminology like orientation, texture, motion, face selectivity, among others, the chapter emphasizes the latest trend in a renewed interest of the field (based on progress in AI and big data) to focus on building stimuli-computable models that can provide falsifiable predictions.

If we consider what we know about the retinal cells, thalamic, and early visuocortical neurons, the description of what these cells encode is often based on human-understandable reduced properties of images like contrast (Albrecht & Hamilton, 1982; Shapley & Victor, 1978), orientation (Hubel & Wiesel, 1959), motion direction (Albright, 1984), among others. Under the assumption that the complexity in describing the stimuli preference of neurons increases along the visuocortical hierarchy, it becomes harder and harder to articulate the nature of that stimulus space beyond empty words like tuning to *complex shapes, textures, and mixed-selectivity*. Thus the traditional approach yields no further success and guidance. At this point, the following quote from P. W. Anderson is rather relevant,

*"The ability to reduce everything to simple fundamental laws does not imply the ability to start from those laws and reconstruct the universe."*

This void in our so-called understanding of the encoding properties of visuocortical neurons makes the advent of convolutional neural networks (see later) all the more relevant. As discussed later, beyond providing the basis for explaining the complex representational properties of high-order cortical neurons (Bashivan, Kar, & DiCarlo, 2019; Yamins et al., 2014), convolutional neural networks have become the leading models for explaining the previously well-studied responses of early visuocortical V1 neurons (Cadena et al., 2019) and retinal ganglion cells (McIntosh, Maheswaranathan, Nayebi, Ganguli, & Baccus, 2016).

## 3.1 The first generation of neural network models

While it is certainly impossible to provide here a comprehensive history of the developments in the artificial neural network literature, I have provided later a brief synopsis of some of the most relevant achievements with regard to the rest of the chapter. The very influential perceptron model was proposed by Frank Rosenblatt in 1958 (Rosenblatt, 1958). This was further refined and analyzed by Minsky and Papert (1969). Convolutional neural networks (CNNs) were first introduced in the 1980s by Yann LeCun and colleagues, building upon the work done by Kunihiko Fukushima, who invented the neocognitron (Fukushima & Miyake, 1982), a very basic image recognition neural network. As the name suggests, the most important operation in these networks is "convolution." This involves a point-by-point multiplication of a smaller pixel pattern (filter) with the original image in steps across the entire image and adding up the resulting maps. It is also often argued that the idea of convolutions as applied to these models has been inspired by the notion of spatial receptive fields in the seminal work of Hubel and Wiesel (1959). The promise of CNN (see Fig. 1 for an example CNN) was evident in early successes like the demonstration of handwritten digit classification using supervised backpropagation algorithms (LeCun et al., 1989), by a network called LeNet (after LeCun). By the 1990s, many hierarchical models of the visual system were being explored. One of the most prominent ones was HMAX (which roughly stands for "Hierarchical Model And X," where X is a highly nonlinear maXimum operation). The HMAX model was initially built to explain a previously collected dataset (Logothetis, Pauls, & Poggio, 1995) involving intracortical recordings in the inferior temporal cortex of rhesus macaques. However, over the years, this model has undergone multiple revisions (Serre & Riesenhuber, 2004; Serre, Wolf, Bileschi, Riesenhuber, & Poggio, 2007). It is often considered that the pivotal point in the trajectory of

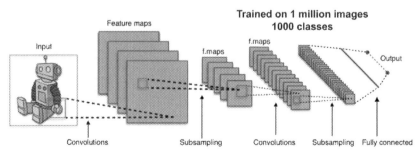

**FIG. 1** Example of a convolutional neural network. A CNN architecture is formed by a stack of distinct layers that transform the input volume into an output volume (e.g., holding the class scores) through a differentiable function. These layers include but are not limited to convolutional layers, pooling layers, ReLU (nonlinearity) layers, and fully connected layers. The convolutional layer is the core building block of a CNN. *(The figure was retrieved from Wikimedia Commons, the free media repository.)*

Visual neuroscience in the age of big data Chapter | 14  **291**

CNN-related research came with the development of a hierarchical 8-layer network, commonly referred to as "AlexNet," by Krizhevsky et al. in 2012. The term "deep-net" often refers to the fact that these networks have multiple computational layers, hierarchically stacked, one feeding into another, making the networks "deep." This network was also trained with backpropagation and became the state of the art in terms of its performance on the ImageNet challenge (Russakovsky et al., 2015). The ImageNet dataset is composed of over a million real-world images. The ImageNet challenge requires the models to classify an image into one of the thousand object categories. This large dataset-based competition and the approach of training networks with supervised learning algorithms and a huge amount of data is arguably a key moment in the history of visual neuroscience.

Around the same time, James DiCarlo and colleagues at MIT were working on similar problems related to visual object recognition in primates. Their approach, different from AlexNet, depended on a performance optimization-based architectural search strategy, yielded similar hierarchical deep convolutional networks as the best models of the primate ventral stream (Yamins, Hong, Cadieu, & DiCarlo, 2013; Yamins et al., 2014). The hierarchical modular optimization (HMO) model proposed by Yamins et al. (2013) showed very high performance in image classification. In addition, remarkably, the internal layers of the models matched the internals of the primate brain (areas V4 and IT), beyond the prediction strength of previous models in the field. It was later shown by the same group that the models emerging from the computer vision field yield very similar results (Cadieu et al., 2014). Some of these models even exceed HMO's performance and prediction capabilities (see following for further details). In the years since these results were published, many different deep network architectures have been explored, with respect to primates' behavior and neural representations (Schrimpf et al., 2018). This same type of relationship between CNN model features and the neurophysiological measurements has also been observed in human fMRI (Guclu & van Gerven, 2015) and MEG (Seeliger et al., 2018).

## 3.2 Next generation of neural network models

Although most of the recent successes are based on models that are predominantly feedforward in nature, many studies (Joukes, Yu, Victor, & Krekelberg, 2017; Kar, Kubilius, Schmidt, Issa, & DiCarlo, 2019; Kietzmann et al., 2019; Quiroga, Morris, & Krekelberg, 2019; Spoerer, McClure, & Kriegeskorte, 2017; Tang et al., 2018) have demonstrated the shortcomings of feedforward models and the need to incorporate recurrent computations in vision models. This has led to the development of deep recurrent neural networks (Kubilius et al., 2019; Nayebi et al., 2018; Zamir et al., 2017) that are still mostly "work in progress" in terms of their performance and predictive powers. Some of these networks (e.g., CORnet-S; Kubilius et al., 2019) indeed

**292** Big data in psychiatry and neurology

predict neural benchmarks much better than their feedforward predecessors. There are many open questions that are very important to address as researchers further develop such models. For instance, computer vision models that are relevant to the primate object recognition literature have heavily relied on training based on the ImageNet database and optimized for image classification tasks. It is not clear whether such images and tasks are the best ones for training a deep recurrent neural network. Therefore researchers might expand the training diet and task regimes to more challenging as well as dynamic regimes. In addition, experimental neuroscientists need to narrow down the specific recurrent circuit motifs that are most relevant for specific tasks, making the development of recurrent loops in the model more tractable. Following I have listed a few of such questions:

1. *What training tasks and image-sets are best suited for recurrent neural networks?*
2. *Is backpropagation through time the best learning algorithm for such networks?*
3. *What are the most critical recurrent pathways (functionally) for a given task?*
4. *What unique experimental predictions can discriminate between different recurrent networks?*

### 3.3 Experiments to falsify and improve models

As demonstrated before, deep convolutional neural networks (DCNNs) have revolutionized the modeling of visual systems. However, despite being good starting points, they do not fully explain brain representations and are not like a brain in many ways. DCNNs differ from the brain with respect to their anatomy, task optimization, learning rules, etc. For instance, most DCNNs lack recurrent connections that are ubiquitous in the primate brain (Felleman & Van Essen, 1991; Rockland, Saleem, & Tanaka, 1994). DCNNs are trained using supervised learning algorithms like backpropagation (Kelley, 1960) that is unlikely to be the exact learning mechanism in biological systems. DCNNs currently do not have brain-like topography (e.g., face patches, orientation columns, etc.). Therefore it can be speculated that the next generation of models could benefit by incorporating some critical brain-inspired constraints. It is also important to acknowledge though that the model predictions need to be experimentally validated and falsified. Not only that, but future experiments should also be specifically designed such that the data collected can be directly used to improve these existing models. A common trend is that experimental data collection and modeling efforts are executed somewhat independently, resulting in very little model falsification, and thus no measurable progress. The synergy and collaboration between computational modeling and experiments are critical to achieving success. The approach needs to be a closed-loop. Models can

Visual neuroscience in the age of big data **Chapter | 14 293**

predict experimental outcomes. Experiments done therefore can falsify models and better models with further experimentally derived constraints can be built.

Following, I have listed two different approaches that are typically used to test the current models of vision—prediction and control. Both of these methods are often based on advanced machine learning algorithms and rely on large-scale datasets.

## 4 How to evaluate the current models of vision?

*"It doesn't matter how beautiful your theory is, it doesn't matter how smart you are. If it doesn't agree with experiment, it's wrong."*

—Richard Feynman

Let us begin by thinking about a scenario where we have some intuition of what the evaluation of a model might look like. During the FIFA world cups, as a spectator, given our "understanding" (aka internal model) of how well the two soccer teams or individual players have performed during the rest of the tournament, we can predict what the score might be in the finals. Depending on the result of the final, and performances of the players, we can then determine the strength of our predictions. The better our predictions are over multiple such games, the more of an expert (ones with the best internal models) we are considered. So when someone asks, "Do you understand the game?" the answer is often based on how good one's predictions have been about the game in the past (beyond the interpretation or knowledge of the rules of the game). This concept of using "prediction" and "understanding" interchangeably, although ubiquitously present in many aspect of our lives, is somehow surprisingly still debated and discussed at length when it comes to defining scientific understanding. It is important to note, however, that irrespective of the strength of our prediction in the soccer matches, we most likely lacked any "control" over the outcome. Studies in vision deal with developing an understanding (aka predictive, falsifiable models) of the related brain processes, and also using that understanding to treat neurobiological disorders. While the progress in the understanding of brain processes can be evaluated by testing the predictions of vision models, the latter requires guidance from such models to efficiently modulate (aka control) the brain processes that they attempt to approximate. The current high-performing computational models in visual neuroscience (enabled by progress in AI and big data) offer both prediction and control capabilities.

## 4.1 Prediction

One way to measure the goodness of any model of a neuron is to estimate how well it can predict the responses of the neuron for novel stimuli. Similarly, for models of behavior, one needs to estimate the goodness of behavioral

**294** Big data in psychiatry and neurology

predictions (for example, see Rajalingham et al., 2018). However, what is meant by novel stimuli is somewhat debatable. For the extent of this chapter, let's assume that any image which was not used to build (e.g., train the parameters of) the model can be considered a novel test stimulus.

Two parallel procedures (brain and model measurements) are involved while assessing the prediction power of any model of vision. On the one hand, direct brain measurements need to be performed. For instance, researchers have recorded from neural sites in different brain areas (e.g., V1, V4, IT), while estimating how well these data can be predicted by current CNNs. For the purposes of illustration, let us consider a neural site recorded in the inferior temporal cortex of a rhesus macaque (as shown in Fig. 2A). Thousands of images (with multiple interleaved repetitions) were shown to the monkey while neural measurements across time were performed at this site. The goal of the prediction metric is to estimate how well the model can predict this data.

In the seminal work of Yamins et al. (2014) and subsequent developments (Kar et al., 2019; Schrimpf et al., 2018), the researchers modeled each such IT neural site (as shown in Fig. 2A; top right panel) as a linear combination of the CNN model features. They first extracted the model features per image, from the DCNNs' layers. Given that these models are image computable, one can plug in any image (provided that the image respects the input size of the model) and ask what is the response of each node (often referred to as feature) in each layer (note: the term layer in a CNN model is not analogous to layers of the cortex) of the model? This is equivalent to showing an image to the human or nonhuman primate subjects and measuring neurophysiological responses from different brain areas. For example, one can extract the features from Alex-Net's (Krizhevsky, Sutskever, & Hinton, 2012) "fc7" layer (as shown in Fig. 2A). Using a 50%/50% train/test split of the images, one can then estimate the regression weights (i.e., how we can linearly combine the model features to predict the neural site's responses) using a partial least squares regression procedure. The neural responses used for training ($R^{TRAIN}$) and testing ($R^{TEST}$) the encoding models were averaged firing rates (measured at the specific site) within the time window considered. This time window is a free parameter. For each set of regression weights ($w$) estimated on the training image responses ($R^{TRAIN}$), we generated the output of that 'synthetic neuron' for the held-out test set ($M^{PRED}$) as,

$$M^{PRED} = \left( w * F^{TEST} \right) + \beta,$$

where $w$ and $\beta$ are estimated via the PLS regression and $F^{TEST}$ are the model activation features for the test image-set.

The percentage of explained variance, *IT predictivity* for that neural site, was then computed by normalizing the $r^2$ prediction value for that site by the self-consistency of the test image responses ($\rho^{R^{TEST}}$) for that site and the

Visual neuroscience in the age of big data Chapter | 14 **295**

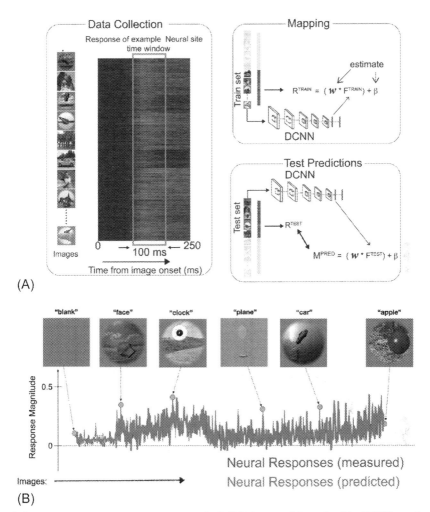

**FIG. 2** (A) Predicting IT neural responses with DCNN features. Schematic of the DCNN neural fitting and prediction testing procedure. This includes three main steps. Data collection: neural responses are collected for each image, e.g., example neural site is shown, a 100 ms time window for estimating the neural response vector (across images) is demarcated. Mapping: We divide the images and the corresponding neural features ($R^{TRAIN}$) into a 50–50 train–test split (shown for demonstration). For the train images, we compute the image evoked activations ($F^{TRAIN}$) of the DCNN model from a specific layer. We then use partial least square regression to estimate the set of weights (w) and biases ($\beta$) that allow us to best predict $R^{TRAIN}$ from $F^{TRAIN}$. Test Predictions: Once we have the best set of weights (w) and biases ($\beta$), we generate the predictions ($M^{PRED}$) from this synthetic neuron for the test image evoked activations of the model $F^{TEST}$. We then compare these predictions with the held-out test image evoked neural features ($R^{TEST}$) to compute the IT predictivity of the model. (B) Comparison of the measured (in *blue* (black in the print version)) neural firing rate per image (x-axis) with the predicted (in *red* (dark gray in the print version)) firing rate per image. The plot demonstrates how closely the model predictions can match the real data.

self-consistency of the regression model predictions ($\rho^{M^{PRED}}$) for that site (estimated by a Spearman Brown corrected trial-split correlation score).

$$\text{IT predictivity} = \left( \frac{corr(R^{TEST}, M^{PRED})}{\sqrt{\rho^{R^{TEST}} * \rho^{M^{PRED}}}} \right)^2$$

To achieve accurate cross-validation results, the prediction of the model needs to be tested on held-out image responses. But to make sure we have exposed the mapping procedure (mapping the model features on to individual IT neural sites) to images from the same full generative space of the image-set, we randomly subsampled image responses from the entire image-set (measured at that specific time window). Fig. 2B shows an example of how well current CNNs can predict real neural responses. The percentage of explained variance can be ~50%, much higher than any previous models (as demonstrated in Yamins et al., 2014).

There are many other ways to compare representational spaces across models and neurophysiological data. For instance, a very popular method, known as representational similarity analysis (RSA), has been pioneered by Kriegeskorte and colleagues (Kriegeskorte et al., 2008). This method (for details of the Method, see Nili et al., 2014) essentially relies on creating a square symmetric matrix indexed by the stimuli horizontally and vertically (in the same order). Each element (row "I," column "j") represents the dissimilarity (distances in the multivariate response space) between the activity patterns associated with two different stimuli (e.g., images). The diagonal entries reflect comparisons between identical stimuli and are 0, by definition. For any given species or measurement type (e.g., human fMRI, CNN model activations, etc.) with a common stimulus set, such matrices can be computed and then compared to assess similarity.

Prediction-related experimental work need not only be limited to behavioral and neural recordings. Model-based predictions should be expanded into the domain of causal perturbation experiments as well. The debate over correlation vs causation with respect to brain and behavior is often settled by causal perturbation experiments (e.g., pharmacological, optogenetic, chemogenetic, and electrical interventions). However, given the limited and arbitrary levels of control that current causal tools offer ( Jazayeri & Afraz, 2017; Wolff & Olveczky, 2018), inferences drawn from such experiments often only confirm prior correlative intuitions, while stronger claims about causal circuit motifs remain challenged. The specific mapping between brain tissue and the computational models allows us to employ causal perturbation techniques to serve as a tool for screening among models. We can now directly test whether in-silico perturbations in models can predict outcomes of in-vivo perturbations. Two recent studies (Kar & DiCarlo, 2021; Kar et al., 2019) demonstrate how models can drive behavioral and neural experiments and can be further improved by performing causal perturbations of specific brain circuit motifs.

## 4.2 Control

Recently, the value of predictive models has been augmented by goal-directed stimulus synthesis of images and sounds. Stimuli are generated that achieve a specified objective in a candidate model, and neural or behavioral activity is measured in response to these synthetic stimuli to test if the model accurately captures sensory representations. These stimuli synthesis procedures can become a new tool to elucidate model successes and failures. Unlike prior attempts in which relatively simple synthetic stimuli were designed based on the experimenters' intuition, current approaches directly take advantage of the DCNN model representations to generate perceptually rich images and sounds. Combined with the advent of easy-to-use tools to implement these algorithms, there is growing interest to begin probing sensory systems using such an approach. For instance, Bashivan et al. (2019) recently demonstrated (see Fig. 3) that by using a CNN-based model of V4 neurons, they can generate synthetic stimuli that drive neurons in this area beyond what could be achieved by the previously known "preferred-stimuli" for the region. They also showed that this approach can be used to target the population of neurons and set them at the desired activation states. A somewhat far-fetched idea but given that specific mental states are derived from specific patterns of neural activity, this same technique can be used to modulate mental states by presenting reverse-engineered patterns of pixels that should appropriately shape neural activity patterns. Indeed, the authors found that models that predict better also generate more accurate controller images. Thus the control methods are directly relevant to translational work providing a direct link between basic science and clinical approaches.

Similar attempts have been made for other areas of the brain. For instance, (Ponce et al. (2019) demonstrated that they can synthesize "super-stimuli" for inferior temporal neurons that drive the activity of these cells beyond their usual response range. In fact, these results often challenge the common terminology in the field given that the super-stimuli for a classical "face-patch" neuron does not resemble a typical human face.

## 5 The vision community is coming together to combine data and models

The ImageNet challenge (Russakovsky et al., 2015) arguably played a big role in the success of computer vision and is still considered a gold standard for judging the goodness of computer vision models. Similarly, the importance of large-scale datasets, open access to some of them, as well as rigorous testing and subsequent ranking of current computational models of vision (that can predict and explain these data) has started to receive an acknowledgment within the scientific visual neuroscience community. Following, I have discussed three collaborative efforts that are leading the way toward this goal. It is by no means an extensive list. And, it is also expected that in the future some of these groups

**298** Big data in psychiatry and neurology

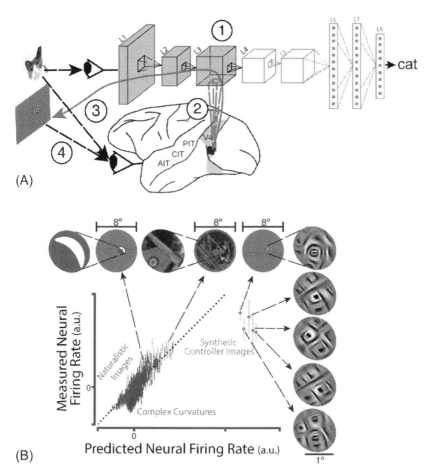

**FIG. 3** (A) The neural control experiments are done in four steps: (1) Parameters of the neural network are optimized by training on a large set of labeled natural images [Imagenet] and then held constant thereafter. (2) ANN "neurons" are mapped to each recorded V4 neural site. The mapping function constitutes an image-computable predictive model of the activity of each of these V4 sites. (3) The resulting differentiable model is then used to synthesize "controller" images for either single-site or population control. (4) The luminous power patterns specified by these images are then applied by the experimenter to the subject's retinae, and the degree of control of the neural sites is measured. AIT, anterior inferior temporal cortex; CIT, central inferior temporal cortex; PIT, posterior inferior temporal cortex. (B) Results for example successful stretch control test. The normalized activity level of the target V4 neural sites is shown for all of the naturalistic images (*blue dots* (black in the print version)), complex-curvature stimuli (*purple dots* (gray in the print version)), and five synthetic stretch controller images (*red dots* (dark gray in the print version); see methods). Best driving images within each category and a zoomed view of the receptive field are shown at the top. *(Adapted from Bashivan, P., Kar, K., & DiCarlo, J. J. (2019). Neural population control via deep image synthesis. Science, 364(6439).)*

Visual neuroscience in the age of big data **Chapter | 14** **299**

might merge or many more such collaborations will be formed. One key thing to emphasize is that each of these collaborations requires an extensive interdisciplinary approach. We need experimental neuroscientists of various kinds to come together with computational neuroscientists, software developers, and a well-equipped management team to successfully run such an effort. This simply cannot be achieved at the level of an individual or an individual laboratory.

## 5.1 Allen brain observatory

The Allen Institute has pioneered the art of industry-grade scientific data collection with standardized protocols and providing excellent interfaces to interact with their databases. The Allen Brain Observatory (2016) offers publicly available tools to access highly standardized measurements of neuronal activity in the mouse visual system. The freely available data and analysis tools in this resource empower researchers around the globe to examine how different circuit motifs in the mouse brain respond during different visual tasks.

According to their website, "The Allen Brain Observatory presents the first standardized in vivo survey of physiological activity in the mouse visual cortex, featuring representations of visually evoked calcium responses from GCaMP6-expressing neurons in selected cortical layers, visual areas and Cre lines.

- Searchable data from hundreds of two-photon calcium imaging sessions across multiple visual areas and depths in the visual cortex
- A variety of data visualization summaries capturing visual coding properties of single cell and cell population responses to sensory stimuli
- Standardized spatial mapping of cellular responses to five types of rich visual stimuli, surveyed from transgenic mouse lines
- Raw data and analysis modules available for download via the Allen Brain Atlas application program interface (API) and Allen Software Development Kit (SDK)"

## 5.2 Brain-score

Despite the plethora of discoveries in visual neuroscience, it remains unclear how much is left to uncover about the visual brain. How does one put everything together? One way to operationalize the knowledge is to build predictive models of the visual system and ask how much variance do these models explain at all current possible levels of investigation. To answer this question, progress in neuroscience needs to be made quantifiable. The development of the platform Brain-score (Schrimpf et al., 2018; also see http://www.brain-score.org) is one such heroic attempt to bring together computational modeling efforts and neural and behavioral data (see Fig. 4) from different research groups across the fields of system neuroscience, psychology, computational neuroscience, and machine learning. According to their website,

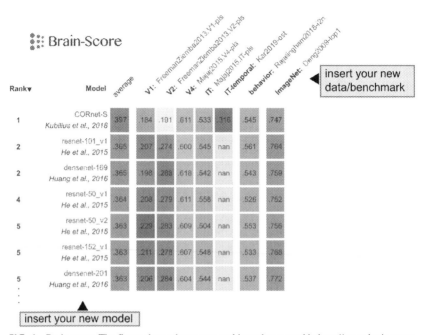

FIG. 4 Brain-score. The figure shows the current ranking scheme used in http://www.brain-score. org. The models are shown on the rows. The different neural and behavioral benchmarks are shown in the columns. The goal is to rank the existing models according to their predictive powers for various benchmarks. Researchers can currently participate in two ways. First, they can evaluate their own models (which will be then added to the rows). Second, they can also contribute their data or benchmarks (which will be then added to the columns).

"The Brain-Score platform aims to yield strong computational models of the ventral stream. We enable researchers to quickly get a sense of how their model scores against standardized brain benchmarks on multiple dimensions and facilitate comparisons to other state-of-the-art models. At the same time, new brain data can quickly be tested against a wide range of models to determine how well existing models explain the data. Brain-Score is organized by the Brain-Score team in collaboration with researchers and labs worldwide. We are working towards an easy-to-use platform where a model can easily be submitted to yield its scores on a range of brain benchmarks and new benchmarks can be incorporated to challenge the models. This quantified approach lets us keep track of how close our models are to the brain on a range of experiments (data) using different evaluation techniques (metrics)."

### 5.3 The Algonauts project

The Algonauts challenge (Cichy et al., 2019) invites the entire computational neuroscience community to compete in producing the best predictive models for two specific types (fMRI and MEG) of the brain measurement dataset. According to their website,

"The primary target of the 2019 challenge is the visual brain: the part of the brain that is responsible for seeing. Currently, particular deep neural networks trained on object recognition (Cichy, Khosla, Pantazis, Torralba, & Oliva, 2016; Khaligh-Razavi & Kriegeskorte, 2014; Yamins et al., 2014 etc.) do best in explaining brain activity. Can your model do better? ... The main goal of the 2019 Algonauts challenge is to predict brain activity from two sources—fMRI data from two brain regions (Track 1), or MEG data from two temporal windows (Track 2)—using computational models. The brain activity is in response to viewing sets of images; for each image set, fMRI and MEG data are collected from the *same* 15 human subjects."

## 6 Conclusion

Data-driven models of vision are undoubtedly part of a huge success story, enabling unprecedented levels of prediction and control of visual processing. As the name suggests, they require a large amount of data to be built. Future experimental efforts in vision should align their goals with what could optimally help improve these models. However, one must exercise caution when testing these models with additional "novel" data. Given the very large number of parameters that most of these models contain, it can become tricky to accurately test and compare these models in the fairest way. Therefore how much and what kind of data these models should be allowed to be fed in their training diet is an important factor to consider. Data sharing with models needs to be therefore carefully managed. The unsolved issue is that it is not clear how much data our brain, with clearly more parameters than the current computational models, have utilized while its own training during evolution and development.

## References

Albrecht, D. G., & Hamilton, D. B. (1982). Striate cortex of monkey and cat: Contrast response function. *Journal of Neurophysiology, 48*, 217–237.

Albright, T. D. (1984). Direction and orientation selectivity of neurons in visual area MT of the macaque. *Journal of Neurophysiology, 52*, 1106–1130.

Allen Brain Observatory (2016).

Bashivan, P., Kar, K., & DiCarlo, J. J. (2019). Neural population control via deep image synthesis. *Science, 364*(6439).

Britten, K. H., Newsome, W. T., Shadlen, M. N., Celebrini, S., & Movshon, J. A. (1996). A relationship between behavioral choice and the visual responses of neurons in macaque MT. *Visual Neuroscience, 13*, 87–100.

Cadena, S. A., Denfield, G. H., Walker, E. Y., Gatys, L. A., Tolias, A. S., Bethge, M., et al. (2019). Deep convolutional models improve predictions of macaque V1 responses to natural images. *PLoS Computational Biology, 15*, e1006897.

**302** Big data in psychiatry and neurology

Cadieu, C. F., Hong, H., Yamins, D. L., Pinto, N., Ardila, D., Solomon, E. A., et al. (2014). Deep neural networks rival the representation of primate IT cortex for core visual object recognition. *PLoS Computational Biology, 10,* e1003963.

Cichy, R. M., Khosla, A., Pantazis, D., Torralba, A., & Oliva, A. (2016). Comparison of deep neural networks to spatio-temporal cortical dynamics of human visual object recognition reveals hierarchical correspondence. *Scientific Reports, 6,* 27755.

Cichy, R. M., Roig, G., Andonian, A., Dwivedi, K., Lahner, B., Lascelles, A., et al. (2019). *The algonauts project: A platform for communication between the sciences of biological and artificial intelligence.* arXiv preprint arXiv:190505675.

Felleman, D. J., & Van Essen, D. C. (1991). Distributed hierarchical processing in the primate cerebral cortex. *Cerebral Cortex, 1,* 1–47.

Fukushima, K., & Miyake, S. (1982). Neocognitron: A self-organizing neural network model for a mechanism of visual pattern recognition. In *Competition and cooperation in neural nets* (pp. 267–285). Berlin, Heidelberg: Springer.

Guclu, U., & van Gerven, M. A. (2015). Deep neural networks reveal a gradient in the complexity of neural representations across the ventral stream. *The Journal of Neuroscience, 35,* 10005–10014.

Hubel, D. H., & Wiesel, T. N. (1959). Receptive fields of single neurones in the cat's striate cortex. *The Journal of Physiology, 148,* 574–591.

Jazayeri, M., & Afraz, A. (2017). Navigating the neural space in search of the neural code. *Neuron, 93,* 1003–1014.

Joukes, J., Yu, Y., Victor, J. D., & Krekelberg, B. (2017). Recurrent network dynamics; a link between form and motion. *Frontiers in Systems Neuroscience, 11,* 12.

Kar, K., & DiCarlo, J. J. (2021). Fast recurrent processing via ventrolateral prefrontal cortex is needed by the primate ventral stream for robust core visual object recognition. *Neuron.* https://doi.org/10.1016/j.neuron.2020.09.035.

Kar, K., Kubilius, J., Schmidt, K., Issa, E. B., & DiCarlo, J. J. (2019). Evidence that recurrent circuits are critical to the ventral stream's execution of core object recognition behavior. *Nature Neuroscience, 22,* 974–983.

Kelley, H. J. (1960). Gradient theory of optimal flight paths. *Aerospace Research Central Journal, 30,* 947–954.

Khaligh-Razavi, S. M., & Kriegeskorte, N. (2014). Deep supervised, but not unsupervised, models may explain IT cortical representation. *PLoS Computational Biology, 10*(11), e1003915.

Kietzmann, T. C., Spoerer, C. J., Sorensen, L. K. A., Cichy, R. M., Hauk, O., & Kriegeskorte, N. (2019). Recurrence is required to capture the representational dynamics of the human visual system. *Proceedings of the National Academy of Sciences of the United States of America, 116,* 21854–21863.

Kriegeskorte, N., Mur, M., Ruff, D. A., Kiani, R., Bodurka, J., Esteky, H., et al. (2008). Matching categorical object representations in inferior temporal cortex of man and monkey. *Neuron, 60,* 1126–1141.

Krizhevsky, A., Sutskever, I., & Hinton, G. E. (2012). ImageNet classification with deep convolutional neural networks. In *Vol. 1. Proceedings of the 25th international conference on neural information processing systems* (pp. 1097–1105). Lake Tahoe, Nevada: Curran Associates Inc.

Kubilius, J., Schrimpf, M., Kar, K., Rajalingham, R., Hong, H., Majaj, N., et al. (2019). Brain-like object recognition with high-performing shallow recurrent ANNs. In *Advances in neural information processing systems* (pp. 12805–12816).

LeCun, Y., Boser, B., Denker, J. S., Henderson, D., Howard, R. E., Hubbard, W., et al. (1989). Backpropagation applied to handwritten zip code recognition. *Neural Computation, 1,* 541–551.

Logothetis, N. K., Pauls, J., & Poggio, T. (1995). Shape representation in the inferior temporal cortex of monkeys. *Current Biology, 5,* 552–563.

Majaj, N. J., Hong, H., Solomon, E. A., & DiCarlo, J. J. (2015). Simple learned weighted sums of inferior temporal neuronal firing rates accurately predict human core object recognition performance. *The Journal of Neuroscience, 35,* 13402–13418.

McIntosh, L., Maheswaranathan, N., Nayebi, A., Ganguli, S., & Baccus, S. (2016). Deep learning models of the retinal response to natural scenes. In *Advances in neural information processing systems* (pp. 1369–1377).

Minsky, M., & Papert, S. (1969). *An introduction to computational geometry.* Cambridge tiass, HIT.

Nauhaus, I., Benucci, A., Carandini, M., & Ringach, D. L. (2008). Neuronal selectivity and local map structure in visual cortex. *Neuron, 57,* 673–679.

Nayebi, A., Bear, D., Kubilius, J., Kar, K., Ganguli, S., Sussillo, D., et al. (2018). Task-driven convolutional recurrent models of the visual system. In *Advances in neural information processing systems* (pp. 5290–5301).

Nili, H., Wingfield, C., Walther, A., Su, L., Marslen-Wilson, W., & Kriegeskorte, N. (2014). A toolbox for representational similarity analysis. *PLoS Computational Biology, 10,* e1003553.

Ponce, C. R., Xiao, W., Schade, P. F., Hartmann, T. S., Kreiman, G., & Livingstone, M. S. (2019). Evolving images for visual neurons using a deep generative network reveals coding principles and neuronal preferences. *Cell, 177*(4), 999–1009.

Quiroga, M. D. M., Morris, A. P., & Krekelberg, B. (2019). Short-term attractive tilt aftereffects predicted by a recurrent network model of primary visual cortex. *Frontiers in Systems Neuroscience, 13,* 67.

Rajalingham, R., Issa, E. B., Bashivan, P., Kar, K., Schmidt, K., & DiCarlo, J. J. (2018). Large-scale, high-resolution comparison of the core visual object recognition behavior of humans, monkeys, and state-of-the-art deep artificial neural networks. *Journal of Neuroscience, 38*(33), 7255–7269.

Richards, B. A., Lillicrap, T. P., Beaudoin, P., Bengio, Y., Bogacz, R., Christensen, A., et al. (2019). A deep learning framework for neuroscience. *Nature Neuroscience, 22,* 1761–1770.

Riesenhuber, M., & Poggio, T. (2000). Models of object recognition. *Nature Neuroscience, 3* (Suppl), 1199–1204.

Rockland, K. S., Saleem, K. S., & Tanaka, K. (1994). Divergent feedback connections from areas V4 and TEO in the macaque. *Visual Neuroscience, 11,* 579–600.

Rosenblatt, F. (1958). The perceptron—A probabilistic model for information-storage and organization in the Brain. *Psychological Review, 65,* 386–408.

Russakovsky, O., Deng, J., Su, H., Krause, J., Satheesh, S., Ma, S., et al. (2015). Imagenet large scale visual recognition challenge. *International Journal of Computer Vision, 115,* 211–252.

Schrimpf, M., Kubilius, J., Hong, H., Majaj, N. J., Rajalingham, R., Issa, E. B., et al. (2018). Brainscore: Which artificial neural network for object recognition is most brain-like? *BioRxiv,* 407007.

Seeliger, K., Fritsche, M., Guclu, U., Schoenmakers, S., Schoffelen, J. M., Bosch, S. E., et al. (2018). Convolutional neural network-based encoding and decoding of visual object recognition in space and time. *NeuroImage, 180,* 253–266.

Serre, T., & Riesenhuber, M. (2004). *Realistic modeling of simple and complex cell tuning in the HMAX model, and implications for invariant object recognition in cortex.* Massachusetts Inst of Tech Cambridge Computer Science and Artificial.

Serre, T., Wolf, L., Bileschi, S., Riesenhuber, M., & Poggio, T. (2007). Robust object recognition with cortex-like mechanisms. *IEEE Transactions on Pattern Analysis and Machine Intelligence, 29,* 411–426.

**304** Big data in psychiatry and neurology

Shapley, R. M., & Victor, J. D. (1978). The effect of contrast on the transfer properties of cat retinal ganglion cells. *The Journal of Physiology, 285*, 275–298.

Spoerer, C. J., McClure, P., & Kriegeskorte, N. (2017). Recurrent convolutional neural networks: A better model of biological object recognition. *Frontiers in Psychology, 8*, 1551.

Tang, H., Schrimpf, M., Lotter, W., Moerman, C., Paredes, A., Ortega Caro, J., et al. (2018). Recurrent computations for visual pattern completion. *Proceedings of the National Academy of Sciences of the United States of America, 115*, 8835–8840.

Tsao, D. Y., Freiwald, W. A., Knutsen, T. A., Mandeville, J. B., & Tootell, R. B. (2003). Faces and objects in macaque cerebral cortex. *Nature Neuroscience, 6*, 989–995.

Wolff, S. B., & Olveczky, B. P. (2018). The promise and perils of causal circuit manipulations. *Current Opinion in Neurobiology, 49*, 84–94.

Yamins, D. L., Hong, H., Cadieu, C., & DiCarlo, J. J. (2013). Hierarchical modular optimization of convolutional networks achieves representations similar to macaque IT and human ventral stream. In *Advances in neural information processing systems* (pp. 3093–3101).

Yamins, D. L., Hong, H., Cadieu, C. F., Solomon, E. A., Seibert, D., & DiCarlo, J. J. (2014). Performance-optimized hierarchical models predict neural responses in higher visual cortex. *Proceedings of the National Academy of Sciences of the United States of America, 111*, 8619–8624.

Zamir, A. R., Wu, T.-L., Sun, L., Shen, W. B., Shi, B. E., Malik, J., et al. (2017). Feedback networks. In *Proceedings of the IEEE conference on computer vision and pattern recognition* (pp. 1308–1317).

# Chapter 15

# Application of big data and artificial intelligence approaches in diagnosis and treatment of neuropsychiatric diseases

Qiurong Song[a], Tianhui Huang[a], Xinyue Wang[a], Jingxiao Niu[a], Wang Zhao[b], Haiqing Xu[c], and Long Lu[a]

[a]*School of Information Management, Wuhan University, Wuhan, PR China,* [b]*Suzhou Zealikon Healthcare Co., Suzhou, PR China,* [c]*Child Health Division, Department of Maternal and Child Health, Maternal and Child Hospital of Hubei Province, Tongji Medical College, Huazhong University of Science and Technology, Wuhan, PR China*

## 1 Introduction

In recent years, big data technology has been increasingly applied to biomedical and medical information research, and large amounts of biological and clinical data have been generated and collected at an unprecedented speed and scale. For example, the next-generation sequencing (NGS) technologies enable the processing of billions of nucleotides data per day, and the application of electronic health records (EHRs) is documenting huge amounts of patient data. With the emergence of data capture and data generating technologies, and the development of hardware and software for parallel computing, the cost of obtaining and analyzing biomedical data is expected to be significantly reduced.

The diagnosis and treatment of neuropsychiatric diseases has received increasing attention and has become one of the most difficult challenges in modern medicine. According to a recent report by the World Health Organization (WHO), 50 million people suffer from epilepsy and 24 million from Alzheimer's disease and other dementias. As a discipline to study and prevent mental disorders, brain disorder has long relied on subjective observations for research. Unlike many traditional "physical" diseases, mental illnesses usually have more obvious physiological characteristics and clinical manifestations, but due to the complexity of brain and mind, there are greater difficulties in the assessment and treatment of mental illnesses. Until now, most

**306** Big data in psychiatry and neurology

of the diagnosis of mental illness has been made on the basis of clinical observations and statistical analysis of symptoms and treatment responses. However, in recent years, with the rise of computational psychiatry, problems such as inaccurate diagnosis (other clinical issues include automatic diagnosis, prediction of treatment outcomes, longitudinal disease progression, and treatment options) may be easier to solve in clinical settings (Qiao, Lin, Cao, & Wang, 2015; Shimada, Shiina, & Saito, 2000). Computational psychiatry uses powerful data analysis, machine learning, and artificial intelligence to detect extreme and abnormal behaviors. On one hand, applying machine learning methods to higher-dimensional data through a data-driven approach can improve disease classification, treatment outcomes prediction, or treatment options (Huys, Maia, & Frank, 2016). On the other hand, the application of artificial intelligence (AI) to the prevention, early screening, diagnosis, treatment, and rehabilitation of mental illnesses can assist doctors, significantly improving the work efficiency of psychiatrists, reducing costs, and complementing the disadvantages of artificial diagnosis and treatment. Furthermore, it can also avoid the subjectivity and bias of psychiatrists. For example, machines can learn directly from medical data to avoid clinical errors caused by human cognitive biases so that a positive impact on patients' diagnosis and treatment could be brought. Today, modern technology and systems enable neurologists to offer appropriate neurological care. Current diagnostic techniques, such as magnetic resonance imaging (MRI), electroencephalography (EEG), generate large amounts of data (size and dimension) for the detection, monitoring, and treatment of neuropsychiatric diseases. A recent study showed that using deep neural networks (DNN) as a computational framework, the AI-driven automatic classification system MNet successfully classified healthy subjects and subjects with two neurological diseases with high accuracy (Aoe et al., 2019). The technique will also be used in the evaluation of the diagnosis, prognosis, and therapeutic effect of various neuropsychiatric diseases.

The development of AI technology and big data enables us to shift from studying biomarkers at the group level to using brain imaging to predict results at the individual level, which means that the diagnostic methods can be improved by the large amount of studying on single patient's data, different from the extract of covert and common biomarkers in various images. Machine learning can objectively and accurately model massive and multidimensional imaging data, so as to quantify the degree of abnormal brain anatomy and function caused by neuropsychiatric diseases, which is conducive to the diagnosis and prognosis biology of mental diseases and neurological diseases.

The development of relevant optimization techniques also improves the accuracy of machine learning results. By applying nonlinear hierarchy structures (trees or graphs), some methods involving AI can establish very complex data models. Therefore, in terms of brain image-based disease classification, AI methods can greatly improve the accuracy, sensitivity, and specificity of computer-aided diagnosis (CAD) systems. With the rapid development of medical technology, most diseases can be diagnosed by advanced equipment. Driven by big data technology, this cloud computing-based "AI" will be more useful through the "cloud computing" method of in-depth excavation of medical pathological images and continuous depiction of the pathological regularity "Facebook." Scientific and reasonable results with high feasibility can further promote the accuracy and authority of diagnosis results.

Moreover, the use of neural network algorithms in the diagnosis of mental illnesses and neurological diseases has also become much more prevalent in the last few years. Neural network algorithms are featured by autonomous learning and complex feature analysis. By using neural network technology to analyze medical data sets, highly abstract diagnostic feature sets can be obtained. In addition, by using multiple data processing techniques in the generated diagnostic feature sets to build neural network models that combined with the patient's vital sign data set including images and other diagnostic criteria, the autonomous learning of the electronic medical record system (EMR) and medical imaging system can be gradually realized, and the diagnostic accuracy of doctors can also be improved.

This chapter focuses on the introduction of clinical data sources like genomics, EEG signals, eye movement data, and neuroimaging, and some particular devices like wearable equipment would be firstly introduced in the third part. Main algorithms used in the diagnosis and prognosis of different brain disorders would also be referred in the third part, including linear and nonlinear machine learning methods and neural networks. The following section provides an in-depth look at the data types and specific algorithms mentioned before, which are used in areas such as autism spectrum disorders (ASD) and Alzheimer's disease. Finally, we will depict the challenges and promising solutions of big data and AI in neuropsychiatric diseases.

## 2 Main data sources

### 2.1 Genomics

Genomic data refer to the genome and DNA data of an organism. They are used in bioinformatics for collecting, storing, and processing the genomes of living things. Genomic data are primarily used in big data processing and analysis techniques, which generally require a large amount of storage and purpose-built software to analyze. Such data are gathered by a bioinformatics system or a genomic data processing software. Typically, genomic data are processed through various data analysis and bioinformatics techniques to find and analyze genome structures and other genomic parameters. Sequencing data analysis techniques and variation analysis are common processes performed on genomic data (Langmead & Nellore, 2018). And genetic counseling is important for the diagnosis of diseases and usually has therapeutic significance. In addition, the related research of genomics can not only be used in the development of related

**308** Big data in psychiatry and neurology

drugs, but also further improve the psychotherapy of mental diseases, including various psychotherapies (such as cognitive behavioral therapy and dialectical behavioral therapy), light therapy, deep-brain stimulation, electroconvulsive therapy, and transcranial magnetic stimulation (TMS), and it will optimize the individualized nursing of patients and improve the prognosis of mental illness (Kumsta, 2019).

## 2.2 EEG signals

Electroencephalography (EEG) is an electrophysiological monitoring method that records the electrical activity of the brain. It is usually noninvasive, with electrodes placed along the scalp, although invasive electrodes are sometimes used, as in the electrocortical cortex. EEG measures the electrical currents from the generated voltage fluctuations of the ions inside the neuron's brain, which is a spontaneous, rhythmic potential change produced by cortical neurons. Clinically, EEG is the recording of spontaneous electrical activity in the brain over a period of time by means of multiple electrodes placed on the scalp. Nowadays, routine EEG examination in clinic has been very popular. With the rapid development of electronic technology, the application of EEG has become an important index to evaluate the state of brain function and is widely used in the diagnosis of mental diseases (Buettner, Rieg, & Frick, 2020). EEG is most commonly used to diagnose epilepsy, which causes abnormalities in EEG readings. It is also used to diagnose sleep disorders, coma, encephalopathies, and brain death (Sazgar & Young, 2019).

## 2.3 Eye movement data

Eye movement technology extracts fixation point, fixation time, eye-jumping distance, pupil size, and other data from eye-movement track records. These data can be used to study the internal cognitive processes of individuals. Nowadays, eye movement research has become an important tool for basic neuroscience research. Measures of eye movement have been used for higher brain functions, such as cognition, social behavior, and higher levels of decision-making. With the development of eye tracker, more and more studies have applied eye movement data to mental disorders and found that the basic eye movement characteristics of patients with mental disorders are different from those of healthy controls. Currently, there are many ways to track eye movements, among which the most common noninvasive method is to obtain the position of eyes through video shooting equipment. In the clinical diagnosis of schizophrenia, ASD, and other mental diseases, eye movement data, including the asymmetry of the speed of visually guided saccades Index, plays a significant part (Shiino et al., 2020).

## 2.4 Neuroimaging data

Medical imaging has long been a significant part of medical big data. Among them, neuroimaging plays an important role in the diagnosis and prognosis of patients with brain diseases, providing more data supports for personalized medical treatment. Neuroimaging or brain imaging is the application of various techniques to either directly or indirectly image the structure, function, or pharmacology of the nervous system (Bzdok, Schulz, & Lindquist, 2019). In particular, imaging genetics, which uses imaging as a quantitative biological phenotype, combines genetics, psychiatry, and neuroscience. Neuroimaging genetics can be used to correlate genetic changes with measurable and repeatable results of brain structure or function. Two main methods of neuroimaging genetics are: (1) to identify the imaging changes in the population with definite genetic diseases; (2) to verify the impact of specific genetic changes. A particularly useful aspect of using imaging to define a phenotype is the ability to determine the distance between a diagnosed disease and a subjective self-report inconsistency to track a patient's prognosis. Most importantly, neuroimaging can better understand the pathology of common mental diseases and provide personalized diagnosis and treatment programs by evaluating the relationship between image phenotype and genetic variation (Chiesa, Cavedo, Lista, Thompson, & Hampel, 2017; Uddén et al., 2019).

## 2.5 Wearable equipment data

The Internet of things (IOT) extends the independence of human interaction, contribution, and cooperation in things. The Internet of things has gradually developed heterogeneous technologies with complex protocols and algorithms. In the process of personalized prognosis of mental and neurological disorders, wearable or implanted sensors can continuously track the physiological characteristics of discharged patients, such as blood pressure, heart rate, body temperature, stress rate, electrocardiogram (ECG), EEG, and other physiological states, and it can also continuously provide various data to the cloud, offering more personalized prognosis for mental patients to support patients in emergency decision-making and emergency contact (McGinnis et al., 2018). Physiological information is monitored by human body sensors and biosensors, such as body temperature, diabetes, blood pressure, and heart rate. For example, a smart watch has built-in thermometer, oximeter, accelerometer, and GPS position tracker. Those sensors embedded in clothing track electrocardiograms, EMGs, and patient stress rates. Through the network, the observed data are transmitted to the cloud through network to store and send messages, and then they are transmitted to the client, i.e. doctors' mobile phones, family members, and hospitals (Chen et al., 2017).

# 310 Big data in psychiatry and neurology

## 3 Main algorithms

There are a lot of algorithms used in neuropsychiatric diseases, mainly including machine learning, signal time–frequency analysis, feature extraction, and so on. In this paper, the principle of support vector machine, decision tree, random forest, convolution neural network, wavelet transform algorithm in time–frequency analysis, and common space pattern algorithm in feature processing are described in detail.

Support vector machine (SVM) is a supervised learning algorithm. It is a binary classification algorithm supported by linear classification and nonlinear classification, but after evolution, it also supports multiclassification problem and can also be applied to regression problem. SVM is based on the VC dimension theory of statistical learning theory and the principle of structural risk minimization. According to the limited sample information, it seeks the best compromise between the complexity of the model (that is, the learning accuracy of a specific training sample, accuracy) and the learning ability (that is, the ability to identify any sample without error), in order to obtain the best generalization ability (or generalization ability) (Guyon, Weston, Barnhill, & Vapnik, 2002).

Decision tree is also a supervised machine learning method, whose generation algorithms include ID3, C4.5, and C5.0. It is a tree structure, in which each internal node represents a judgment on attribute, each branch represents an output of judgment result, and finally each leaf node represents a classification result. Decision tree can be divided into classification tree and regression tree. Classification tree makes decision tree for discrete variables while regression tree makes it for continuous variables (Pradhan, 2013).

Random forest is a classifier with multiple decision trees, and the output category is determined by the mode of the output category of the individual tree. Each decision tree is a classifier, so for an input sample, n trees will have n classification results. Random forest integrates all the classification voting results and specifies the category with the most voting times as the final output. It is not sensitive to multivariate common linear, and the results are more robust to missing data and unbalanced data. Furthermore, it can predict the role of up to thousands of explanatory variables well and improve the prediction accuracy without significant improvement in the calculation (Diaz-Uriarte & de Andres, 2006).

Convolution neural network is made up of three parts. The first part is the input layer. And the second one is composed of several convolution layers and pooling layers. Then the third one consists of a fully connected multilayer perceptron classifier. In the convolution layer of the convolution neural network, a neuron is only connected with some neighboring neurons and usually contains several feature planes. Each feature plane is comprised of some neurons arranged in a rectangle. The neurons in the same feature plane share the weight, which is the convolution kernel. Convolution kernel can reduce the connection between all layers of the network and the risk of overfitting (Lessmann et al., 2018).

Application of big data and artificial intelligence Chapter | 15 **311**

In wavelet transform, the infinite trigonometric function basis in Fourier transform is replaced by the finite attenuation wavelet basis. In this way, not only the frequency can be obtained, but also the time can be located. Wavelet transform has two variables: scale a and translation $\tau$. Scale a controls the expansion and contraction of wavelet function, while translation $\tau$ controls the translation of wavelet function. The scale corresponds to the frequency (inverse ratio), while the translation $\tau$ corresponds to the time (Adeli, Zhou, & Dadmehr, 2003).

Common spatial pattern (CSP) is a spatial filtering feature extraction algorithm for two classification tasks, which can extract spatial distribution components of each category from multichannel BCI data. The basic principle of the common spatial pattern algorithm is to find a group of optimal spatial filters to project by using the diagonalization of the matrix, so as to maximize the difference between the variance values of the two types of signals, thus obtaining the eigenvectors with high differentiation (Ang, Chin, Zhang, & Guan, 2008).

## 4 Applications

Medical data types and machine learning algorithms commonly used in the field of neuropsychiatry are briefly introduced before. This section will describe the application of big data and AI algorithms in neuropsychiatric diseases from the perspectives of early warning, diagnosis, and prognosis. The major brain disorders involved include ASD, Alzheimer's disease, depressive disorders, and schizophrenia.

### 4.1 Early warning

Most mental diseases are caused by a variety of causes, and their clinical characteristics are diverse. The onset of diseases is often sudden and uncertain, and the effectiveness of treatment varies from person to person. Therefore early warning of diseases through the collection of physiological data can systematically monitor the patient's condition and directly provide a basis for the follow-up nursing decisions, which can buy more time for the treatment of doctors. In that case, the incidence and mortality of mental diseases can be reduced, and the harm caused by them can be greatly moderated. In order to achieve the accurate effect of early warning, we need to rely on a large amount of information collected by various tools. Only through further analysis of big data can we get specific early warning results. Early warning of epilepsy is the most popular application in the early warning through physiological data.

Epilepsy is a chronic brain dysfunction syndrome caused by various causes. Abnormal discharge of brain nerve cells leads to brain dysfunction and thus epilepsy, which is the clinical manifestation of paroxysmal abnormal hypersynchronous electrical activity of brain nerve cells. It is characterized by repeatability, abruptness, and temporality. Nearly 25% of the world's 50 million

## 312 Big data in psychiatry and neurology

people with epilepsy still have no effective treatments. Due to the sudden and irregular nature of epileptic seizures, early warning of epileptic seizures is the key to break the bottleneck of epileptic treatment, as it can buy patients and doctors time to take appropriate protective measures to avoid accidental injuries caused by sudden seizures.

For collection of patients with intracranial EEG, cortex EEG signals, such as power spectrum data index, epilepsy early warning system uses the machine learning algorithms such as convolution algorithm of neural network to distinguish between the normal EEG and epileptic seizure EEG, and between epileptic seizure phase and epileptic seizure EEG, allowing doctors to access and track patients' data at any time through the system and then deal with the treatment of late decisions. It not only helps to reduce the rate of seizures and mortality, but also helps to further reveal the regularity and physiological basis of seizures.

## 4.2 Diagnosis

### 4.2.1 EEG signals used in the diagnosis of autism

Brain electrical activity is a spontaneous and rhythmic potential change generated by cortical neurons. Nowadays, routine clinical electroencephalography has long been pretty popular. And with the rapid development of electronic technology, EEG has become increasingly widely applied. It has become an important index to evaluate the state of brain function and is widely used in the diagnosis of mental diseases. The data output of this kind of data is very large, which provides a sufficient data source for the diagnosis of mental diseases. Among them, it is most commonly used in the field of autism.

Autism is a highly inherited neurological disorder characterized by psychological symptoms, including repetitive and stereotypical behaviors, as well as a lack of emotional expression, verbal, and social communication skills. It is a mental illness that contains a range of complex neurodevelopmental disorders. Due to the heterogeneity and clinical diversity of autism, the diagnosis method mainly relies on the diagnostic medical staff's judgment of the patient's emotion, language, and social behavior, which is a kind of subjective analysis. However, since its pathophysiology has not yet been established and the existing diagnostic methods are highly subjective, medical staffs will have to work harder to improve the diagnostic accuracy.

EEG signals can detect the electrical activity of neurons in the brain by traversing the electrodes on the scalp, which has the advantages of no damage, high temporal resolution, and real-time monitoring (Wang et al., 2013). As the real-time data collection of brain electrical signal is so large that it is impossible to accurately analyze all the data according to the subjective judgment of medical staffs, so it is an inevitable trend to apply artificial intelligence technology to data analysis. In addition, existing studies have found that the EEG phase of autistic patients is different from that of normal people. The EEG power spectrum of autistic patients is different between groups in the cerebral hemisphere and between hemispheres. The EEG signals of autistic patients are located in the frontal lobe, significantly decreased in the brain area of the back, reflecting the cognitive decline of autistic patients. The EEG signals of autistic patients are located in the frontal lobe (Daoust, Limoges, Bolduc, Mottron, & Godbout, 2004), which are significantly increased in the parietal lobe area and significantly weakened (Burnette et al., 2011) in the back area of the brain, reflecting the cognitive decline of autistic patients. The characteristic values of children's EEG signals can be extracted by artificial intelligence technology and compared with those of autistic patients. In this way, the high-precision diagnosis of autism can be realized and the work burden of medical workers can be effectively reduced.

Due to certain behaviors in social situations, it is difficult for autistic patients to keep quiet in the process of data collection. However, the acquisition of resting EEG signals has a high tolerance for the activities of the subjects and can well adapt to the behavioral changes of autistic patients without causing too many errors. Up to now, the diagnostic accuracy of EEG has reached 80%.

### 4.2.2 Electroencephalogram and MRI used in diagnosis of Alzheimer's disease

Alzheimer's Disease (AD) is an irreversible degenerative neuropsychiatric disease. Because the etiology of Alzheimer's disease has not been fully elucidated and there is no effective treatment, early diagnosis and intervention of AD quite essential. At present, as a clinical precursory stage, the early diagnosis of Mild Cognitive Impairment (CMI) has also attracted more attention in the academic community. The common clinical diagnosis of AD and MCI on account of neuroimaging is based on, for example, MRI and Positron Emission Computed Tomography (PET). And common analysis methods for neuroimaging are mainly based upon image feature markers (such as hippocampal volume, cortical thickness, etc.), or using diagnostic methods on the basis of machine learning and deep learning to construct diagnostic algorithms.

At present, the AI diagnosis process for AD includes finding participants, cleaning image data, standardizing classifiers and classification criteria, and extracting as well as recognizing biological features, so as to achieve the classification of different types of AD and MCI. The basic approaches involve SVM, whose aim is to find a predictive model which is able to perform binary group separation. Some researches also use Naive Bayes or Random Forest for pattern recognition models, in terms of Voxel-based diagnosis on functional magnetic resonance imaging (fMRI) (Armananzas, Iglesias, Morales, & Alonso-Nanclares, 2017). Although there are minimal differences when using different voxel subset selection and classification paradigms, the quantitative results suggest that classification models are highly applicable in clinical scenarios.

**314** Big data in psychiatry and neurology

In this field, the most widely studied and promising method is prediction based on Convolutional Neural Networks (CNN). CNN is a kind of feedforward neural network that shows excellent performance in large-scale image processing, including 1D CNN, 2D CNN, and 3D CNN. Traditional machine learning methods usually need to manually check out areas of the brain where AD patients have more obvious changes, such as hippocampus, amygdala, and other regions of interest (ROI). However, CNN and other Deep Learning methods can achieve autonomous learning of feature representations to manage image classification, and then diagnose and classify patients with Alzheimer's disease or mild cognitive impairment. For example, Suk, Lee, and Shen (2014) used pairs of MRI and PET images from ADNI to execute Patch-Level Deep Feature Learning based on Deep Boltzmann Machine (DBM). Increasing evidences have proved that biomarkers from various modalities can provide more complementary information in AD/MCI diagnosis, so multimodal deep feature fusion for different kinds of images is carried out under the joint efforts of Deep Learning methods.

In addition, deep machine learning algorithms such as Stacked Autoencoders, Noise Reduction Autoencoders, and Deep Belief Networks (DBN) are also gradually applied to the clinical diagnosis of Alzheimer's disease, which usually has high robustness and accuracy (Armananzas et al., 2017). Although most of the related Artificial Intelligence methods require a large amount of data and images for training, they can effectively avoid physicians' personal decision errors and limitations, and multimodal analysis of multiple imaging data can also greatly improve the accuracy and reliability of diagnosis results.

### 4.2.3 Eye movement data used in diagnosis of schizophrenia

Eye movement technology is to extract such data as fixation point, fixation time, eye-hop distance, pupil size, and so on from the records of eye movement trajectory, so as to study the internal cognitive process of individuals. Nowadays, eye movement data is increasingly used in detection and classification of mental disorders, combined with AI.

Schizophrenia is a relatively common psychiatric disease, which is clinically manifested as a syndrome of multiple symptoms. As early as the 1970s, relevant researchers have found that eye movements of patients with schizophrenia are significantly different from normal people. Therefore, in the clinical diagnosis of schizophrenia, eye movement data (including the asymmetry of the speed of visually guided saccades Index, and so on) have become important monitoring indicators. With the widespread application of AI and Big Data technologies in the field of mental illness, eye movement data processing in the diagnosis of schizophrenia has gradually been automatized and intelligentized, and outstanding results have been achieved.

In the clinical diagnosis of schizophrenia, eye movement data is often directly obtained from medical equipment such as electro-optic eye tracker,

Application of big data and artificial intelligence Chapter | 15 **315**

corneal reflection eye tracker, or pupil-corneal reflection vector method medical eye tracker. Data diagnosis mostly plays a role in auxiliary research. Li Yu from Shanghai Jiao Tong University has used principal component analysis in machine learning combined with support vector machine (PCA-SVM) algorithm to establish a diagnostic discrimination model, aiming at using saccade behavior and reverse saccade behavior amplitude in eye movement data to treat schizophrenia patients. The accuracy of classification, diagnosis, and treatment of patients with latency can reach more than 80% (Yu, 2017).

In addition, the integration of eye movement data and the chronological comparison before and after the illness greatly improves the accuracy of diagnosis. In specific researches, eye movement data can also be combined with EEG data to construct a brain function network based on multilayered complex neural networks, which plays an essential role in the pathological investigation and prognosis of schizophrenia.

Big Data relevant methods have been widely used in eye movement data analysis. Particular machine-learned computational models (MLCMs) can be used in eye tracking to model reading comprehension (D'Mello & Southwell, 2020), and also, huge amount of eye movement data can assist deep learning models to establish visual sensory–perceptual–cognitive dynamical systems (Assadi et al., 2018). Although big data methods have not yet been applied to diagnosis or early warning of neuropsychiatric diseases detected by eye movement data, eye movement still enjoys significance and intuition as parameter for schizophrenia and depression, etc. It is believed that big data methods would be used in the analysis of eye movement data and give assistance to clinical neuropsychiatric diseases.

In general, big data or AI which uses eye movement data in the diagnosis of schizophrenia mostly involve saccade research, but do not go deep into smooth tracking or other aspects, so there is still a lot of space for development. For example, the results of exploratory eye movement monitoring can be used as both characteristics and state markers, and the response scores can be used as important biological markers for the diagnosis of schizophrenia.

### 4.2.4 Genomics data used in diagnosis of autism

Recent studies have revealed that genetic neurological channelopathies can lead to many different neurological diseases. In general, neurofunctional diseases are inherited in an autosomal dominant manner and cause paroxysmal disorders of neurologic function. These diseases are rare in some cases, but accurate diagnosis is important because they have the function of genetic counseling and usually have therapeutic significance (Spillane, Kullmann, & Hanna, 2016).

Autism spectrum disorder (ASD) is a complex neurodevelopmental disorder with a strong genetic basis. Evidence accumulated over recent years has shown the correlation of hundreds of gene variants in autism. Yet, only a small fraction of potentially causal genes—about 65 genes out of an estimated several

316  Big data in psychiatry and neurology

hundred—are known with strong genetic evidence from sequencing studies. The effects of gene variants are highly variable, and they are usually related to other disorders besides autism. These findings suggest that genetic heterogeneity, mutation penetrance, and genetic pleiotropy are common features of autism.

Researchers have developed an evidence-weighted machine learning approach on the basis of a human brain-specific gene network to present a genome-wide prediction of autism risk genes and identified likely pathogenic genes within frequent autism-associated copy-number variants (Krishnan et al., 2016).

In recent years, SNV identification has also made great progress, so that the whole-genome sequencing (WGS) and the whole-exome sequencing (WES) have become feasible alternative methods for selective genotyping. Through the identification of novel and rare variants, it has the significance of risk prediction and diagnosis, and is applied to the treatment of autism and other neuropsychiatric diseases (Carter & Scherer, 2013).

## 4.3  Prognosis

The central theme of personalized medicine is to adopt specific therapies according to the unique physiological characteristics of individuals. Such characteristics mainly include: genetic alterations and epigenetic modifications, clinical symptomatology, biomarker changes, and environmental factors (Ozomaro, Wahlestedt, & Nemeroff, 2013). Therefore the primary objective of personalized medicine is to predict the individual's susceptibility to disease, to achieve accurate diagnosis, and to optimize the most effective and beneficial response to treatment. The realization of personalized medicine in psychiatry can promote a significant reduction in incidence rate and mortality rate.

Pharmacogenomics involves the customization of therapies through individual genetic makeup, rational drug development, and drug reuse. Several large research teams related to electronic health record (EHR) data have developed gradually. Like new genetic variants that predict drug action have been discovered in many ways, supporting Mendelian randomized trials, showing drug efficacy, and suggesting new indications for existing drugs. Big data approaches may also help identify subtypes of disease, which could suggest differential treatment and prognoses.

Evidence from recent studies indicates that genetic targets often indicate effective drug targets. The subject of most of these studies was genome-wide association analysis (GWAS), representing drug targets with significant impact on the disease or trait. Electronic health record (EHR) is an effective tool for evaluating the efficacy of drugs. Pharmacogenomics research using EHR data can more efficiently obtain the "actual" situation, complications, related drugs, and long-term therapeutic effect of drug use in patients, so as to better discover the clinical effect of pharmacogenomic interactions and make better use of patient data. Networks including the eMERGE network have established a strong track record for accurately identifying disease and drug response phenotypes from EHRs, and a published algorithm with a median positive predictive value (PPV) of >95% (Kirby et al., 2016).

Depressive disorders (DDs) are a common disease in mental disease pathology. About 350 million people worldwide are affected by the condition, according to the World Health Organization. Genetic factors contribute significantly to the risk of DDs and more than 20 genes are linked to episodes of depression (Schosser & Kasper, 2009). Through genome-wide association analysis (GWAS) and electronic case data, a large number of samples of patients and healthy people were used in the algorithm to obtain statistically significant results (Cai et al., 2015; Sullivan et al., 2009). Research has shown that we are able to assess substantial pharmacogenetic traits with accurate performance even if there is not manual review, just as for algorithms of pharmacogenomics. Similarly, the response of antidepressants to major depression also varies with specific genetic changes in certain candidate genes (Shadrina, Bondarenko, & Slominsky, 2018). The associated genetic changes affect not only MDD susceptibility, but also the effectiveness of antidepressant treatment. Thus multiple linear regression was used to generate the algorithm by using the stable dosage of the drugs for depressed disease as the dependent variable and the clinical and genotypic variables as the independent variables. These new applications of biomedical informatics, machine learning, and statistical knowledge have improved the ability to interpret clinical information to identify patients' complex phenotypes and subsets, and to tailor medication and other treatment regimens, accelerating the development of sophisticated medicine.

This approach has also been shown in personalized medicine for other diseases. For example, coumarins, the most widely used anticoagulants for the treatment and prevention of thromboembolism, vary greatly from individual to individual in the dose required to achieve stable anticoagulation. The multiple linear regression algorithm can be used to predict effectively and improve the dose selection of acenocoumarol (Tong et al., 2016) (Table 1).

# 5 Challenges and promising solutions

## 5.1 Privacy and security of patient information

When we get a large amount of data to help patients with personalized prognosis, the privacy of patients including personal details, physiological information, diagnosis reports, and treatment plans will be easily exposed. The storage and use of these data in the cloud may cause the leakage of patient's privacy and bring unnecessary trouble for patients. The security of data is based on the previously mentioned data privacy. When the user's personal privacy is disclosed, the hacker can modify the sensitive information, including health information, so as to cause misdiagnosis or incorrect disease evaluation, further

# 318 Big data in psychiatry and neurology

**TABLE 1** Briefly summarizes the use of big data and artificial intelligence technologies in early warning, diagnosis, and prognosis, listing the types of data commonly used and the main algorithms used in studies of related diseases, respectively, and the studies of neuropsychiatric diseases listed are presented specifically in the previous paper.

| Disease | Data type | Main algorithm | Application |
| --- | --- | --- | --- |
| Epilepsy | Electroencephalogram | Convolutional neural networks | Early warning |
| Autism | Electroencephalogram | Wavelet transform, common spatial pattern, autoregressive model | Diagnose |
| Autism | Genomics | Evidence-weighted machine learning approach | Diagnose |
| Depressive disorders (DDS) | Genomics, electronic health record | Multiple linear regression | Prognosis |
| Alzheimer's disease (AD) | Electroencephalogram and MRI | Support Vector Machine, Convolutional Neural Network, Deep Belief Network | Diagnose |
| Schizophrenia | Eye movement | Neural Network, Principal Component Analysis | Diagnose |

leading to improper treatment or prognosis, thus increasing the death rate. The data of patients discharged from hospital monitored by Internet of things devices are prone to generate security risks in the transmission process. It can be said that data security is one of the most important challenges in the personalized prognosis of mental diseases or other diseases. In the process of tracking patients' prognosis, hardware and software are also vulnerable to security threats. In order to ensure the secure communication between Internet of things devices, SSL/TLS protocol is widely used. Encryption technology can be used to provide secure communication, but it also needs more storage and processing capacity, which is a major challenge for wireless sensors and computing devices used in healthcare applications. Developing algorithms and solutions to protect health-related information from unauthorized users is a challenging issue, as design and implementation phase rationale errors can also lead to security risks.

To solve this problem, malicious users should not be miscalculated. This requires the establishment of an authentication mechanism for IoT sensor data on the server side to prevent malicious access by third-party users and monitor the application's end-to-end delay in synchronization of data packets. In the case of patients moving independently while wearing sensors, dynamic network topology can quickly adapt to the security switching mechanism from one network to another. At the same time, in order to achieve high throughput of Internet of Things health monitoring and efficient sensing, both communication costs and computation should be cost effective.

## 5.2 Information island

Information island means the problem that information is not shared with each other and information is disconnected from business processes and applications. It is also an important issue that plagues the prognosis of AI and big data in psychiatric disorders. Data islands mainly refer to that medical data cannot be exchanged and shared among hospitals, regions, and departments. Medical activities will generate a large amount of data, but cannot be fully utilized. There are very few data used to guide medical activities, and the overall planning of resources cannot be achieved, which brings great challenges to the integration of medical big data. At present, big data and AI have brought new medical models, but because of the differences in the nature of ownership, the closeness of cooperation, and benefit orientation, among others, it is easy to create new data "islands." On the one hand, different hospitals, medical enterprises, among others, have their own established information systems, where they store massive diagnostic and prognostic information. Therefore, in the process of medical data collection, the lack of unified standards in medical data means that the majority of medical data cannot be shared and used, resulting in information islands; on the other hand, because medical data inevitably contains personal privacy, it is difficult to achieve full sharing in a real sense. At this stage, a large proportion of hospitals maintain a negative attitude toward data sharing, and it is hard to achieve information exchange between hospitals.

## 5.3 Storage and analysis capabilities

In the process of individualized prognosis, storage and analysis capabilities are the major challenges. The storage problem is mainly caused by the number, speed, and various forms of big data analysis. Big data analysis is difficult to use in traditional storage system directly. Therefore cloud storage systems like Amazon S2, Amazon EC2, and elastic block store provide solutions for big data analysis by providing infinite storage systems with fault tolerance. In addition to storage, another challenge is how to move big data to cloud services at lower cost and faster speed. Most of the work in the cloud relies on Amazon Elastic Compute Cloud (Amazon EC2) to provide scalable computing power in the

320   Big data in psychiatry and neurology

cloud. Nevertheless, cloud platform cannot be effectively used to meet the rapid growth demand for big data analysis and healthcare information.

## 5.4   Lack of specialized personnel

Similarly, based on the vigorous development of personalized medical field, the scarcity of professionals has become one of the challenges facing the construction of medical big data. More specialized operation puts forward more requirements for high-quality compound talents, who not only need to grasp certain information technology ability and data analysis ability, but also need solid medical basic knowledge to better participate in the processing of medical big data. Faced with the huge amount of medical data resources, the value of big medical data cannot be fully realized without professional personnel dealing with it. At present, most hospitals are seriously short of high-quality composite talents, which seriously hinder the development of big medical data. To this end, more support is needed for the training of composite talents, and more excellent talents are guided to join this field.

## 6   Conclusions

In recent years, big data and AI have been widely used in the field of psychiatry. In this paper, the characteristics of biomedical big data itself are analyzed from genetic data, EEG signals, eye movement data, electroencephalogram data, and wearable device data. The main algorithms, such as support vector machine, decision tree, random forest, convolution neural network, wavelet transform algorithm in time–frequency analysis, and common space pattern algorithm in feature processing, are briefly introduced. Due to the rapid development of this field, biomedical big data is widely used in the early warning, diagnosis, and prognosis of mental diseases. And the major brain disorders involved include autism, Alzheimer's disease, and schizophrenia. Finally, we analyzed the challenges encountered in the process of data application, such as possible patient data security issues, information islands, and other issues, and proposed our solutions.

The future direction of work is mainly to pay more attention to the protection of patients' privacy, while relying on emerging technologies, including Internet of things technology, knowledge mapping technology, etc. to support doctors' diagnosis, treatment, and prognosis. Moreover, we need to accelerate the integration of existing data resources to build a reliable and accurate model for clinical practice; to exchange and share medical data among hospitals, regions, and departments; and to store and use data in a unified way. At the same time, we also need to enhance the ability of data storage and analysis to meet the growing demand for data, as well as to strengthen the training of professionals, which is also the focus of attention in the future.

# References

Adeli, H., Zhou, Z., & Dadmehr, N. (2003). Analysis of EEG records in an epileptic patient using wavelet transform. *Journal of Neuroscience Methods, 123*(1), 69–87. https://doi.org/10.1016/s0165-0270(02)00340-0.

Ang, K. K., Chin, Z. Y., Zhang, H. H., & Guan, C. T. (2008). Filter bank common spatial pattern (FBCSP) in brain-computer interface. In (Vols. 1–8). *2008 IEEE international joint conference on neural networks* (pp. 2390–2397). IEEE: New York.

Aoe, J., Fukuma, R., Yanagisawa, T., Harada, T., Tanaka, M., Kobayashi, M., ... Kishima, H. (2019). Automatic diagnosis of neurological diseases using MEG signals with a deep neural network. *Scientific Reports, 9*(5057). https://doi.org/10.1038/s41598-019-41500-x.

Armananzas, R., Iglesias, M., Morales, D. A., & Alonso-Nanclares, L. (2017). Voxel-based diagnosis of Alzheimer's disease using classifier ensembles. *IEEE Journal of Biomedical and Health Informatics, 21*(3), 778–784. https://doi.org/10.1109/JBHI.2016.2538559.

Assadi, A. H., et al. (2018). Deep learning models for visual sensory-perceptual cognitive dynamical systems from eye movement data and categories of natural images. *Investigative Ophthalmology & Visual Science, 59*(9).

Buettner, R., Rieg, T., & Frick, J. (2020). Machine learning based diagnosis of diseases using the unfolded EEG spectra: Towards an intelligent software sensor. In *Paper presented at the information systems and neuroscience, Cham.*

Burnette, C. P., Henderson, H. A., Inge, A. P., Zahka, N. E., Schwartz, C. B., & Mundy, P. C. (2011). Anterior EEG asymmetry and the modifier model of autism. *Journal of Autism and Developmental Disorders, 41*(8), 1113–1124. https://doi.org/10.1007/s10803-010-1138-0.

Bzdok, D., Schulz, M.-A., & Lindquist, M. (2019). Emerging shifts in neuroimaging data analysis in the era of "big data". In I. C. Passos, B. Mwangi, & F. Kapczinski (Eds.), *Personalized psychiatry: Big data analytics in mental health* (pp. 99–118). Cham: Springer International Publishing.

Cai, N., Bigdeli, T. B., Kretzschmar, W., Li, Y., Liang, J., Song, L., ... Consortium, C. (2015). Sparse whole-genome sequencing identifies two loci for major depressive disorder. *Nature, 523*(7562), 588–591. https://doi.org/10.1038/nature14659.

Carter, M., & Scherer, S. (2013). Autism spectrum disorder in the genetics clinic: A review. *Clinical Genetics, 83*. https://doi.org/10.1111/cge.12101.

Chen, M., Ma, Y., Li, Y., Wu, D., Zhang, Y., & Youn, C. (2017). Wearable 2.0: Enabling human-cloud integration in next generation healthcare systems. *IEEE Communications Magazine, 55*(1), 54–61. https://doi.org/10.1109/MCOM.2017.1600410CM.

Chiesa, P. A., Cavedo, E., Lista, S., Thompson, P. M., & Hampel, H. (2017). Revolution of resting-state functional neuroimaging genetics in Alzheimer's disease. *Trends in Neurosciences, 40*(8), 469–480. https://doi.org/10.1016/j.tins.2017.06.002.

Daoust, A.-M., Limoges, É., Bolduc, C., Mottron, L., & Godbout, R. (2004). EEG spectral analysis of wakefulness and REM sleep in high functioning autistic spectrum disorders. *Clinical Neurophysiology, 115*(6), 1368–1373. https://doi.org/10.1016/j.clinph.2004.01.011.

Diaz-Uriarte, R., & de Andres, S. A. (2006). Gene selection and classification of microarray data using random forest. *BMC Bioinformatics, 7*, 13. https://doi.org/10.1186/1471-2105-7-3.

D'Mello, S. K., & Southwell, R. (2020). Machine-learned computational models can enhance the study of text and discourse: A case study using eye tracking to model reading comprehension. *Discourse Processes, 57*(5–6), 420–440.

Guyon, I., Weston, J., Barnhill, S., & Vapnik, V. (2002). Gene selection for cancer classification using support vector machines. *Machine Learning, 46*(1–3), 389–422. https://doi.org/10.1023/a:1012487302797.

Huys, Q. J. M., Maia, T. V., & Frank, M. J. (2016). Computational psychiatry as a bridge from neuroscience to clinical applications. *Nature Neuroscience, 19*(3), 404–413. https://doi.org/10.1038/nn.4238.

Kirby, J. C., Speltz, P., Rasmussen, L. V., Basford, M., Gottesman, O., Peissig, P. L., … Denny, J. C. (2016). PheKB: A catalog and workflow for creating electronic phenotype algorithms for transportability. *Journal of the American Medical Informatics Association, 23*(6), 1046–1052. https://doi.org/10.1093/jamia/ocv202.

Krishnan, A., Zhang, R., Yao, V., Theesfeld, C. L., Wong, A. K., Tadych, A., … Troyanskaya, O. G. (2016). Genome-wide prediction and functional characterization of the genetic basis of autism spectrum disorder. *Nature Neuroscience, 19*(11), 1454–1462. https://doi.org/10.1038/nn.4353.

Kumsta, R. (2019). The role of epigenetics for understanding mental health difficulties and its implications for psychotherapy research. *Psychology and Psychotherapy: Theory, Research and Practice, 92*(2), 190–207. https://doi.org/10.1111/papt.12227.

Langmead, B., & Nellore, A. (2018). Cloud computing for genomic data analysis and collaboration. *Nature Reviews Genetics, 19*(4), 208–219. https://doi.org/10.1038/nrg.2017.113.

Lessmann, N., van Ginneken, B., Zreik, M., de Jong, P. A., de Vos, B. D., Viergever, M. A., & Isgum, I. (2018). Automatic calcium scoring in low-dose chest CT using deep neural networks with dilated convolutions. *IEEE Transactions on Medical Imaging, 37*(2), 615–625. https://doi.org/10.1109/tmi.2017.2769839.

McGinnis, R. S., McGinnis, E. W., Hruschak, J., Lopez-Duran, N. L., Fitzgerald, K., Rosenblum, K. L., & Muzik, M. (2018). Wearable sensors and machine learning diagnose anxiety and depression in young children. In *Paper presented at the 2018 IEEE EMBS international conference on biomedical & health informatics (BHI).* 4–7 March 2018.

Ozomaro, U., Wahlestedt, C., & Nemeroff, C. B. (2013). Personalized medicine in psychiatry: Problems and promises. *BMC Medicine, 11*(1), 132. https://doi.org/10.1186/1741-7015-11-132.

Pradhan, B. (2013). A comparative study on the predictive ability of the decision tree, support vector machine and neuro-fuzzy models in landslide susceptibility mapping using GIS. *Computers & Geosciences, 51*, 350–365. https://doi.org/10.1016/j.cageo.2012.08.023.

Qiao, C., Lin, D. D., Cao, S. L., & Wang, Y. P. (2015). The effective diagnosis of schizophrenia by using multi-layer RBMs deep networks. In J. Huan, S. Miyano, A. Shehu, X. Hu, B. Ma, S. Rajasekaran, V. K. Gombar, I. M. Schapranow, I. H. Yoo, J. Y. Zhou, B. Chen, V.Pai & B. Pierce (Eds.), *IEEE international conference on bioinformatics and biomedicine – BIBM* (pp. 603–606).

Sazgar, M., & Young, M. G. (2019). Encephalopathies, brain death, and EEG. In M. Sazgar, & M. G. Young (Eds.), *Absolute epilepsy and EEG rotation review: Essentials for trainees* (pp. 183–198). Cham: Springer International Publishing.

Schosser, A., & Kasper, S. (2009). The role of pharmacogenetics in the treatment of depression and anxiety disorders. *International Clinical Psychopharmacology, 24*(6), 277–288. https://doi.org/10.1097/yic.0b013e3283306a2f.

Shadrina, M., Bondarenko, E. A., & Slominsky, P. A. (2018). Genetics factors in major depression disease. *Frontiers in Psychiatry, 9*, 334. https://doi.org/10.3389/fpsyt.2018.00334.

Shiino, T., Miura, K., Fujimoto, M., Kudo, N., Yamamori, H., Yasuda, Y., … Hashimoto, R. (2020). Comparison of eye movements in schizophrenia and autism spectrum disorder. *Neuropsychopharmacology Reports, 40*(1), 92–95. https://doi.org/10.1002/npr2.12085.

Shimada, T., Shiina, T., & Saito, Y. (2000). Detection of characteristic waves of sleep EEG by neural network analysis. *IEEE Transactions on Biomedical Engineering, 47*(3), 369–379. https://doi.org/10.1109/10.827301.

Application of big data and artificial intelligence **Chapter | 15 323**

Spillane, J., Kullmann, D. M., & Hanna, M. G. (2016). Genetic neurological channelopathies: Molecular genetics and clinical phenotypes. *Journal of Neurology, Neurosurgery & Psychiatry, 87*(1), 37. https://doi.org/10.1136/jnnp-2015-311233.

Suk, H., Lee, S., & Shen, D. (2014). Hierarchical feature representation and multimodal fusion with deep learning for AD/MCI diagnosis. *NeuroImage, 101*, 569–582. https://doi.org/10.1016/j.neuroimage.2014.06.077.

Sullivan, P. F., de Geus, E. J. C., Willemsen, G., James, M. R., Smit, J. H., Zandbelt, T., ... Penninx, B. W. J. H. (2009). Genome-wide association for major depressive disorder: A possible role for the presynaptic protein piccolo. *Molecular Psychiatry, 14*(4), 359–375. https://doi.org/10.1038/mp.2008.125.

Tong, H. Y., Dávila-Fajardo, C. L., Borobia, A. M., Martínez-González, L. J., Lubomirov, R., Perea León, L. M., ... Group, P.-A. I. (2016). A new pharmacogenetic algorithm to predict the most appropriate dosage of acenocoumarol for stable anticoagulation in a mixed Spanish population. *PLoS One, 11*(3), e0150456. https://doi.org/10.1371/journal.pone.0150456.

Uddén, J., Hultén, A., Bendtz, K., Mineroff, Z., Kucera, K. S., Vino, A., ... Fisher, S. E. (2019). Toward robust functional neuroimaging genetics of cognition. *The Journal of Neuroscience, 39*(44), 8778. https://doi.org/10.1523/JNEUROSCI.0888-19.2019.

Wang, J., Barstein, J., Ethridge, L. E., Mosconi, M. W., Takarae, Y., & Sweeney, J. A. (2013). Resting state EEG abnormalities in autism spectrum disorders. *Journal of Neurodevelopmental Disorders, 5*.

Yu, L. (2017). *Characteristics of eye-movement in first-episode schizophrenia and first-episode depression and modelling diagnosis and differential diagnosis for schizophrenia.* Master Degree, Shanghai Jiaotong University.

# Chapter 16

# Harnessing big data to strengthen evidence-informed precise public health response

## G.V. Asokan[a] and Mohammed Yousif Mohammed[b]

[a]*Public Health Program, College of Health and Sport Sciences, University of Bahrain, Salmanya Medical Complex, Manama, Kingdom of Bahrain,* [b]*Medical Laboratory Sciences Program, College of Health and Sport Sciences, University of Bahrain, Salmanya Medical Complex, Manama, Kingdom of Bahrain*

## 1 Public health

Sir Donald Acheson (1988) defined public health as "the art and science of preventing disease, prolonging life and promoting health through the organized efforts of society." It is identified as both a science and an art—a blend of knowledge and action. Through organized efforts public health focuses on population health (public health paradigm) and not on an individual's health (clinical medicine paradigm) by invigorating public health capacities to serve the population and assure conditions under which people can stay healthy. In brief, the Institute of Medicine (1988) defined public health as "what we, as a society, do collectively to assure the conditions for people to be healthy." Further, the mission of public health is to provide access to quality health care and apply diverse disease prevention strategies by focusing on modifiable risk factors of morbidity and mortality while reducing health disparities. In addition to identifying social, environmental, and biological factors, the dynamics of public health allow to embrace contributions from health informatics, big data, and genomics to prevent disease and improve health care. Data on disease burden, research on intervention effectiveness, and estimates of the resultant health benefits for the population are key to inform public health response.

Evidence-based public health (EBPH) focuses on informed, explicit, and judicious use of current best evidence that are derived from various scientific disciplines and evaluation process (Brownson, Fielding, & Maylahn, 2009, p. 175). According to a perspective (Bayer & Galea, 2015, p. 499), to make informed decisions, EBPH utilize the best available scientific evidence, capture

Big Data in Psychiatry and Neurology. https://doi.org/10.1016/B978-0-12-822884-5.00003-9
Copyright © 2021 Elsevier Inc. All rights reserved.

**326** Big data in psychiatry and neurology

data through information systems systematically, adopt program frameworks, engage the community in decision making, conduct evaluations, and disseminate the outcomes. "The transition from evidence-based approach in public health to genomic approach to individuals with a paradigm shift from a "reactive" medicine to a more "proactive and personalized healthcare" may look extraordinary; nevertheless, the success of precision medicine relies on population perspective" (Vaithinathan & Asokan, 2017, p. 79).

In a correspondence by the public health leadership society (Institute of Medicine, 2003), to strengthen public health, delivery of 10 essential public health operations was listed as necessary. They are as follows:

1. surveillance and assessment of the population's health and well-being
2. identification of health problems and health hazards in the community
3. health protection services (environment, occupational, food safety)
4. preparedness for and planning of public health emergencies
5. disease prevention operations
6. health promotion operations
7. evaluation of quality and effectiveness of personal and community health services
8. assurance of a competent public health and personal health care workforce
9. leadership, governance and the initiation, development and planning of public health policy
10. health-related research.

Among the 10 essential public health services, monitor the health status to identify community health problems, and diagnose and investigate health problems and health hazards in the community are data driven. The public health services leading to EBPH are dependent on the data generated. Moreover, the public health policy makers are inspired by the data-driven transformation occurring rapidly in the digital world. The underlying principles to achieve the operations for evidence-informed public health policy are gathering, collating, analyzing, and responding of the data.

## 2 Global burden of disease

Most metrics of the global burden of disease (GBD) are available from 2000 onward in 5-year increments from the World Health Organization (WHO) Global Health Observatory (2019a). The Institute of Health Metrics and Evaluation (2020) publishes GBD from 1990 onward. From the estimates of 2017, more than 60% of the global burden of disease was attributable to non-communicable diseases (NCDs); 28% of communicable, maternal, neonatal, and nutritional diseases; and around 10% of injuries. There is a significant shift in the burden toward NCDs by a reduction in communicable and preventable disease with a rise in income and living standards. In high-income economies, NCDs account for more than 80% of the disease burden contrasting the communicable diseases that are as low as less than 5%. In contrast, communicable disease contributes more than 60% across many low-income economies.

## 2.1 Noncommunicable diseases

We struggle to cope with the growing burden of lifestyle diseases alongside the complex and growing health needs of aging populations that challenge and stress the health systems (World Bank, 2020). The NCDs are moving rapidly along the trend of epidemiological transition and are the predominant causes of morbidity and mortality than all other causes combined. Of the 56.9 million deaths reported worldwide in 2016, 54% was accounted for the top 10 causes; the leading killers were ischemic heart disease and stroke causing 15.2 million deaths, and continues to be the leading cause of global deaths in the last two decades (World Health Organization, 2018a). They were attributed to "metabolic risk factors"—a term that denotes the "combination of an increase in blood pressure, blood glucose and blood lipids, as well as obesity." Among the major burden of NCD deaths, cardiovascular diseases account for most NCD deaths, or 17.9 million people annually, followed by cancers (9.0 million), respiratory diseases (3.9million), and diabetes (1.6 million). These 4 groups of diseases account for over 80% of all premature NCD deaths. Each year, between the ages of 30 and 69 years the NCDs cause 15 million deaths (World Health Organization, 2018b). Detection, screening, and treatment of NCDs, as well as palliative care, are key components of the response to NCDs.

## 2.2 Infectious diseases

Infectious diseases have wrecked us right through our existence, decimated whole populations as well as ended lineage. Casualties due to infectious diseases were higher than wars and had historical significance. In 2016 the WHO (World Health Organization, 2018a) reported three infectious diseases among the top 10 causes of death worldwide. They are lower respiratory infections (3.0 million deaths), diarrheal diseases (1.4 million deaths), and tuberculosis (1.3 million deaths). Even as the incidence of infectious diseases such as HIV (1.0 million deaths in 2016 compared with 1.5 million in 2000), tuberculosis, and malaria has reduced in developed economies, they are still a leading cause of death in developing economies.

Taylor et al.(2001, p. 985) identified 1415 species of infectious organisms known to be pathogenic to humans, of which 217 are viruses and prions, 538 are bacteria and rickettsia, 307 are fungi, 66 are protozoa, and 287 are helminths. Zoonoses constitute 61% of all known infectious diseases, while humans serve as the primary reservoir for just 3% of the total. Of the 175 emerging infectious species, 75% are identified as zoonotic. The sources for almost 90% of all human food-borne illnesses have been traced to foods of animal origin. Wildlife serves as the reservoir of unknown microorganisms from which unfamiliar

**328** Big data in psychiatry and neurology

pathogens continue to emerge and cause havoc (Brackett, Medellin, Caceres, & Mainka, 2004). We could document only 1/5 to 1/50 of microbial species so far; therefore, the reservoir of potential zoonotic pathogens is enormous and alarming.

Improvement in sanitation and advent of vaccines had effectively controlled major infectious diseases in the mid-20th century. However, in the last three decades, we witnessed the epidemics and pandemics of infectious diseases due to mobile, interdependence, and interconnectedness of the world. The frequency of emerging and reemerging pathogens appears more quickly than before and the rapid global transmission causing epidemics such as the emergence of pneumonic plague in India (1994), panzootic H5N1 influenza in Hong Kong (1997), Nipah encephalitis in Malaysia (1999), severe acute respiratory syndrome (SARS CoV1) in China (2003), and pandemic swine origin H1N1 influenza in Mexico (2009). Among the corona viruses, SARS CoV1 quickly spread to 26 countries before being contained after about 4 months. More than 8000 people fell ill from SARS and 774 died. Middle East respiratory syndrome (MERS) also caused by a coronavirus was first reported in Saudi Arabia in September 2012 and has since spread to 27 countries (Centers for Disease Control and Prevention, 2003). Since its emergence until January 2020, WHO confirmed 2519 MERS cases and 866 deaths (34%). When the countdown begun for the new year of 2020, the WHO was alerted on 31 December 2019 to several cases of pneumonia in Wuhan City, Hubei Province of China. It was named SARS CoV2. In consideration of the global threat, following the recommendations of the Emergency Committee of the WHO, a public health emergency of international concern or PHEIC was declared on 30 January 2020 (World Health Organization, 2020a). Globally, as of September 2, 2020, there have been 25,541,380 confirmed cases of COVID-19, including 852,000 deaths, reported to WHO (World Health Organization, 2020b).

## 3   Health systems and public health system

The pivotal goal of a health system is to prolong healthy life into attaining old age gracefully. It is believed that the stronger public health system offers a stronger response to health concerns. The public health system is located within a broader health system. Through a systems approach, public health is concerned with the whole system and not confined to the mere eradication of a specific disease. System models have mechanisms of positive and negative feedback and reflect the fact that diseases emerge within complex molecular, biological, and social systems (Ness, Koopman, & Roberts, 2007, p. 565). Broadly, the public health system can be divided into public health activities primarily located within a health system and those related to the wider health system through an intersectoral coordination with other sectors. Collaboration is a key element to augment public health response. Fundamentally, the organized efforts/activities are normally intertwined with the framework of

Harnessing big data Chapter | 16 **329**

government, nongovernmental organizations, private organizations, and individuals. An effective network of collaboration and intersectoral coordination can adequately respond to outbreaks of disease, disseminate knowledge, improve standards of living, and support research and treatment options. The core activities of the networking efforts listed in a correspondence by the public health leadership society (2002) include data collection and monitoring (surveillance), health improvement, health protection, and improvement in quality of health services and are reflected in capacity requirements for a public health workforce.

The public health system is constantly expanding. Big data and digital transformation of public health systems form the basis for research and action by the policy makers, for example Evidence-informed Policy Network (EVIPNet) is a network established by the WHO to promote the systematic use of research evidence in health policy-making in order to strengthen health systems to reduce GBD by offering the right programs and services to whom it is required (World Health Organization, 2019a).

## 3.1 Public health surveillance system

Public health surveillance continues to evolve. Public health surveillance is an ongoing systematic, frequent, and regular data collection for analysis and interpretation, thus, providing the platform essential to the planning, implementation, and evaluation of public health practice linked to prevention and control, and dissemination to those who need to know (Soucie, 2012). The surveillance data include demographic, socioeconomic, and clinical characteristics of the population; data on the main outcomes of morbidity and mortality; and data on mitigating behaviors or risk factors. A good surveillance data can define the problem, estimate the magnitude of the problem, determine the geographical location of the problem, describe the natural history of a disease, alert epidemics, generate hypotheses, monitor alterations in infectious agents, detect changes in health practices, and support emergency planning.

The aim of improving public health surveillance methodology is to promote system designs that encompass all the activities of data collection, filtering, analysis, visualization, and response that account for timely, simple, representative, flexible, rapid, valid, reliable, and cost-effective data. "Surveillance system design may thus be viewed as an optimization problem, suggesting enlistment of operations research analysts and sampling statisticians" (Burkom, 2017, p. 848). An efficient electronic surveillance system requires three categories of expertise:

- domain specific (medical, epidemiological, environmental—drivers of the problems)
- technological (e.g., database, network, programming, visualization)
- analytical (e.g., mathematics, statistics, machine learning).

**330** Big data in psychiatry and neurology

Sometimes it is challenging to achieve a stronger collaboration among domain experts and technology developers. The barriers are varied professional cultures, and the experts in specific domains may not possess up-to-date expertise in all three categories. Thus a beneficial collaboration must ensure repurposing the information suiting the requirements of surveillance from electronic health records and other sources in a technology-driven big data environment (Burkom, 2017, p. 848).

The existing traditional types of surveillance follow a descriptive epidemiology of time, place, and person that include active (most complete and accurate), passive (regular reporting to the health department by physicians and laboratories), sentinel (data about a particular disease that cannot be obtained through a passive system), syndromic (application of real-time indicators for disease), risk factor (indicator based), outbreak (event based). For early warning and preparedness, risk surveillance and syndromic surveillance are used. Each type of surveillance has their strengths and limitations. Most surveillance data originate from vital records (birth, death, hospital records), surveys of institutions (school) and individuals (physicians), monitoring systems (environmental), and animal health surveillance (zoonoses). Some of the international public health response systems are listed here:

- The Global Outbreak Alert Response Network (GOARN) of WHO ensures quick and appropriate technical support to populations affected by human disease epidemics on a national, regional, or international level. The GOARN focuses on technical and operational resources from scientific institutions of member states, medical and surveillance initiatives, regional technical networks, networks of laboratories, United Nations Organizations (e.g., UNICEF, UNHCR), the Red Cross, Red Crescent Societies, national societies and international humanitarian nongovernmental organizations (e.g., Médecins sans Frontières, International Rescue Committee, Merlin and Epicenter).
- The WHO and Food and Agriculture Organization (FAO) host International Food Safety Authorities Network (INFOSAN), which alerts national focal points on the occurrence of regional or global concerns for a food safety event. 'INFOSAN Emergency' partners with the GOARN. The INFOSAN systematically monitors potential international food safety-related events in addition to receiving information through INFOSAN emergency contact points.
- The FAO—World Organization for Animal Health (OIE)—WHO Global Early Warning and Response System (GLEWS) for major animal diseases, including zoonoses, combines the alert and response mechanisms of the three organizations to avoid duplication, and coordinate verification processes.
- The World Animal Health Information System (WAHIS) provides countries with a simple and rapid method of sending notifications, and reports on animal disease information.

- The FAO and OIE have developed a joint Network of Expertise on Animal Influenza (AI-OFFLU) to support international efforts to monitor and control infections of AI.
- The Emergency Prevention System (EMPRES) for Transboundary Animal and Plant Pests, and Diseases was established by FAO in 1994, which provides information, training, and emergency assistance to countries to prevent, contain, and control the world's most serious livestock diseases, while also surveying for emerging pathogens. EMPRES Global Animal Disease Information System (EMPRES-i) is a web-based application that has been designed to support veterinary services by facilitating regional and global disease information.
- The Global Framework for the Control of Transboundary Animal Diseases (GF-TAD) provides a clear vision and addresses endemic and emerging infectious diseases, including zoonoses.
- The Global Virus Network (GVN) is an international group of leading virologists and medical researchers tasked with providing scientific expertise to government agencies in the face of emerging infectious viral outbreak.
- The Program for Monitoring Emerging Diseases (ProMED) mail is a reporting system dedicated to rapid global dissemination of information on outbreaks of infectious diseases, and acute exposures to toxins that affect human health, animal health, and plants grown for food or animal feed.
- Morbidity and Mortality Weekly Report (MMWR) of Centers for Disease Control (CDC)
- Global Public Health Information Network (GPHIN) of Health Canada
- UK public health network
- European Centre for Disease Prevention and Control (ECDC)
- PACNET in the Pacific region
- Sentiweb in France
- Mekong Basin Disease Surveillance Network (MBDS) for China, Cambodia, Vietnam, Thailand, Myanmar, and Lao PDR.
- Middle East Consortium on Infectious Disease Surveillance (MECIDS) for Israel, Jordan, and the Palestinian Territory
- South-eastern European Health Network (SEEHN) for Albania, Bosnia and Herzegovina, Bulgaria, Croatia, Montenegro, the Republic of Moldova, Romania, Serbia, and the former Yugoslav Republic of Macedonia (FYROM)
- Asian Partnership on Emerging Infectious Disease Research (APIER) for Cambodia, China, Lao PDR, Indonesia, Thailand, and Vietnam
- South African Centre for Infectious Disease Surveillance (SACIDS) for DR Congo, Mozambique, South Africa, Zambia, and Tanzania
- East African Integrated Disease Surveillance Network (EAIDSNet) for Kenya, Tanzania, and Uganda

The purpose of a public health surveillance system must ensure a secure database with proper data management and quality control procedures. The progress in public health surveillance has often depended on adequate funding, continuity, and strategic consistency. Surveillance systems are not always perfect and that can compromise its usefulness. Surveillance systems do have limitations, mostly due to underreporting, lack of representativeness, and lack of timeliness. Certain measures can be implemented to overcome these limitations. Studies have shown that only a fraction of the cases of the notifiable diseases under surveillance are reported. Such an underreporting delay effective action of prevention and control (Doyle, 2002, p. 873). The big data revolution in the last two decades is envisaged to overcome the barriers in the existing traditional surveillance systems. "It would improve the granularity and timeliness of available epidemiological information, with hybrid systems augmenting rather than supplanting traditional surveillance systems, and better prospects for accurate diseases models and forecasts" (Bansal, Chowell, Simonsen, Vespignani, & Viboud, 2016).

## 4 Big data in precision public health

Precision public health is technology driven with more precise descriptions, and analyses of individuals and population groups for improving the overall health of populations (Baynam et al., 2017). It aims to improve the ability of preventing disease, promote health, and reduce health disparities in populations by applying novel emerging methods and technologies for measuring disease, identifying pathogens, exposures, behaviors, and susceptibility in populations (Khoury & Galea, 2016, p. 1357). Succinctly, precision public health is about accuracy and granularity in delivering many types of targeted interventions in the defined population through evidence-informed policies.

In the digital era, public health is amid a data revolution. Data generation occurs at a staggering and ever-increasing rate. Data that is rapid, complete, reliable, and abundant generates useful health information. Big data is the structured and unstructured data characterized by volume, velocity, complexity, diversity, and timeliness. Public health specialty areas that harness big data include community health, epidemiology, environmental health science, infectious diseases and NCDs, maternal and child health, occupational health and safety, food hygiene and nutrition. Dealing with massive multinomics of big data that is not apriori can be challenging; nevertheless, the benefits in the public health sector are numerous. Consistently, it has provided opportunities to understand public health in depth and helped achieve the goals of precision public health. In public health research, big data were utilized in identifying efficacious treatment interventions in pediatric asthma, pediatric obesity, diarrhea, Hepatitis C, HIV, injectable drug use, malaria, misuse of opioid medication, use of smokeless tobacco, and the Zika virus. In public health surveillance, big data advances real-time tracking of diseases and predicting disease outbreaks. Precision real-time signaling using big data has shown efficacy in monitoring air pollution, antibiotic resistance and drug safety, cholera, dengue, malaria,

Influenza A H1N1, Lyme disease, Ebola, HIV, drowning, exposure to electromagnetic field, monitoring food intake, and whooping cough (Dolley, 2018; Asokan & Asokan, 2015, p. 311). In a study from Mooney and Pejaver (2018, p. 108) it has been described that the networks of big data work on a single task simultaneously by handling various resources with resiliency, consistency, storage, and analysis to provide a robust evidence-informed precise public health policies by squeezing more actionable information.

Big data sets in public health commonly include one or more of the following descriptions (Heitmueller et al., 2014, p. 1524):

- measures of genomic or metabolomic data sets of the participants
- geospatial analyses of the participants through global positioning system (GPS)
- medical record data that enhances sample size compared to a study limited to primary data collection
- measures from the data effluent created by life in a digital world, such as web search term records, social media, or mobile phone records.

However, according to Lazer et al. (2014, p. 1203), the risks of the big data were realized after the failure of Google Flu Trends and the flaws on methods to predict risk using the internet and social media were recognized. This alerted to make a more cautious approach to combine big data, with nonsocial media data sources to avoid overfitting models that has few cases.

On the other hand, the utility of predictive computing tools in health care worldwide has brought noteworthy transformations in public health and is expected to continue improving precision public health. The volume of health data with the availability of various technologies like big data, machine learning, internet of things (IoT), and artificial intelligence (AI) has augmented predictive modeling capacities (Allam, Tegally, & Thondoo, 2019, p. 264). Since the beginning of this decade, mobile technologies have been in use to track the transmission during outbreaks. To track the transmission of flu a mobile app, FluPhone, was introduced in 2011 by Cambridge University, UK. Likewise, the mobile phone data and applications were used in Africa to fight the Ebola outbreak between 2014 and 2016 (Lin & Hou, 2020).

## 5 Case studies

### 5.1 Noncommunicable diseases

Neurological disorders affect millions of people. The Centers for Disease Control report (2020a) on plans to utilize public health data modernization and the new world of public health data (Table 1) to actively engage in surveillance for a range of neurological conditions. The augmented approach to surveillance being developed is expected to use state-of-the-art tools and robust analytic methods as part of a systems approach to data collection, surveillance

## 334 Big data in psychiatry and neurology

**TABLE 1** The new world of public health data.

| The reality | The opportunity |
| --- | --- |
| *Reacting*: Always behind when epidemics occur | *Predicting*: Getting ahead of epidemics to stop them quickly |
| *Counting*: Collecting data without the ability to rapidly analyze it | *Understanding*: Rapid data analysis to gain real-time insights |
| *Storing separately*: Siloed systems that restrict data sharing | *Sharing effectively*: Interoperable, accessible data for action |
| *Moving slowly*: Outdated, paper-based systems with multiple points of data transfer | *Moving fast*: A true digital highway to automate transfer of critical data in real time |
| *Using resources inefficiently*: New resources always required to do new data collection | *Connecting resources*: Leveraging existing resources and making common investments for the future |

to derive actionable and timely information. In the first stage of National Neurological Conditions Surveillance System strategy of the Centers for Disease Control (2020b), embarked on how to derive best estimate rates of disease for two conditions: multiple sclerosis and Parkinson's disease, as both involve significant morbidity and place a substantial burden on patients and their families.

### 5.2 Infectious disease: COVID-19

Although Coronavirus disease (COVID-19) caused by SARS, CoV2 is not the first global health crisis for which mobile technology and big data have been used to fight epidemics, AI modeling developed by companies such as BlueDot and Metabiota was able to predict the COVID-19 in China before its alarming spread globally, that is, a week before the WHO released an official notice of the outbreak (Lin & Hou, 2020). The BlueDot scoured data from news reports, airline ticketing, and animal disease outbreaks to predict regions that would be at risk for the outbreak of SARS CoV2 from the epicenters of China. The Metabiota used the similar approach through Big Data analytics to track flight data to accurately anticipate that countries such as Japan, Thailand, Taiwan, and South Korea would be at risk to the SARS CoV2 outbreak well before an index case was reported in those countries. Thus big data and AI-based infectious disease surveillance algorithms captured and analyzed in real time, aid and signal early infectious disease detection. In contact tracing, big data and AI are useful in modeling and predicts the course of a pandemic for public health response strategies. Vast amounts of data from security camera footage, credit card usage records and GPS data from cars, railway and mobile phones were used to trace the movement of individuals with COVID-19. Mainland China, Hong Kong, Taiwan, and South Korea use big data and AI for strict enforcement in dynamic epidemic risk management of COVID-19 (Gaille, 2019; Heilweil, 2020; Lin & Hou, 2020). The Executive Director of the Communicable Diseases Cluster, WHO, David Heymann who led the international response to SARS quoted "By monitoring social media, newsfeeds, or airline ticketing systems, we can warn if there's something wrong" and that requires further exploration (McCall, 2020).

## References

Acheson, D. (1988). *Public health in England. The report of the committee of inquiry into the future development of the public health function (Cm 289).* London: HMSO.

Allam, Z., Tegally, H., & Thondoo, M. (2019). Redefining the use of big data in urban health for increased liveability in smart cities. *Smart Cities, 2*(2), 259–268. https://doi.org/10.3390/smartcities2020017.

Asokan, G. V., & Asokan, V. (2015). Leveraging "big data" to enhance the effectiveness of "one health" in an era of health informatics. *Journal of Epidemiology and Global Health, 5*(4), 311. https://doi.org/10.1016/j.jegh.2015.02.001.

Bansal, S., Chowell, G., Simonsen, L., Vespignani, A., & Viboud, C. (2016). Big data for infectious disease surveillance and modeling. *Journal of Infectious Diseases, 214*(suppl 4), S375–S379. https://doi.org/10.1093/infdis/jiw400.

Bayer, R., & Galea, S. (2015). Public health in the precision-medicine era. *New England Journal of Medicine, 373*(6), 499–501. https://doi.org/10.1056/nejmp1506241.

Baynam, G., Bauskis, A., Pachter, N., Schofield, L., Verhoef, H., Palmer, R. L., et al. (2017). 3-Dimensional facial analysis—Facing precision public health. *Frontiers in Public Health, 5,* 31. https://doi.org/10.3389/fpubh.2017.0003.

Brackett, D., Medellin, R. A., Caceres, C., & Mainka, S. (2004). *United Nations. Commissioned issue paper of the United Nations millennium project task force on environmental sustainability.* Retrieved from www.unmillenniumproject.org/documents/.TF6.IP1.Biodiversity.pdf.

Brownson, R. C., Fielding, J. E., & Maylahn, C. M. (2009). Evidence-based public health: A fundamental concept for public health practice. *Annual Review of Public Health, 30*(1), 175–201. https://doi.org/10.1146/annurev.publhealth.031308.100134.

Burkom, H. S. (2017). Evolution of public health surveillance: Status and recommendations. *American Journal of Public Health, 107*(6), 848–850. https://doi.org/10.2105/ajph.2017.303801.

Centers for Disease Control and Prevention. (2003). The global threat of new and Reemerging infectious diseases: Reconciling U.S. National Security and Public Health Policy. *Emerging Infectious Diseases, 9*(9). https://doi.org/10.3201/eid0909.030442.

Centers for Disease Control and Prevention. (2020, August 5). *Public health data modernization initiative: Harnessing the power of data to save lives.* Retrieved from https://www.cdc.gov/surveillance/pdfs/Data-and-IT-Transformation-IB-508.pdf.

Centers for Disease Control and Prevention. (2020, May 13). *NNCSS.* Centers for Disease Control and Prevention. Retrieved from https://www.cdc.gov/surveillance/neurology/what_is_the_national_neurological_surveillance_system.html.

Dolley, S. (2018). Big data's role in precision public health. *Frontiers in Public Health, 6.* https://doi.org/10.3389/fpubh.2018.00068.

## 336 Big data in psychiatry and neurology

Doyle, T. J. (2002). Completeness of notifiable infectious disease reporting in the United States: An analytical literature review. *American Journal of Epidemiology*, *155*(9), 866–874. https://doi.org/10.1093/aje/155.9.866.

Gaille, B. (2019). 29 wearable technology industry statistics. In *Trends & analysis* BrandonGaille. Com. February 19 https://brandongaille.com/29-wearable-technology-industry-statistics-trends-analysis/.

Heilweil, R. (2020). *How AI is battling the coronavirus outbreak*. Retrieved from https://www.vox.com/recode/2020/1/28/21110902/artificial-intelligence-ai-coronavirus-wuhan.

Heitmueller, A., Henderson, S., Warburton, W., Elmagarmid, A., Pentland, A. S., & Darzi, A. (2014). Developing public policy to advance the use of big data in health care. *Health Affairs*, *33*(9), 1523–1530. https://doi.org/10.1377/hlthaff.2014.0771.

Institute of Medicine. (1988). *The future of public health*. Washington, DC: The National Academies Press. https://doi.org/10.17226/1091.

Institute of Medicine. (2003). *The future of the Public's health in the 21st century*. Washington, DC: The National Academies Press. https://doi.org/10.17226/10548.

Khoury, M. J., & Galea, S. (2016). Will precision medicine improve population health? *Journal of American Medical Association*, *316*(13), 1357. https://doi.org/10.1001/jama.2016.12260.

Lazer, D., Kennedy, R., King, G., & Vespignani, A. (2014). The parable of Google flu: Traps in big data analysis. *Science*, *343*(6176), 1203–1205. https://doi.org/10.1126/science.1248506.

Lin, L., & Hou, Z. (2020). Combat COVID-19 with artificial intelligence and big data. *Journal of Travel Medicine*, *27*(5). https://doi.org/10.1093/jtm/taaa080.

McCall, B. (2020). COVID-19 and artificial intelligence: Protecting health-care workers and curbing the spread. *The Lancet Digital Health*, *2*(4), e166–e167. https://doi.org/10.1016/s2589-7500(20)30054-6.

Mooney, S. J., & Pejaver, V. (2018). Big data in public health: Terminology, machine learning, and privacy. *Annual Review of Public Health*, *39*(1), 95–112. https://doi.org/10.1146/annurev-publhealth-040617-014208.

Ness, R. B., Koopman, J. S., & Roberts, M. S. (2007). Causal system modeling in chronic disease epidemiology: A proposal. *Annals of Epidemiology*, *17*(7), 564–568. https://doi.org/10.1016/j.annepidem.2006.10.014.

Soucie, J. M. (2012). Public health surveillance and data collection: General principles and impact on hemophilia care. *Hematology*, *17*(sup1), s144–s146. https://doi.org/10.1179/102453312x13336169156537.

Taylor, L. H., Latham, S. M., & woolhouse, M. E. J. (2001). Risk factors for human disease emergence. *Philosophical Transactions of the Royal Society of London. Series B: Biological Sciences*, *356*(1411), 983–989. https://doi.org/10.1098/rstb.2001.0888.

The Institute for Health Metrics and Evaluation. (2020, September 2). *GBD results tool*. Retrieved from http://ghdx.healthdata.org/gbd-results-tool.

The World Bank. (2020, April 2). *Health overview*. Retrieved from https://www.worldbank.org/en/topic/health/overview.

Vaithinathan, A. G., & Asokan, V. (2017). Public health and precision medicine share a goal. *Journal of Evidence-Based Medicine*, *10*(2), 76–80. https://doi.org/10.1111/jebm.12239.

World Health Organization. (2018, My 24). *The top 10 causes of death*. World Health Organization. Retrieved from https://www.who.int/news-room/fact-sheets/detail/the-top-10-causes-of-death.

World Health Organization. (2018, June 1). *Non communicable diseases*. World Health Organization. Retrieved from https://www.who.int/news-room/fact-sheets/detail/noncommunicable-diseases.

World Health Organization. (2019, Mrch 26). *Disease burden and mortality estimates.* World Health Organization. Retrieved from https://www.who.int/healthinfo/global_burden_disease/estimates/en/index1.html.

World Health Organization. (2020, Februry 12). *COVID-19 public health emergency of international concern (PHEIC) global research and innovation forum.* World Health Organization. Retrieved from https://www.who.int/publications/m/item/covid-19-public-health-emergency-of-international-concern-(pheic)-global-research-and-innovation-forum.

World Health Organization. (2020, September 2). *WHO coronavirus disease (COVID-19) dashboard.* World Health Organization. Retrieved from https://covid19.who.int/.

# Chapter 17

# How big data analytics is changing the face of precision medicine in women's health

**Maryam Panahiazar[a,b], Maryam Karimzadehgan[c], Roohallah Alizadehsani[d], Dexter Hadley[e], and Ramin E. Beygui[a]**

[a]*Department of Surgery, Division of Cardiothoracic Surgery, School of Medicine, University of California San Francisco, San Francisco, CA, United States,* [b]*Bakar Computational Health Sciences Institute, School of Medicine, University of California San Francisco, San Francisco, CA, United States,* [c]*Independent Researcher, Google,* [d]*Institute for Intelligent Systems Research and Innovation, Deakin University, Melbourne, VIC, Australia,* [e]*Department of Clinical Sciences, College of Medicine, University of Central Florida, Orlando, FL, United States*

## 1 Introduction

Since the early ages of history, health care providers have been helping their patients with different kinds of treatments, observing the results and improving upon it. Previous medical approaches have been on "one size fits all" applying the same treatments to the patients with the same disease. Having more accurate, impactful results for each individual has always been the goal of physicians. Nowadays, with the advancement of personalized medicine, health care providers are able to take this mission by getting help with electronic records, genetic information, big data analysis, all the necessary tools and information that are required to truly move toward the precise and personalized medicine (National Research Council, 2011). Personalized medicine is one of the promising approaches to tackle different kinds of diseases.

Research in personalized medicine has enabled physicians to have medical treatments that are safe and effective for each individual patient. Physicians can choose a treatment strategy based on the patient's individual information to provide targeted treatments and prevention plans, minimize the side effects of the treatment, and lead to more successful outcomes. Genomic data have enabled researchers to collect large amounts of data from diverse patients. Combining this data with clinical information has enabled physicians to observe patterns for understanding the effectiveness of the treatment which will bring faster and more effective care to more people (Seyhan & Carini, 2019).

Big Data in Psychiatry and Neurology. https://doi.org/10.1016/B978-0-12-822884-5.00001-5
Copyright © 2021 Elsevier Inc. All rights reserved.

**339**

**340** Big data in psychiatry and neurology

While the last decade was dominated by challenges in handling large amounts of patient data, we need to shift the focus on making the use of data for knowledge discovery. Due to the advancement of technology, it is easy now to gather enormous amounts of data generated from patient health records, diagnosis, treatments, genomic information, labs even from fitness apps and smartphones. These datasets are usually heterogeneous (or multiview data) and in unstructured format such as image or text (Peck, 2018). So big data in health care needs to process a large amount of both structured and unstructured data for each patient. As a result, a successful integration of these multiple datasets is a key to personalized medicine to be able to take advantage of available data and use that for new discovery or hypothesis generation.

The main question in personalized medicine then is how to extract knowledge from big data (Hulsen et al., 2019). Nowadays, advances in Machine learning, AI, Natural Language Processing, and Deep learning methods enable faster and better mining of patient data (Mesko, 2017). Technologies targeting big data are playing an important role in health care. Deep learning is suited well for medical data as it can identify patterns in sparse, noisy, and heterogeneous data. Multiview data can be represented as multiple facets of data instances (Li, Wu, & Ngom, 2016). As a result, machine learning methods can be applied to heterogeneous multifaceted data to predict the output of interest. Common machine learning models are Naive Bayes Classifier, ensemble-learning models (random forest) (Su, Peña, Liu, & Levine, 2018), and more recently deep learning models (Abadi et al., 2016; Glorot & Bengio, 2010).

Deep learning models have been applied to different ranges of health data integration and modeling applications. They have been applied for better understanding of the sequence specificities of DNA- and RNA-proteins (Alipanahi, Delong, Weirauch, & Frey, 2015), or predicting the interactions based on sequence-based features (Singh, Yang, Póczos, & Ma, 2019), or for training as generative models for molecular structures (Segler, Kogej, Tyrchan, & Waller, 2018). Deep learning models have also made big improvements to help identify, classify, and quantify patterns in medical images (Shen, Wu, & Suk, 2017). Specifically, due to the great improvements in computer vision, medical image analysis has gained improvements in image fusion (Suk, Lee, & Shen, 2014), computer-aided diagnosis (Suk et al., 2015; Suk, Wee, Lee, & Shen, 2016), lesion detection (Pereira, Pinto, Alves, & Silva, 2016), and microscopic imaging analysis (Cireşan, Giusti, Gambardella, & Schmidhuber, 2013). The top causes of death among adult women in the United States include heart disease and cancer. Women who have mental health issues, such as depression and anxiety, are at higher risk of having conditions such as cardiovascular disease (CVD), and breast cancer. Psychiatric conditions (e.g., depression and stress) also are important factors in the causation of skin disease in women. In the following, we describe the role of big data, data analytics, and deep learning to improve women's health and we explain a few use cases in CVD, breast cancer, and skin disease.

How big data analytics is empowering women's health **Chapter | 17** **341**

## 2 The role of big data and deep learning in personalized medicine to empower women's health

Women experience certain health care challenges and are more likely to be diagnosed with chronic disease and conditions such as heart disease, cancer, and diabetes than men. Women's health refers to the branch of medicine that focuses on the treatment and diagnosis that affect a woman's physical health as well as emotional and mental health. As deep learning techniques have transformed health care and helped in treating diseases, one main question to ask is can we apply deep learning models to women's health?

Breast cancer is a leading cause of cancer death among women in the USA. Mammographic screening significantly reduces the breast cancer deaths in women (Tabar et al., 2003; Tabár et al., 2001). Interpretation of mammograms is still very difficult (Miglioretti et al., 2009) and extensive experience is needed for accurate interpretation. Computer-aided detection (CAD) was among the first few methods developed to assist mammogram interpretation (Giger, Chan, & Boone, 2008). Early studies (Burhenne et al., 2000; Freer & Ulissey, 2001) showed that CAD technique was enough for detecting cancer. For example, the study in (Burhenne et al., 2000) showed that the radiologists had a false-negative rate of 21% and the CAD method helped to reduce this false-negative rate by 77%. The study in (Freer & Ulissey, 2001) used 12,860 screening mammograms; each mammogram was interpreted without the assistance of CAD, followed by a reevaluation of the areas that are marked by the CAD system. Using CAD, the recall rate was improved from 6.5% to 7.7% and 19.5% increase in the number of cancers were detected. However, recently due to large-scale clinical trials, this technique has failed to improve diagnostic performance due to high false-positive rates (Lehman et al., 2015).

There has been new advancement in applying AI and deep learning models to improve the false-positive rates in mammographic interpretation where they can achieve similar performance to experts (Akselrod-Ballin et al., 2019). In this study, the deep learning algorithm was trained on 9611 mammograms of women to make two breast cancer predictions: (1) predict biopsy malignancy and to (2) differentiate normal from abnormal screening examinations. Their algorithm identified 34 of 71 (48%) false-negative cases on mammograms. For the malignancy prediction, their algorithm obtained an area under the curve (AUC) of 0.91.

The recent study in 2020 (McKinney et al., 2020) used large datasets and showed superior performance to radiologists in cancer detection. More specifically, they used a large dataset from the UK (25,000 mammograms collected between 2012 and 2015) and a large enriched dataset from the USA (3000 mammograms collected between 2001 and 2018). They used an ensemble of 3 deep neural networks where the inputs are mammogram images and the output is breast cancer risk score between 0 and 1. The results showed an absolute reduction of 5.7% and 1.2% (USA and UK) in false positives and 9.4% and 2.7% in

**342** Big data in psychiatry and neurology

false negatives. These studies were a few examples to demonstrate how deep learning models have revolutionized women's health.

As personalized medicine is also evolving and researchers are collecting data to describe the staggering complexity of biological processes with the advancements of deep learning models, the question is can we better predict, prevent, and treat diseases that are uniquely affecting women due to their myriad of biological systems? This will help to address conditions that are more common in women and to identify the risk factors. For example, one of the most common applications of personalized medicine is for women with breast cancer. About 30% of breast cancer patients are identified by overexpression of human epidermal growth factor receptor 2 (HER2) (Romond et al., 2005). Standard methods won't work for these women and only an antibody drug— Herceptin—would work. As a result, developing models that target specific needs of an individual woman is becoming supercritical.

In the following sections, we describe a few use cases on skin conditions, CVD, and breast cancer.

## 3  Use case studies

### 3.1  Advanced data analytics on skin conditions from genotype to phenotype

Women and men can have many of the same skin conditions. However, they affect women differently. This could be because of hormones and reproductive differences between the genders. Advanced data analytics using Big Data make it possible to find the common genes and phenotype (e.g. drugs, biomarkers, biological process) for different skin conditions (Panahiazar et al., 2018). We have included hyperpigmentation, postinflammatory hyperpigmentation, melasma, rosacea, actinic keratosis, and pigmentation in this study. These conditions have been selected based on the reasoning of big-scale UCSF patient data of 527,273 females for 8 years and related publications regarding skin conditions. We proposed a novel framework for large-scale available public data to find the common genotypes and phenotypes of different skin conditions.

We extracted a group of genes for each skin condition from Gene expression Omnibus (GEO) from National Center for Biotechnology Information (NCBI) (Barrett et al., 2012). We found over 1000s of genes with the Search Tag Analyze Resource for the GEO (STARGEO) platform (Hadley et al., 2017). STARGEO has been explained in detail (Hadley et al., 2017). GEO is an open database of more than 2 million samples of functional genomics experiments. STARGEO platform allows for a meta-analysis of genomic signatures of disease and tissue through the tagging of biological samples from multiple experiments (Hadley et al., 2017). Then we analyzed the signature (genes) from our STARGEO analyses in Ingenuity Pathway Analysis (IPA) (IPA. QIAGEN Inc. Available online: https://www.qiagenbioinformatics.com/

How big data analytics is empowering women's health Chapter | 17 **343**

products/ingenuity-pathwayanalysis), restricting genes that showed statistical significance ($P < .05$) and an absolute experimental log ratio greater than 0.1 between conditions and control samples. These selected genes have been used for the next step analysis in IPA to extract selected genes, drugs, biological processes, and other factors. First, hundreds of genes were extracted from IPA for the skin conditions as we explained before, including 32 genes for hyperpigmentation, 63 genes for actinic keratosis, 21 genes for melasma, 42 genes for rosacea, 5 genes for postinflammatory, and 87 genes for pigmentation. For each gene, we extracted related information including Symbol (e.g., TYR), Entrez Gene Name (e.g., tyrosinase), Location (e.g., Cytoplasm), Type (e.g., Enzyme), Biomarker Application (e.g., efficacy), drugs (e.g., hydroquinone, azelaic acid), Entrez Gene ID for Human (e.g., 7299). Term frequency-inverse document frequency (TF-IDF) is a numerical statistic that is intended to reflect how important a word is to a document in a collection or corpus. Term frequency-inverse document frequency (TF-IDF) was used to find the more common types and genes in all conditions based on all results. TF-IDF is a numerical statistic that is intended to reflect how important a word is to a document in a collection or corpus. Next, we sought to understand the phenotypes for each condition, such as relevant drugs.

We chose hyperpigmentation, postinflammatory hyperpigmentation, melasma, rosacea, actinic keratosis, and pigmentation as more frequent skin conditions in women. Women are at greater risk for immune system diseases that can cause melanoma as a most dangerous form of skin cancer. Women are more likely to get rosacea than men, especially during menopause. Melasma is sometimes referred to as the mask of pregnancy because the splotches typically show up during pregnancy in women. After choosing different conditions and meta-analysis for multidimensional gene study and find several genes and corresponding phenotypes, we hypothesize that by finding the overlap in genotype and phenotype between different skin conditions, we can suggest a drug that is used in one condition for treatment in another condition that has similar genes or other common phenotypes. For example, Azelatic Acid is a drug used to treat rosacea and acne, kills acne bacteria, and reduces the production of keratin, reduces inflammation, reduces the synthesis of melanin, and is used for the treatment of skin pigmentation. Thioredoxin reductase and TYR genes are affected by Azelaic Acid and these two genes are presented as extracted genes in all these conditions. The overlap of these genes across conditions suggests the hypothesis of using this drug for the treatment of all conditions and not just for rosacea and/or postinflammatory as it says in current guidelines.

The outcome of this study based on Advance Data Analytics provides information on skin conditions and their treatments to the research community and introduces new hypotheses for possible genotype and phenotype targets. The novelty of this work is a metaanalysis of different features on different skin conditions on big-scale data. Instead of looking at individual conditions with one or two features, which is how most of the previous works are conducted, we looked

**344** Big data in psychiatry and neurology

at several conditions with different features to find the common factors between them. This work has implications for discovery and new hypotheses to improve women's health quality in most popular skin conditions in women and is geared toward making Big Data useful. More outcomes of this study can be found in our paper (Panahiazar et al., 2018).

## 3.2 Big data platform to use machine learning on EHR data for personalized medicine in heart failure survival analysis and patient similarity

Cardiovascular diseases (CVD) encompass a broad range of conditions. According to the AHA 2020 Heart Disease Statistics, gender, ethnic, racial, and age discrepancies within diagnosis and treatment exist and have been well reported. When it comes to cardiovascular disease, worldwide, women continue to have poorer outcomes than men. The causes of these discrepancies have yet to be fully addressed. Precision Cardiovascular Medicine (PCM) refers to the customization of CVD care for patients to maximize the success of preventive and therapeutic interventions by using patient-specific information such as demographics and genetic data. In most patients diagnosed with CVD, aggressive medical therapy is the cornerstone of treatment, while the responses to these therapies vary significantly among individual patients. One of the most common complications of CVD is Heart Failure (HF). We developed a big data platform at Mayo Clinic to study survival analysis and patient similarity (Panahiazar, Taslimitehrani, Pereira, & Pathak, 2015a, 2015b) that lead to individualized and personalized treatment based on patient characteristics. From an architectural perspective, we used Hadoop in our applications as a complement to existing data systems. It offers an open-source technology designed to run on large numbers of connected servers to scale-up data storage and processing at a very low cost and is proven to scale to the needs of the largest web properties in the world. We designed the Hortonworks Data Platform (HDP) in Mayo's health care systems. HDP is powered by Open Source Apache Hadoop that provides all of the Apache Hadoop projects necessary to integrate Hadoop as part of a Modern Data Architecture.

Based on our architecture, we store our datasets from different resources including EHRs, Genomics, and Medical Imaging into the Hortonworks repository and then use scripting tools like Pig and Hive to clean and prepare our data. One of the applications of an implementation of this architecture at Mayo Clinic is data retrieval and cohort creation. There are many data sources available in different departments of Mayo Clinic and each one includes millions of EHR data and creating cohorts is one of the main steps in each project. Using spreadsheets for extracting records from millions of records of EHR data based on the cohort criteria is time consuming. Pig is one of the big data tools that produce a sequence of MapReduce programs to run complex tasks which consists of multiple interrelated transformations. In one of our projects about the integration of

How big data analytics is empowering women's health Chapter | 17 **345**

different data sources such as lab results, medications, and patient demographics to predict survival score of each heart failure patients, our cohort was patients with heart failure diagnosis event with at least one EF (Ejection Fraction) value within 3 months of the heart failure diagnosis date. To create our cohort, we extracted our desirable records from the aggregation of four large datasets including one heart failure clinical trial and three EHR datasets from different Mayo's clinical systems. Using any spreadsheet-based tool or even SQL to retrieve data from these datasets is almost impossible.

We implemented our cohort criteria in the form of pig queries in three steps: we filter all patients with heart failure ICD9 code, then in the second step, we joined the results of the first query with the patient's Ejection Fraction (EF) records, and finally, the results of the second query are being filtered based on the time intervals defined in the cohort by domain experts and clinicians. Pig translates our queries to a sequence of MapReduce jobs and the jobs were sent to the servers sequentially. Using pig to create our cohorts was faster and easier than any other tools. Our dataset included more than 150 million patient records that require the usage of parallel querying and computation, and this highlights the use of big-scale data analysis.

We assessed the performance of the survival model (e.g., SHFM) applying Mayo Clinic's Big scale EHR dataset. Our results demonstrate an improvement in accuracy as compared to the standard SHFM and also suggest the applicability of our model to standard clinical care in the community. We also incorporated additional predictor variables that included the patient's characteristics including 26 comorbidities into our models that led to further improvement in the prognostic predictive accuracy. Finally, we built a heart failure risk prediction model using a series of machine learning techniques and observed that logistic regression and random forest return more accurate models compared to other classifiers. Since models that are built based on EHRs are more accurate (11% improvement in AUC) and are applicable in standard routine care, it is imperative to leverage EHR data for survival analysis and prediction modeling in HF and other chronic conditions. Find out more about this research in our original paper (Panahiazar et al., 2015a).

In Precision Medicine, the ability to match the right drug with the right dose to the right patient at the right time is vital. We developed a multidimensional patient similarity assessment technique that leverages multiple types of information from the EHR and predicts a medication plan for each new patient based on the prior knowledge (e.g., demographic information such as gender, patient history) and data from similar patients. In our algorithm, patients have been clustered into different groups using a hierarchical clustering approach and subsequently have been assigned a medication plan based on the similarity index to the overall patient population. We evaluated the performance of our approach on a cohort of heart failure patients identified from EHR data at Mayo Clinic and achieved an AUC of 0.74. Our results suggest that it is possible to harness population-based information from EHRs for an individual

**346** Big data in psychiatry and neurology

patient-specific assessment. We found and introduced different medication plans for individual patients based on their characteristics.

Our objective in this study was to propose an approach to use patient similarity techniques to determine the medication plan for a new patient based on the EHR data. To this end, we defined a patient similarity framework, allowing us to exploit the similarity-based medication recommendation. We calculated the distribution of medication plans in our cohort. Fifty-seven percent of the patients responded to HF therapy and their EF measurements increased by at least 10% after 6 months from the first EF measurement and initiation of HF therapy. In our cohort, we detected 28 different medication plans as a combination of 5 medication classes. The results show that the combination of Angiotensin-converting enzyme inhibitors (ACEIs), Beta blockers (BBs), and Statins is the most popular medication plan in our cohort with 17% of the patients being prescribed this combination therapy, and with more than 50% demonstrating an improvement in EF by at least 10%. The next common plan is ACEIs and BBs. More than 12% of the patients were prescribed ACEIs and BBs and 51% of the patients demonstrated a good response to therapy. Note that statins and BBs are commonly prescribed to HF patients, which affirm the clinical practice guidelines. In one other study we showed that there is gender. See more detail in our original paper (Panahiazar et al., 2015a, 2015b). For more cases in cardiovascular disease and women health, please refer to our systematic review of gender-based studies of diagnosis and treatment of cardiovascular disease in the last 20 years (Panahiazar, Alizadehsani, Bishara, Chern, & Hadley, 2019). According to our systematic review, there are significant findings around different risk factors and outcomes in CVD for both men and women. There are gender-based discrepancies in diagnosis and treatment of CVD. But the causes of these discrepancies have yet to be fully addressed and require further research in this field to overcome poor outcomes in women.

### 3.3 Large-scale labeling of free-text pathology report for deep learning to improve women's health in breast cancer

As explained earlier, screening mammography is effective in reducing mortality but has a high rate of unnecessary recalls and biopsies. While deep learning can be applied to mammography, large-scale labeled datasets, which are difficult to obtain, are required. In our previous study (Panahiazar et al., 2019), we demonstrate the feasibility of creating a hybrid framework using traditional Machine learning models and cloud-based automated solutions from IBM Watson (Panahiazar et al., 2019) to derive pathologic diagnosis from free-text breast pathology reports. About 7000 women (mean age = 51.8 years), ~10,000 reports were extracted from an in-house pathology database at UCSF. The "final diagnosis" section was considered in the analysis.

Both traditional machine learning models and Watson's Natural Language Classifier (NLC) performed well for cases under 1024 characters with weighted

How big data analytics is empowering women's health **Chapter | 17** **347**

average F-measures above 0.96 across all classes. Performance of traditional NLP was lower for cases over 1024 characters with an F-measure of 0.83. We demonstrate a hybrid framework using traditional models combined with IBM Watson to annotate over 10,000 breast pathology reports for the development of a large-scale database to be used for deep learning in mammography.

The pathology report text was converted to the ARFF file format and imported to WEKA—a machine learning framework that can be used for NLP. Each report was first tokenized. In our case, each token was an N-gram (a set of co-occurring words) in lengths of one to three words. For example, the phrase "no ductal carcinoma" will result in the following tokens: "no," "ductal," "carcinoma," "no ductal," "ductal carcinoma," and "no ductal carcinoma." Each pathology report was tokenized into a vector of N-grams to serve as the input for TF-IDF. Following the construction of the TF-IDF matrix, seven supervised machine learning algorithms were tested to determine the best performing classifier for predicting the label for each report: PART, decision tables, AdaBoost, Naive Bayes, multiclass logistic regression (one vs. all), support vector machine (SVM), and majority vote classifier (ZeroR). For our dataset, logistic regression outperformed all other classifiers. In future we will focus on expanding this process to other medical records such as radiology reports and clinical notes as well as testing other automated solutions from Facebook, Google, Amazon, and Microsoft. We hope to design an automated pipeline for large-scale clinical data annotation so that existing clinical records can be efficiently utilized for the development of deep learning algorithms. More details about this study can be found in our published paper (Trivedi et al., 2019).

# 4 Conclusion

This chapter reviews recent advances in applying big data analytics, machine learning, text mining, and deep learning in personalized medicine to empower women's health for a few important diseases in women that psychiatric conditions (e.g., depression and stress) are important factors in the causation of these diseases in women. Big data has already started to demonstrate its clinical value in the field of personalized medicine since a decade ago. However, to realize Big Data's full potential, we proposed "big data analytics" as a requirement to enable downstream analysis and extraction of meaningful information from data. We described the related works and explained the real use cases based on our research studies in Dermatology, Cardiovascular Disease, and Breast Cancer. For more details about each research study, a reader should refer to the original papers. Our long-term goal is to use the Personalized Medicine platform to translate the multidimensional big data (e.g., EHR, direct genetic information, GEO, and OMICS). This will improve and assist the clinical care decision making that will ultimately improve the outcome for individual patients and as a result reduces medical cost.

**348** Big data in psychiatry and neurology

Moreover, the completion of this chapter makes the foundation to understand the importance of big data analytics and personalized medicine for a better outcome. This will help to develop novel translational interventions through powerful big data-driven analytics that leverage the wide availability of data. As an implementation of clinical care, this study's goal is to improve precision diagnosis and ultimately the management of diseases for both early detection and identification of patients at risk for rapid progression of diseases impacting women's health.

## Acknowledgements

We thank all our collaborators from University of California San Francisco, University of California Berkeley, and Mayo Clinic who were listed as authors of the published research studies that we refer to as use cases in this chapter.

## Author contribution

M.P. wrote the use case research studies, abstract, and conclusion; M.K. wrote the introduction and outlined the deep learning in the personalized medicine and women health section; M.P., M.K., R.A., D.H., and R.B. read and revised the article.

## References

Abadi, M., Agarwal, A., Barham, P., Brevdo, E., Chen, Z., Citro, C., … Devin, M. (2016). Tensorflow: Large-scale machine learning on heterogeneous distributed systems. *arXiv preprint arXiv.* 1603.04467.

Akselrod-Ballin, A., Chorev, M., Shoshan, Y., Spiro, A., Hazan, A., Melamed, R., … Guindy, M. (2019). Predicting breast cancer by applying deep learning to linked health records and mammograms. *Radiology, 292*(2), 331–342. https://doi.org/10.1148/radiol.2019182622.

Alipanahi, B., Delong, A., Weirauch, M. T., & Frey, B. J. (2015). Predicting the sequence specificities of DNA- and RNA-binding proteins by deep learning. *Nature Biotechnology, 33*(8), 831–838. https://doi.org/10.1038/nbt.3300.

Barrett, T., Wilhite, S. E., Ledoux, P., Evangelista, C., Kim, I. F., Tomashevsky, M., … Soboleva, A. (2012). NCBI GEO: Archive for functional genomics data sets—Update. *Nucleic Acids Research, 41*(D1), D991–D995. https://doi.org/10.1093/nar/gks1193.

Burhenne, L. J. W., Wood, S. A., D'Orsi, C. J., Feig, S. A., Kopans, D. B., O'Shaughnessy, K. F., … Castellino, R. A. (2000). Potential contribution of computer-aided detection to the sensitivity of screening mammography. *Radiology, 215*(2), 554–562. https://doi.org/10.1148/radiology.215.2.r00ma15554.

Cireşan, D. C., Giusti, A., Gambardella, L. M., & Schmidhuber, J. (2013). In Mitosis detection in breast cancer histology images with deep neural networks Paper presented at the medical image computing and computer-assisted intervention—MICCAI Berlin, Heidelberg.

Freer, T. W., & Ulissey, M. J. (2001). Screening mammography with computer-aided detection: Prospective study of 12,860 patients in a community breast center. *Radiology, 220*(3), 781–786. https://doi.org/10.1148/radiol.2203001282.

Giger, M. L., Chan, H.-P., & Boone, J. (2008). Anniversary paper: History and status of CAD and quantitative image analysis: The role of medical physics and AAPM. *Medical Physics, 35*(12), 5799–5820. https://doi.org/10.1118/1.3013555.

How big data analytics is empowering women's health **Chapter | 17** **349**

Glorot, X., & Bengio, Y. (2010). Understanding the difficulty of training deep feedforward neural networks. In Paper presented at the proceedings of the thirteenth international conference on artificial intelligence and statistics.

Hadley, D., Pan, J., El-Sayed, O., Aljabban, J., Aljabban, I., Azad, T. D., ... Butte, A. J. (2017). Precision annotation of digital samples in NCBI's gene expression omnibus. *Scientific Data*, *4*(1), 170125. https://doi.org/10.1038/sdata.2017.125.

Hulsen, T., Jamuar, S. S., Moody, A. R., Karnes, J. H., Varga, O., Hedensted, S., ... McKinney, E. F. (2019). From big data to precision medicine. *Frontiers in Medicine*, *6*(34). https://doi.org/10.3389/fmed.2019.00034.

Lehman, C. D., Wellman, R. D., Buist, D. S. M., Kerlikowske, K., Tosteson, A. N. A., Miglioretti, D. L., & Breast Cancer Surveillance Consortium. (2015). Diagnostic accuracy of digital screening mammography with and without computer-aided detection. *JAMA Internal Medicine*, *175*(11), 1828–1837. https://doi.org/10.1001/jamainternmed.2015.5231.

Li, Y., Wu, F.-X., & Ngom, A. (2016). A review on machine learning principles for multi-view biological data integration. *Briefings in Bioinformatics*, *19*(2), 325–340. https://doi.org/10.1093/bib/bbw113.

McKinney, S. M., Sieniek, M., Godbole, V., Godwin, J., Antropova, N., Ashrafian, H., ... Shetty, S. (2020). International evaluation of an AI system for breast cancer screening. *Nature*, *577*(7788), 89–94. https://doi.org/10.1038/s41586-019-1799-6.

Mesko, B. (2017). The role of artificial intelligence in precision medicine. *Expert Review of Precision Medicine and Drug Development*, *2*(5), 239–241. https://doi.org/10.1080/23808993.2017.1380516.

Miglioretti, D. L., Gard, C. C., Carney, P. A., Onega, T. L., Buist, D. S. M., Sickles, E. A., ... Elmore, J. G. (2009). When radiologists perform best: The learning curve in screening mammogram interpretation. *Radiology*, *253*(3), 632–640. https://doi.org/10.1148/radiol.2533090070.

National Research Council. (2011). *Toward precision medicine: Building a knowledge network for biomedical research and a new taxonomy of disease*. National Academies Press.

Panahiazar, M., Alizadehsani, R., Bishara, A. M., Chern, Y., & Hadley, D. (2019). Systematic review of gender based studies of diagnosis and treatment of cardiovascular disease in last 20 years. *Advancements in Cardiovascular Research*, *2*(4), 192–194. https://doi.org/10.32474/ACR.2019.02.000143.

Panahiazar, M., Fadavi, D., Aljabban, J., Safeer, L., Aljabban, I., & Hadley, D. (2018). Large scale advanced data analytics on skin conditions from genotype to phenotype. In Paper presented at the Informatics.

Panahiazar, M., Taslimitehrani, V., Pereira, N., & Pathak, J. (2015a). Using EHRs and machine learning for heart failure survival analysis. *Studies in Health Technology and Informatics*, *216*, 40–44.

Panahiazar, M., Taslimitehrani, V., Pereira, N. L., & Pathak, J. (2015b). Using EHRs for heart failure therapy recommendation using multidimensional patient similarity analytics. *Studies in Health Technology and Informatics*, *210*, 369–373.

Peck, R. W. (2018). Precision medicine is not just genomics: The right dose for every patient. *Annual Review of Pharmacology and Toxicology*, *58*(1), 105–122. https://doi.org/10.1146/annurev-pharmtox-010617-052446.

Pereira, S., Pinto, A., Alves, V., & Silva, C. A. (2016). Brain tumor segmentation using convolutional neural networks in MRI images. *IEEE Transactions on Medical Imaging*, *35*(5), 1240–1251. https://doi.org/10.1109/TMI.2016.2538465.

Romond, E. H., Perez, E. A., Bryant, J., Suman, V. J., Geyer, C. E., Davidson, N. E., ... Wolmark, N. (2005). Trastuzumab plus adjuvant chemotherapy for operable HER2-positive breast cancer.

**350** Big data in psychiatry and neurology

*New England Journal of Medicine, 353*(16), 1673–1684. https://doi.org/10.1056/NEJMoa052122.

Segler, M. H. S., Kogej, T., Tyrchan, C., & Waller, M. P. (2018). Generating focused molecule libraries for drug discovery with recurrent neural networks. *ACS Central Science, 4*(1), 120–131. https://doi.org/10.1021/acscentsci.7b00512.

Seyhan, A. A., & Carini, C. (2019). Are innovation and new technologies in precision medicine paving a new era in patients centric care? *Journal of Translational Medicine, 17*(1), 114.

Shen, D., Wu, G., & Suk, H.-I. (2017). Deep learning in medical image analysis. *Annual Review of Biomedical Engineering, 19*(1), 221–248. https://doi.org/10.1146/annurev-bioeng-071516-044442.

Singh, S., Yang, Y., Póczos, B., & Ma, J. (2019). Predicting enhancer-promoter interaction from genomic sequence with deep neural networks. *Quantitative Biology, 7*(2), 122–137. https://doi.org/10.1007/s40484-019-0154-0.

Su, X., Peña, A. T., Liu, L., & Levine, R. A. (2018). Random forests of interaction trees for estimating individualized treatment effects in randomized trials. *Statistics in Medicine, 37*(17), 2547–2560. https://doi.org/10.1002/sim.7660.

Suk, H.-I., Lee, S.-W., & Shen, D. (2014). Hierarchical feature representation and multimodal fusion with deep learning for AD/MCI diagnosis. *NeuroImage, 101*, 569–582. https://doi.org/10.1016/j.neuroimage.2014.06.077.

Suk, H.-I., Lee, S.-W., Shen, D., & The Alzheimer's Disease Neuroimaging Initiative. (2015). Latent feature representation with stacked auto-encoder for AD/MCI diagnosis. *Brain Structure and Function, 220*(2), 841–859. https://doi.org/10.1007/s00429-013-0687-3.

Suk, H.-I., Wee, C.-Y., Lee, S.-W., & Shen, D. (2016). State-space model with deep learning for functional dynamics estimation in resting-state fMRI. *NeuroImage, 129*, 292–307. https://doi.org/10.1016/j.neuroimage.2016.01.005.

Tabár, L., Vitak, B., Chen, H.-H. T., Yen, M.-F., Duffy, S. W., & Smith, R. A. (2001). Beyond randomized controlled trials. *Cancer, 91*(9), 1724–1731. https://doi.org/10.1002/1097-0142(20010501)91:9<1724::Aid-cncr1190>3.0.Co;2-v.

Tabar, L., Yen, M.-F., Vitak, B., Chen, H.-H. T., Smith, R. A., & Duffy, S. W. (2003). Mammography service screening and mortality in breast cancer patients: 20-year follow-up before and after introduction of screening. *The Lancet, 361*(9367), 1405–1410. https://doi.org/10.1016/S0140-6736(03)13143-1.

Trivedi, H. M., Panahiazar, M., Liang, A., Lituiev, D., Chang, P., Sohn, J. H., ... Hadley, D. (2019). Large scale semi-automated labeling of routine free-text clinical records for deep learning. *Journal of Digital Imaging, 32*(1), 30–37. https://doi.org/10.1007/s10278-018-0105-8.

# Index

Note: Page numbers followed by *f* indicate figures, *t* indicate tables, and *b* indicate boxes.

## A

Abstraction studies, administrative databases, 160
Accelerometer, 222
Access to care, 70
Accuracy of diagnostic and procedural codes, 159
Acute care, 70
Administrative databases (AD), 156–157
pros and cons of, 157–162
Advanced data analytics, on skin conditions, 342–344
AI. *See* Artificial intelligence (AI)
AIC. *See* Akaike's information criterion (AIC)
Akaike's information criterion (AIC), 10
AlexNet, 290–291
Algonauts project, 300–301
Allen Brain Observatory, 299
Alzheimer's disease (AD), 151
diagnosis of, 313–314
Alzheimer's Disease Neuroimaging Initiative (ADNI) database, 190
Amazon Elastic Compute Cloud (Amazon EC2), 319–320
Android Wear OS 2.0, 221
Angiotensin-converting enzyme inhibitors (ACEIs), 346
ANNs. *See* Artificial neural networks (ANNs)
Apache Spark, 39
Artificial intelligence (AI), 26–28, 70, 87–88
Artificial neural networks (ANNs), 26–28, 27*f*, 227–228
Assumption of homoscedasticity, 6, 7*f*
Assumption of independence, 6
Assumption of linearity, 6
unrelated features, 7–8
Assumption of normality, 6–7
Autism diagnosis
EEG signals used in, 312–313
genomics data used in, 315–316
Autoencoder (AE), 248, 268–269

Automated machine learning (AutoML), 226–228, 234–235*f*
AWS Simple Storage Service (S3), 222–223
Azelaic acid, 343

## B

Balanced accuracy, 17–18
Balancing datasets, 24
Batch processing, 38
Bayesian information criterion (BIC), 10
Beta blockers (BBs), 346
Bias, 19–20
BIC. *See* Bayesian information criterion (BIC)
Big data, 69, 138*f*, 155–157
analytics, 44
in personalized healthcare, 42–46
characteristics of, 36–37, 36*t*
complex networks, 141–145, 142*f*
data sharing and integration
data integration, 81–84
data ownership, 77
data sharing initiatives, 79–81
support for, 78–79
definition, 36–37, 137
diagnose neurodegenerative diseases and anomalous aging, 150–152
in healthcare area, 39–41
learning from data, 145–150
and MRI analyses, 137–141
in personalized healthcare, 42–46
pictorial representation of, 138*f*
in precision public health, 332–333
privacy and ethics
explicit consent, 85
in industry, 85–86
stricter regulations, 84–85
standardization
interoperability and reusability, 71–72
standards for bio sample data, 76–77
standards for clinical data, 72–74
standards for imaging data, 75–76

351

**352** Index

Big data *(Continued)*
  standards for -omics data, 74–75
  teaching data science
    available courses in clinical data science, 88
    to medical students, 87–88
    need for more training, 87
  technologies, 38, 38*t*
    for neuroimaging, 41, 42*t*
    use cases in healthcare, 41, 42*t*
    *vs.* traditional research methods, 37, 37*t*
Big data and artificial intelligence approaches
  applications, 311
    diagnosis (*see* Diagnosis)
    early warning, 311–312
    prognosis, 316–317
  challenges and promising solutions
    information island, 319
    patient information, privacy and security of, 317–319
    specialized personnel, lack of, 320
    storage and analysis capabilities, 319–320
  data sources
    electroencephalography (EEG) signals, 308
    eye movement data, 308
    genomics, 307–308
    neuroimaging data, 309
    wearable equipment data, 309
  main algorithms, 310–311
Binary classification between early and late classes, 279–281
Biobank Sample Communication Protocol (BioSCOOP), 76
Biosample data, standards for, 76–77
  Biobank Sample Communication Protocol (BioSCOOP), 76
  Minimum Information About BIobank data Sharing (MIABIS), 76
  Ontology for BIoBanking (OBIB), 76
  Ontology for Biomedical Investigations (OBI), 77
Box plots of segmentation, 203–204, 203*f*
Brain-score, 299–300, 300*f*
Breast cancer predictions, 341

## C

California Consumer Privacy Act (CCPA), 84–85
Cambridgeshire and Peterborough NHS Foundation Trust (CPFT), 103

Cancer care, 70
Cardiovascular diseases (CVD), 98–99, 120, 344, 346
Cascade prediction, 243–244, 249–252
  background, 243–245
  long short-term memory (LSTM), 251, 252*f*
  method
    dataset, 245
    evaluation metrics, 253
    generating graph embeddings, 245–249
    model calibration, 251–252
    multilayer perceptron (MLP), 250–251, 250*f*
    using LSTM, 253–254
    using MLP, 253
cBioPortal, 84
CCPA. *See* California Consumer Privacy Act (CCPA)
CDL. *See* Clinical Data Lake (CDL)
CDSS. *See* Clinical decision support system (CDSS)
Cellular component, Gene Ontology, 74
CFA. *See* Confirmatory factor analysis (CFA)
China Brain Project, 35
Classical neuroscience approach, 288
Classification phase, 125
Clinical data, standards for, 72–74
  Clinical Data Interchange Standards Consortium (CDISC) Operational Data Model (ODM), 73
  Fast Healthcare Interoperability Resources (FHIR), 73–74
  Health Level Seven International (HL7), 73–74
  Logical Observation Identifiers Names and Codes (LOINC), 73
  National Cancer Institute Thesaurus (NCIt), 73
  project-specific codebooks, 74
  Systematized Nomenclature of Medicine Clinical Terms (SNOMED CT), 72
Clinical Data Interchange Standards Consortium (CDISC) Operational Data Model (ODM), 73
Clinical Data Lake (CDL), 82–83
Clinical decision support system (CDSS), 106
Clinical decision support (CDS) tool, 87–88
Clinical Practice Research Datalink (CPRD), 97–98
Clinical Record Interactive Search (CRIS) database, 101–106
Clinical Records Anonymization and Text Extraction (CRATE) database, 101–106

Index **353**

Clinical significance *vs.* statistical significance, 161
Cloud computing, 222
  method, 306–307
Cloud services, 222
Clustered data and study outcomes, 161
CNNs. *See* Convolutional neural networks (CNNs)
Common spatial pattern (CSP), 311
Complex networks, big data, 141–145, 142*f*
Computational pathology, 70
Computer-aided detection (CAD), 341
Computer vision-based medication adherence monitoring, 219
Confirmatory factor analysis (CFA), 56
Construction of rule-based systems, 119–120
Conventional learning-based methods, 184
Convolutional neural networks (CNNs), 264, 289–291, 290*f*, 295*f*, 314
Convolution neural network, 310
Coronavirus disease (COVID-19), 334–335
Covariance structure, 64
COVID-19. *See* Coronavirus disease (COVID-19)
CPFT. *See* Cambridgeshire and Peterborough NHS Foundation Trust (CPFT)
CPRD. *See* Clinical Practice Research Datalink (CPRD)
CVD. *See* Cardiovascular disease (CVD)

**D**
DAG. *See* Directed acyclic graph (DAG)
DAGs. *See* Dynamic directed acyclic graphs (DAGs)
Data base, 120
Data cleansing, 37
Data formatting, 2–4
Data integration, 45–46, 71, 81–84
  cBioPortal, 84
  Clinical Data Lake (CDL), 82–83
  Informatics for Integrating Biology and the Bedside (i2b2), 82
  tranSMART, 82
Data leakage, 23–24
Data linkages in epidemiology
  linking local, 96–99
  linking routinely- and non-routinely-collected data, 99–100
  linking structured and unstructured routinely collected data, 100–109
  national routinely-collected data, 96–99

Data ownership, 77
Data privacy, 45–46
Data processing, 38
Data Processor Agreement (DPA), 84–85
Data reduction and scaling, 23
Data science, teaching
  available courses, 88
  to medical students, 87–88
  need for more training, 87
Data security, 45–46
Datasets, 78, 126–127, 131–132, 245
Data sharing, 45–46, 78
  initiatives, 79–81
    DataSHIELD, 81
    DistibutedLearning.ai, 80–81
    GIFT-Cloud, 79
    Personal Health Train (PHT), 81
    Personalized Consent Flow, 79–80
    Sync for Science (S4S), 80
  and integration
    data integration, 81–84
    data ownership, 77
    data sharing initiatives, 79–81
    support for data sharing, 78–79
  support for, 78–79
    Open Science, 78
    sharing with industry, 78–79
DataSHIELD, 81
Data standardization, 139–140
Data storage, 222–223
DCNNs. *See* Deep convolutional neural networks (DCNNs)
Decision trees, 25–26, 310
Decoding problem, 289
DeepCas, 244
Deep convolutional neural networks (DCNNs), 292–294, 295*f*, 297
Deep learning, 26–28, 39, 184, 340
Deep learning-based methods for Hippocampus segmentation, 194–209
Deep machine learning algorithms, 314
Deep neural network (DNN) models, 194–195
DeepWalk, 246
Defuzzifier, 120
Dementia with Lewy bodies (DLB), 106–109
Deneutrosophication, 124
Depressive disorders (DDs), 317
Description of datasets, 159
Detail *vs.* approximation coefficients, 273–274, 274*f*
Diabetes, 129

**354** Index

Diabetics using NRCS, 129–130
  diabetic predictive using the NRCS, 129–130
  PIMA dataset, 129, 129$t$, 130–131$f$
Diagnosis
  Alzheimer's Disease (AD)
    electroencephalogram, 313–314
    magnetic resonance imaging (MRI),
      313–314
  autism diagnosis
    EEG signals used in, 312–313
    genomics data used in, 315–316
  schizophrenia diagnosis
    eye movement data used in, 314–315
Diagnostic algorithms, development of
  structured data, 98–99
  unstructured data, 106–109
DICOM. *See* Digital Imaging and
    Communications (DICOM)
Different patch-based multiatlas labeling
    methods, 190–194
Diffusion weighted imaging (DWI), 143–144
Digital Imaging and Communications
    (DICOM), 75
Digital Twin, 89
Dilated convolutional kernels, 206, 207$f$
Dilated dense network, 206–207, 207$f$
Dilated residual dense U-net, 205–207
Dimensionality, 13
  reduction, 268, 268$f$, 274–277, 275$f$
Directed acyclic graph (DAG), 223
Direct monitoring, medication adherence, 219
DistibutedLearning.ai, 80–81
Distributed data processing, 223
Distributed preprocessing, 224–226
DLB. *See* Dementia with Lewy bodies (DLB)
DPA. *See* Data Processor Agreement (DPA)
DWI. *See* Diffusion weighted imaging (DWI)
Dynamic directed acyclic graphs (DAGs), 242

**E**

Early warning, 311–312
EBPH. *See* Evidence-based public health
    (EBPH)
Ecological studies, administrative databases,
    160
EFA. *See* Exploratory factor analysis (EFA)
Electroencephalogram, 313–314
Electroencephalography (EEG), 144–145,
    263–264, 308, 312–313
Electronic health record (EHR), 39, 316–317
Electronic medical records (EMRs), 39, 95,
    100–101, 156–157

Electronic surveillance system, 329–330
Embedded-IC algorithm, 244
Embedded metadata, 2–4
Emergency prevention system (EMPRES), 331
EMRs. *See* Electronic medical records (EMRs)
Encoding problem, 289
End-to-end deep learning methods, 207–209
End-to-end dilated residual dense U-net,
    205–209
  dilated residual dense U-net, 205–207
  evaluation, 207–209
Epilepsy, 263, 311–312
Evaluation metrics, 253
Evidence-based public health (EBPH),
    325–326
Evidence-informed Policy Network (EVIPNet),
    329
Expert system, 119
Explicit consent, 85
Exploratory factor analysis (EFA), 55
Extreme gradient boosting (XGBoost), 26, 227
Eye contact, 156–157
Eye movement technology, 308

**F**

FA. *See* Factor analysis (FA)
Factor analysis (FA), 55–58, 57$f$
Fast Healthcare Interoperability Resources
    (FHIR), 73–74
FCN-based confidence estimation, 196–199,
    198$f$
FCNs. *See* Fully convolutional networks
    (FCNs)
Fertility dataset, 131–132, 132$f$
FHIR. *See* Fast Healthcare Interoperability
    Resources (FHIR)
Food and Agriculture Organization (FAO), 330
FSN. *See* Fully Specified Name (FSN)
Fully convolutional networks (FCNs), 184
Fully Specified Name (FSN), 72
Functional magnetic resonance imaging
    (fMRI), 300–301
Fuzzifier, 120
Fuzzy rule- based system, 120
Fuzzy set (FS) theory, 122

**G**

Gated Recurrent Unit (GRU), 244
GBD. *See* Global burden of disease (GBD)
GDPR. *See* General Data Protection Regulation
    (GDPR)

Index **355**

Gene expression Omnibus (GEO), 342–343
Gene Ontology (GO), 74
General Data Protection Regulation (GDPR), 84–85
Generalized linear models (GLM), 227
Genome-wide association analysis (GWAS), 316–317
Genomics, 307–308
GIFT-Cloud, 79
GLM. *See* Generalized linear models (GLM)
Global burden of disease (GBD), 326–328
  infectious diseases, 327–328
  noncommunicable diseases (NCDs), 327
Global Early Warning and Response System (GLEWS), 330
Global Framework for the Control of Transboundary Animal Diseases (GF-TAD), 331
Global Outbreak Alert Response Network (GOARN), 330
Global Virus Network (GVN), 331
GO. *See* Gene Ontology (GO)
Goal-directed stimulus synthesis, 297
Gold-standard studies, administrative databases, 160
Google Trends, 168, 171, 175–176, 175*f*
Gradient boosting, 227
Graph embedding, 243–249
Gyroscope sensors, 222

# H

Hadoop distributed file system (HDFS), 226
HDFS. *See* Hadoop distributed file system (HDFS)
Healthcare data, 39, 40*t*, 96–97
Health Insurance Portability and Accountability Act (HIPAA), 84–85
Health Level Seven International (HL7), 73–74
HealthWise Wales (HWW), 99–100
Heart failure survival analysis, 344–346
Hierarchical Model And X (HMAX), 290–291
Hierarchical modular optimization (HMO) model, 291
High-frequency smart watch sensor data, 218
High-performance computing (HPC), 39
Hippocampus segmentation, 181, 182*f*
  automatic and accurate, 182
  deep learning-based methods for, 194–209
  end-to-end dilated residual dense U-net for, 205–209
  learning-based methods, 184

multiatlas-based deep learning method for, 196–204
multiatlas-based methods, 183–184
patch-based multiatlas labeling for, 184–194
  evaluation of, 190–194
  local learning-based label fusion, 187–188
  supervised metric learning for label fusion, 188–190
  weighted voting label fusion, 185–187
HL7. *See* Health Level Seven International (HL7)
HMO. *See* Hierarchical modular optimization (HMO) model
Homoscedasticity, 6
Hortonworks data platform (HDP), 344
HPC. *See* High-performance computing (HPC)
Human Brain Project, 35
Human Phenotype Ontology (HPO), 75
Human vision, 287
HWW. *See* HealthWise Wales (HWW)

# I

i2b2. *See* Informatics for Integrating Biology and the Bedside (i2b2)
IDI. *See* Integrated Data Infrastructure (IDI)
Image-guided therapy, 70
ImageNet, 290–291, 297–299
Image registration, 196
Imaging data, standards for, 75–76
  Digital Imaging and Communications (DICOM), 75
  Quantitative Imaging Biomarker Ontology (QIBO), 75–76
  RadLex, 75
Independent Subspace Analysis (ISA) network, 195
Indirect monitoring, medication adherence, 219
Infectious diseases, 327–328
Inference engine, 120, 124
Informatics for Integrating Biology and the Bedside (i2b2), 82
Information cascades, 241–242, 255–257
Information extraction phase, 124
Information island, 319
Informed consent, 96–97
In-memory processing, 41
Inpatient treatment, 156
Integrated Data Infrastructure (IDI), 95–96
Interictal and preictal binary classification, 267–269, 272–277

**356** Index

International Food Safety Authorities Network (INFOSAN), 330
Internet of Things (IoT), 40–41, 41*t*, 43, 46, 309
Internet search queries as data, 169–171
Interoperability and reusability, 71–72
IoT. *See* Internet of Things (IoT)
Irritant effect, NRCS algorithm, 127
IT neural predictivity, 294–296, 295*f*

## K

KB. *See* Knowledge base (KB)
K-fold cross validation, 22, 22*f*
K-nearest neighbors (KNN), 270, 277, 278*f*
Knowledge base (KB), 119–120
Krokodil, 170–171

## L

Label fusion with FCN-based confidence estimation, 199–200
Latent curve model (LCM), 51–52, 60
LCM. *See* Latent curve model (LCM)
LDA. *See* Linear discriminant analysis (LDA)
Learning-based methods, 184
Leave-one-out cross validation, 22–23
LG Watch Sport, 221, 222*t*
Life expectancy, 217
Linear discriminant analysis (LDA), 14, 15*f*, 277, 278*f*
Linear regression, 5–6
Linking local, 96–99
Linking routinely- and non-routinely-collected data, 99–100
Linking structured and unstructured routinely collected data, 100–109
   CRIS and CRATE databases, 101–106
   development of diagnostic algorithms, 106–109
Local learning-based label fusion, 187–188
Local weighted voting label fusion methods, 185–186
Logical Observation Identifiers Names and Codes (LOINC), 73
Logistic regression, 24–25
LOINC. *See* Logical Observation Identifiers Names and Codes (LOINC)
Long data, 69
Longitudinal data, 51
Longitudinal measurement invariance, 58–59, 59*f*
Long short-term memory (LSTM), 251, 252*f*, 253–254, 254*f*
LSTM. *See* Long short-term memory (LSTM)

## M

Machine-learned computational models (MLCMs), 315
Machine learning (ML), 1, 28, 38–39, 226–228, 306, 340
Magnetic resonance imaging (MRI), 181–182, 313–314
Magnetoencephalography (MEG), 144–145, 300–301
MapReduce, 223
Matplotlib, 41
Matthew's Correlation Coefficient (MCC), 18–19
Mayo Clinic Platform, 79
MCC. *See* Matthew's Correlation Coefficient (MCC)
MCCV. *See* Monte Carlo cross validation (MCCV)
MCI. *See* Mild cognitive impairment (MCI)
Measurement model, 61
Medical crowdsourcing, 89
Medication adherence, 218–219
   monitoring, 219
Melanoma, 343
Melasma, 343
Memory-based networks, 244–245
Memory optimization, 26
Mental disorders, 156
MERS. *See* Middle East respiratory syndrome (MERS)
Micro-batch processing, 38
Middle East respiratory syndrome (MERS), 328
MILCM. *See* Multiple indicators latent curve model (MILCM)
Mild cognitive impairment (MCI), 151, 313
Minimum Information About BIobank data Sharing (MIABIS), 76
MLP. *See* Multilayer perceptron (MLP)
Model calibration, 249, 251–252
Molecular function, Gene Ontology, 74
Monitoring the future (MTF), 168, 171
Monte Carlo cross validation (MCCV), 21
Movember Global Action Plan 3 (GAP3) database, 82, 83*f*
MR brain images, 182, 183*f*
MTF. *See* Monitoring the future (MTF)
Multiatlas-based deep learning methods, 195–204, 197*f*
   evaluation, 200–204
   FCN-based confidence estimation, 196–199
   label fusion with FCN-based confidence estimation, 199–200
Multiatlas-based methods, 183–184

## Index  357

Multiatlas segmentation method, 184, 185*f*
Multicollinearity, 7–8
Multilayer perceptron (MLP), 250–251, 250*f*, 253, 254*f*
Multiple indicator growth curve models, 60–64
  factor analysis, 57–58
  principal component analysis, 55*f*, 57–58
Multiple indicators latent curve model (MILCM), 53
  equations, 61–63
    measurement model, 61
    structural model, 61–63
  with linear function, 63, 63*f*
  specification details, 63–64
    covariance structure, 64
    identification and scaling, 64
  steps in fitting, 64–66
Multiplex model, for brain connectivity, 150–152
Multivariate dimension reduction techniques
  factor analysis, 55–56, 57*f*
  principal component analysis, 54–55
Mutagenic effect, NRCS algorithm, 127

## N

Naive Bayes, 25
National Cancer Institute Thesaurus (NCIt), 73
National routinely-collected data, 96–99
Natural language processing (NLP) algorithms, 104–106, 109
NCIt. *See* National Cancer Institute Thesaurus (NCIt)
Near-field communication (NFC), 219
Network community, 142
Neural network, 149–150, 149*f*
  algorithms, 307
  architectures, 250–251
  models
    first generation, 290–291
    next generation, 291–292
Neuroimaging data, 309
Neuroscience, 138
Neutrosophication, 123–125
Neutrosophic logic (NL), 121, 123
Neutrosophic rule-based classification system (NRCS), 121, 121*f*, 123–125, 124*f*
  classification phase, 125
  information extraction phase, 124
  medical applications
    for diabetics using NRCS, 129–130
    for seminal quality, 130–132

    for toxicity effects assessment of biotransformed hepatic drugs, 126–128
  neutrosophication phase, 124–125
  neutrosophic logic and neutrosophic set, 122–123
  rules generation phase, 125
Neutrosophic set, 122–123
Neutrosophy, 122
NFC. *See* Near-field communication (NFC)
NL. *See* Neutrosophic logic (NL)
NLP algorithms. *See* Natural language processing (NLP) algorithms
Node2Vec, 246–248, 247*f*
Noncommunicable diseases (NCDs), 326–327, 333–334
Nonlocal weighted voting label fusion methods, 186
Normality, 6–7
Normalized zero-crossings, 279–281, 280*t*
Novel psychoactive substances (NPS), 167
  in Google Trends searches, 171, 171–172*t*
  Internet search queries as data, 169–171
  methods, 171–172
  monitoring the future (MTF), 168, 171, 173*t*
NPS. *See* Novel psychoactive substances (NPS)
NRCS. *See* Neutrosophic rule-based classification system (NRCS)

## O

OBI. *See* Ontology for Biomedical Investigations (OBI)
OBIB. *See* Ontology for BIoBanking (OBIB)
OBO. *See* Open Biological and Biomedical Ontologies (OBO)
Omics data, standards for, 74–75
  Gene Ontology (GO), 74
  Human Phenotype Ontology (HPO), 75
Ontology for BIoBanking (OBIB), 76
Ontology for Biomedical Investigations (OBI), 77
Open Biological and Biomedical Ontologies (OBO), 77
Open Science, 78
Overall accuracy, 16–17

## P

Parallelization, 26
Parsimonious models, choosing, 9–11
Partitioning, 23
Patch-based multiatlas labeling, for hippocampus segmentation, 184–194
Patient identifiers (PI), 96–97
PCA. *See* Principal component analysis (PCA)
Performance metrics, 15–19

**358** Index

Personal Health Train (PHT), 81
Personalized healthcare, 36, 46, 69
Personalized medicine, 43, 339–340
Pharmacogenomics, 45–46, 316–317
Pharmacology, 217
Phenotypic personalized medicine
  (PPM), 45
Philips AI Principles, 85–86, 86*t*
Philips Data Principles, 85–86, 86*t*
PHT. *See* Personal Health Train (PHT)
PhysicianSN dataset, 245
PI. *See* Patient identifiers (PI)
PIMA dataset, 129, 129*t*, 130*f*
  precision of, 131*f*
  sensitivity of, 131*f*
  specificity of, 132*f*
Poor medication adherence, 217–218
Positive predictive value (PPV), 161
PPM. *See* Phenotypic personalized medicine
  (PPM)
PPV. *See* Positive predictive value (PPV)
Precision, 17
  diagnosis, 70
Precision cardiovascular medicine
  (PCM), 344
Precision health, 42–43, 46
Precision healthcare, 36
Precision medicine, 44–45
Precision public health, 332–333
Predicted values, 6
Preictal classification, 269–271
  different selections, 270
  into early and late stages, 277–283
    different classifiers and feature extraction
      methods, 282–283
    different selections, 277–281
    different wavelet functions and
      levels, 281
Principal component analysis (PCA), 13–14,
  13*f*, 55*f*, 57–58, 268–269
Principal components (PCs), 268–269
Privacy and ethics, big data
  explicit consent, 85
  in industry, 85–86
  stricter regulations, 84–85
Program for Monitoring Emerging Diseases
  (ProMED), 331
Project-specific codebooks, 74
Proximity tags, 219
Psychiatry, 156
Public health, 325–326
  big data in, 332–333

cases
  coronavirus disease (COVID-19),
    334–335
  noncommunicable diseases, 333–334
  data, new world of, 334*t*
  health systems and, 328–332
  surveillance system, 329–332
Python Pandas, 41

## Q

Quantitative Imaging Biomarker Ontology
  (QIBO), 75–76
Quantum computing, 39

## R

Radial basis function (RBF), 269
Radiofrequency identification (RFID), 219
RadLex, 75
Random forest classifier, 227, 310
Random Forests, 26, 147–148
RBF. *See* Radial basis function (RBF)
Real-world data, 158
Rectified linear unit (ReLU), 198
Recurrent neural network (RNN), 242
Reduction of data dimensionality, 11–14
  linear discriminant analysis, 14, 15*f*
  principal component analysis, 13–14, 13*f*
  scaling, 12
  variable selection, 12–13
Region-of-Interest (ROI), 140–141
Relationships, 72
Representational learning method, 242
Representational similarity analysis (RSA), 296
Reproductive effect, NRCS algorithm, 127
Resampling methods, 19–23
ResDUNet, 205, 206*f*
ResUNet, 198, 199*f*
RFID. *See* Radiofrequency identification
  (RFID)
RNN. *See* Recurrent neural network (RNN)
Rule base, 120
Rule-based expert system, 119–120
Rules generation phase, 125

## S

Sample size estimation, 8–9
Scalable medication intake monitoring system
  algorithms, 223–228
    automated machine learning (AutoML),
      226–228

distributed preprocessing, 224–226
machine learning, 226–228
experiment results, 228–236
related work, 218–220
system architecture, 220–223
cloud services, 222
data storage, 222–223
distributed data processing, 223
smartwatch application, 221–222
Scaling, reduction of data dimensionality, 12
Schizophrenia diagnosis, eye movement data, 314–315
Scikit-Learn, 41
Screening mammography, 346
SDNE. *See* Structured deep network embedding (SDNE)
Search engines, 169–170
Search tag analyze resource for the GEO (STARGEO), 342–343
Seizure prediction, 264–266, 269
Seminal quality using NRCS, 130–132
classification using the NRCS model, 132
dataset description, 131–132
Sensitivity, 17
Severe acute respiratory syndrome (SARS CoV2), 328
Smartwatches, 218, 221–222
SNA. *See* Social network analysis (SNA)
SNOMED CT. *See* Systematized Nomenclature of Medicine Clinical Terms (SNOMED CT)
Social network analysis (SNA), 243
Social networks, 243
Software as a Service (SaaS), 138–139
Specificity, 17
Spectral power-based method, 272
SSL/TLS protocol, 317–318
Stacked ensemble, 228
Standardization, big data
for biosample data, 76–77
for clinical data, 72–74
for imaging data, 75–76
interoperability and reusability, 71–72
for -omics data, 74–75
Statistical assumptions, 5–8
Stream processing, 38
Stricter regulations, big data, 84–85
Structural connectivity, 143–144
Structural model, 61–63
Structured deep network embedding (SDNE), 248–249
Substance abuse, 167

Supervised machine learning
artificial intelligence, 26–28
best practice 1, 2*b*
statistical assumptions, 5–8
best practice 2
sample size estimation, 8–9
best practice 3, 9*b*
choosing parsimonious models, 9–11
best practice 4, 10*b*
reduction of data dimensionality
(*see* Reduction of data dimensionality)
best practice 5, 13*b*
performance metrics, 15–19
best practice 6, 15*b*
best practice 7, 18*b*
resampling methods, 19–23
best practice 8
data leakage, 23–24
best practice 9, 23*b*
balancing datasets, 24
data reduction and scaling, 23
partitioning, 23
variable selection, 23–24
best practice 10, 28*b*
limitations and future directions, 28
classifiers, 24–26
deep learning, 26–28
Supervised metric learning for label fusion, 188–190
Supervised Naive Bayes, 25
Support vector machines (SVM), 25, 147–148, 310
SVM. *See* Support vector machines (SVM)
Sync for Science (S4S), 80
Synonym, 72
Systematized Nomenclature of Medicine Clinical Terms (SNOMED CT), 72

## T

Term frequency-inverse document frequency (TF-IDF), 342–343
Three-dimensional (3D) printing, 45–46
Tidy datasets, 2–4
Time-dependent nature of variables, 161
Toxicity classification using the NRCS, 127–128
tranSMART, 82
Tumorigenic effect, NRCS algorithm, 127
Two-stage wavelet-based method, 265, 265*f*
comparative analysis methods, 271–272
materials and methods
dataset, 265–266

**360** Index

Two-stage wavelet-based method *(Continued)*
two-stage zero-crossings wavelet-based
framework, 266–271
results, 272–283

## U

Univariate latent curve model, 60, 60*f*

## V

V4 neurons, 297
Variability, 155–156
Variable selection, 12–13, 23–24
Variance, 19–20
Variety, 155–156
VBM. *See* Voxel-Based Morphometry
(VBM)
Velocity, 155–156
Veracity, 155–156
Visual neuroscience
control, 293, 297, 298*f*
evaluation, 293–297
neural network models
first generation of, 290–291
next generation of, 291–292
neural network models, first generation of,
290–291
prediction, 293–296
understanding vision, 288–293
vision community, 297–301
Algonauts project, 300–301
Allen Brain Observatory, 299
brain-score, 299–300, 300*f*
Voxel-Based Morphometry (VBM), 140–141,
140*f*

## W

Wavelet energy, 264
Wavelet energy-based method, 272
Wavelet entropy-based method, 271–272
Wavelet features-based method, 271
Wavelet functions and levels, 281
Wavelet output, 264
Wavelet transform, 264, 267, 311
Weak tree learner, 26
Wearable equipment data, 309
Weighted voting label fusion, 185–187
Whole-exome sequencing (WES), 316
Whole-genome sequencing (WGS), 316
Wide data, 69
Women's health, personalized medicine to
big data and deep learning, 341–342
advanced data analytics on skin conditions,
342–344
cases, 342–347
at Mayo Clinic, 344–346
World Animal Health Information System
(WAHIS), 330

## X

XGBoost. *See* Extreme gradient boosting
(XGBoost)

## Y

Years lost due to disability (YLD), 98–99

## Z

Zero-crossings, 279–281, 280*t*
wavelet-based framework, 266–271, 267*f*

Printed in the United States
by Baker & Taylor Publisher Services